Advances in High-speed Rail Technology

More information about this series at http://www.springer.com/series/13506

Zhang-Dui Zhong · Bo Ai
Gang Zhu · Hao Wu · Lei Xiong
Fang-Gang Wang · Lei Lei
Jian-Wen Ding · Ke Guan
Rui-Si He

Dedicated Mobile Communications for High-speed Railway

北京交通大学出版社
http://press.bjtu.edu.cn

Springer

Zhang-Dui Zhong
Beijing Jiaotong University
Beijing
China

Bo Ai
Beijing Jiaotong University
Beijing
China

Gang Zhu
Beijing Jiaotong University
Beijing
China

Hao Wu
Beijing Jiaotong University
Beijing
China

Lei Xiong
Beijing Jiaotong University
Beijing
China

Fang-Gang Wang
Beijing Jiaotong University
Beijing
China

Lei Lei
Beijing Jiaotong University
Beijing
China

Jian-Wen Ding
Beijing Jiaotong University
Beijing
China

Ke Guan
Beijing Jiaotong University
Beijing
China

Rui-Si He
Beijing Jiaotong University
Beijing
China

ISSN 2363-5010 ISSN 2363-5029 (electronic)
Advances in High-speed Rail Technology
ISBN 978-3-662-54858-5 ISBN 978-3-662-54860-8 (eBook)
DOI 10.1007/978-3-662-54860-8

Jointly published with Beijing Jiaotong University Press, Beijing, China

The print edition is not for sale in China Mainland. Customers from China Mainland please order the print book from: Beijing Jiaotong University Press

Library of Congress Control Number: 2017941485

Printed on acid-free paper

This Springer imprint is published by Springer Nature
The registered company is Springer-Verlag GmbH Germany
The registered company address is: Heidelberger Platz 3, 14197 Berlin, Germany

Contents

1 Review of the Development of Dedicated Mobile Communications for High-Speed Railway 1
1.1 Railway Development in China 1
1.2 High-Speed Railway Development in the World 3
1.2.1 High-Speed Railway Development in China 5
1.3 The Active Role of Mobile Communications for Railway 8
1.4 GSM for Railway 9
1.4.1 The Development of GSM-R 10
1.4.2 GSM-R Key Technology and Engineering Measures 11
1.5 Next-Generation Mobile Communication System for Railway 14
References 17

2 Key Issues for GSM-R and LTE-R 19
2.1 GSM-R Architecture 19
2.1.1 GSM-R Network Composition 19
2.1.2 Mobile Switching Subsystem 19
2.1.3 Mobile Intelligent Network Subsystem 21
2.1.4 General Packet Radio Service (GPRS) Subsystem 22
2.1.5 Base Station Subsystem 23
2.1.6 Operation and Support Subsystem (OSS) 24
2.1.7 Terminal 25
2.2 GSM-R Network Hierarchical Structure 25
2.2.1 Mobile Switching Network 25
2.2.2 Intelligent Network 26
2.2.3 General Packet Radio Service Network 26
2.3 LTE-R Architecture 27
2.4 Key Technologies for GSM-R 37
2.5 Key Technologeis for LTE-R 42
2.5.1 The Application Requirements of the Next-Generation Railway Mobile Communication System 44

2.5.2 The Technology System and Network Architecture of the Next-Generation Railway Mobile Communication System 45
2.5.3 Frequency and Bandwidth Requirements of the Next-Generation Railway Mobile Communication System 46
2.5.4 The Key Technology in the Next-Generation Railway Mobile Communication System 46
2.5.5 Hybrid Networking of GSM-R and the Next-Generation Mobile Communication System 50
2.5.6 The Evaluation and Optimization of High-Speed Railway Wireless Resource Management Mechanism. . . . 51
2.6 Summary 53
References 53

3 Radio Propagation and Wireless Channel for Railway Communications 57
3.1 High-Speed Railway Propagation Scenarios. . . . 57
3.1.1 High-Speed Railway Propagation Scenarios Definition 57
3.1.2 Propagation Scenarios of Wide-Sense Vehicle-to-X Communications 60
3.2 High-Speed Railway Channel Measurements 68
3.2.1 Measurement Methods and System. . . . 68
3.2.2 Measurement Campaign 71
3.3 Narrowband Channel Characterization of High-Speed Railways 75
3.3.1 Path Loss 75
3.3.2 Shadow Fading. . . . 78
3.3.3 Small-Scale Fading. . . . 86
3.4 Wideband Channel Characterization of High-Speed Railways 100
3.4.1 Delay Characteristics 100
3.4.2 Doppler Effect 104
3.4.3 Angular Characteristics. . . . 109
3.5 Summary 116
References 116

4 Cooperation and Cognition for Railway Communications. . . . 125
4.1 Cooperation Scenarios 125
4.1.1 Improved Channel Reliability 129
4.1.2 Improved System Throughput. . . . 130
4.1.3 Seamless Service Provision. . . . 133
4.2 Key Techniques for Cooperation 133
4.2.1 Relay Protocol 133
4.2.2 MIMO and Cooperative Communication 144
4.2.3 Distributed Space–Time Coding 147
4.2.4 Physical Layer Network Coding and Cooperative Communication. . . . 151

4.3 Signal Classification in Cognitive Radio 155
4.3.1 Spectrum Sensing 159
4.3.2 Automatic Modulation Classification 165
4.3.3 Specific Emitter Identification 172
4.4 Cooperation and Cognition for High-Speed Railway 176
4.4.1 Relay Selective Cooperation in Railway Network 178
4.4.2 A Cooperative Handover Scheme for High-Speed Railway 181
4.4.3 Cognition for High-Speed Railway 185
4.5 Summary 190
4.5.1 Cooperative Diversity in Wireless Sensor Networks 194
4.5.2 Cooperative Diversity in Cognitive Radio 194
4.5.3 Summary of Cognitive Radio 198
References 199

5 Resource Management for High-Speed Railway Mobile Communications 205
5.1 Introduction 205
5.2 Overview and Survey 207
5.2.1 Admission Control 207
5.2.2 Level-Based Admission Control 208
5.2.3 Handover-Based Admission Control 209
5.2.4 Priority-Based Admission Control 209
5.2.5 Resource Allocation 210
5.2.6 Interference-Aware Resource Allocation 211
5.2.7 QoS-Aware Resource Allocation 212
5.2.8 Cross-Layer Dynamic Resource Allocation 213
5.2.9 Power Control 214
5.3 Resource Allocation and Power Control 216
5.3.1 System Model 216
5.3.2 Time-Distance Mapping 217
5.3.3 BS-RS Link Capacity 218
5.3.4 Utility-Based Resource Allocation 218
5.3.5 Problem Formulation 219
5.3.6 PAT Problem 220
5.3.7 PAS Problem 221
5.3.8 Problem Transformation 223
5.3.9 The Greedy Algorithm 226
5.3.10 Numerical Results and Discussions 228
5.4 Dynamic Resource Management 235
5.4.1 System Model 235
5.4.2 Problem Formulation 238
5.4.3 Dynamic Resource Management Schemes 240
5.4.4 Lyapunov Drift-Plus-Penalty Approach 241
5.4.5 Dynamic Resource Management Algorithm 242

5.4.6 Dual Optimization Framework ... 245
5.4.7 Simulation Results ... 247
5.5 Challenges and Open Issues ... 251
5.5.1 Location-Aware Resource Management ... 252
5.5.2 Cross-Layer Based Resource Management ... 252
5.5.3 Energy-Efficient Resource Management ... 253
5.5.4 Robust Resource Management ... 253
5.5.5 Resource Management for 5G Communications ... 254
5.6 Summary ... 254
References ... 255

6 LTE-R Network ... 259
6.1 LTE-R Network Services ... 259
6.2 LTE-R Network Architecture ... 262
6.3 LTE-R Network Performance Evaluation ... 265
6.3.1 Queueing Theory ... 265
6.3.2 Petri Nets ... 266
6.3.3 Network Calculus ... 266
6.3.4 System Model ... 270
6.3.5 Stochastic Arrival Curve for Train Control Service ... 277
6.3.6 Stochastic Service Curve for HSR Fading Channel ... 278
6.3.7 Performance Evaluation ... 284
References ... 292

7 Security of Dedicated Mobile Communications for Railway ... 295
7.1 Security Threats of Mobile Communications for Railway ... 295
7.1.1 Security Threats ... 295
7.1.2 Security Issues in GSM-R ... 296
7.1.3 Problems Still Existing in GSM-R ... 300
7.2 Security Enhancement for GSM-R ... 303
7.2.1 Security Measures Taken by GSM-R System ... 303
7.2.2 Bidirectional Authentication ... 307
7.2.3 End-to-End Encryption ... 310
7.2.4 Anti SIM Card Clone ... 312
7.3 Security of Wireless Heterogeneous Networks for Railway ... 315
7.3.1 Fast Re-authentication in Hot Spots ... 316
7.3.2 Wlan and Cellular Authentication ... 323
7.3.3 Relay Security ... 325
7.3.4 Access Authentication for Mobile Trusted Computing ... 329
7.4 Future and Challenges ... 332
References ... 332

8 Channel Simulation Technologies for Railway Broadband Mobile Communication Systems ... 335
8.1 Simulation Approaches ... 335
8.2 Simulation Scenario for Railway ... 336

8.2.1 Scenario 1: Open Space SFN 336
8.2.2 Scenario 2: Tunnel Environment. 337
8.2.3 Scenario 3: Open Space ENB to RP. 338
8.2.4 Scenario 4: Public Network 338
8.3 Channel Model in Simulation 339
8.3.1 Single-Tap HST Channel Model. 339
8.3.2 Two-Tap HST Channel Model 340
8.3.3 WINER Channel Model 342
8.4 Hardware-in-Loop Simulation Testbed. 347
8.4.1 Architecture 347
8.4.2 HIL Simulation Results 348

Chapter 1
Review of the Development of Dedicated Mobile Communications for High-Speed Railway

1.1 Railway Development in China

In recent years, a large number of reforms and improvements in technologies have been achieved in China railway system to promote the development of transportation and communication in railways. These reforms and improvements include the heavy-loan transportation, the electrification upgrade, several-time speed up for all lines, the development of passenger railway lines, and the development of high-speed railways.

In 2010, the investment in the infrastructure construction for railways is 707.459 billion. It is a number of 107.012 billon increase over last year's figure, which equals to a 17.8% increase. At the end of 2010, there are about 91,000 km of railway in service, achieving the target of the "11th Five-Year Plan" for the railway construction. Also in 2010, there are 329 large and medium-sized projects (excluding the local railways) under construction in railway infrastructure network, and the number of newly started projects is 97.

In 2011, the investment in the infrastructure construction for railways is 460.127 billion. Some major projects, such as Beijing–Shanghai high-speed railway and Guangzhou–Shenzhen–Hong Kong express railway line, are in operation in this year. The 3348.9 km new line of track laying, 2174.3 km double-track laying, and 3430.9 km electrified railways have been completed. There are 299 large and medium-sized projects (excluding the local railways) under construction in railway infrastructure network, and the number of newly started projects is 15.

In 2012, the investment in the infrastructure construction for railways is 521.5 billion, with year-on-year growth of 13.3%. The investment is more than 20 billion in total for Shanghai railway administration, Beijing railway administration, Nanning railway administration, Guangzhou railway group corporation, Guiguang railway corporation, and Jinyulu railway corporation. From September to December, the investment in all the railways is 284.8 billion, which increases more than double compared to the same period of the previous year, and the average

Z.-D. Zhong et al., *Dedicated Mobile Communications for High-speed Railway*, Advances in High-speed Rail Technology, DOI 10.1007/978-3-662-54860-8_1 monthly investment is more than 70 billion. Totally 68 projects have been completed, including the Harbin–Dalian express railway, Beijng–Shijiazhuang–Wuhan express railway, Harbin–Hami express railway, and so on. In this year, 5389 km new line of track laying, 4792 km double-track laying, and 6073 km electrified railways have been completed, and among which the new high-speed line of track laying is 2722 km. There are 301 large and medium-sized projects (excluding the local railways) under construction in railway infrastructure network, and the number of newly started projects is 28.

In 2013, more major projects have been completed, including the Nanjing–Hangzhou high-speed railway, Hangzhou–Ningbo high-speed railway, Tianjin–Qinhuangdao high-speed railway, Xiamen–Shenzhen high-speed railway, Xi'an–Baoji high-speed railway, and so on. Forty-nine new projects are under construction, including Chongqing–Guiyang high-speed railway, Hohhot–Zhangjiakou high-speed railway, Jiujiang–Jingdezhen–Quzhou high-speed railway, and so on. The investment in all the railways is 663.8 billion, and the total length of new railways is 5586 km. As a result, a new milestone is reached in the history of the development in China railways: there are more than 100,000 km railways in service, including more than 10,000 km high-speed railways.

In 2014, the construction of railways has been accelerated. The nationwide investment in the infrastructure construction for railways is 808.8 billion, and the total length of new railways in service is 8427 km. Both numbers achieve a highest record in history. There are more than 5000 km high-speed railways in service, including a large number of major projects such as Lanzhou–Xinjiang high-speed railway, Guizhou–Guangzhou high-speed railway, Nanning–Guangzhou high-speed railway, and the Changhuai section in Shanghai–Kunming high-speed railway. Therefore, the conditions are created and the foundations are laid to achieve the target of the "12th Five-Year Plan" for the railway development. By the end of 2014, there are 112,000 km railways in service, including 16,000 km high-speed railways. In order to promote the development of high-speed railway and intercity high-speed railway, the "Design Specification Standards for High-Speed Railway" and the "Design Specification Standards for Intercity High-Speed Railway" have been officially unveiled, which can also provide technical supports to promote the "exportation" of China high-speed railway.

In 2015, the nationwide investment in railways is 823.8 billion, and the total length of new railways in service is 9531 km. There are also 61 new projects under construction.

During the "12th Five-Year Plan" period, the total investment in railways comes to 3580 billion, and the total length of new railways in service is 30500 km. By the end of 2015, there are 121,000 km railways in service, ranking second in the world. Taking up more than 60% of the high-speed railways, there are more than 19,000 km in service in China, ranking first in the world (Fig. 1.1 and Table 1.1).

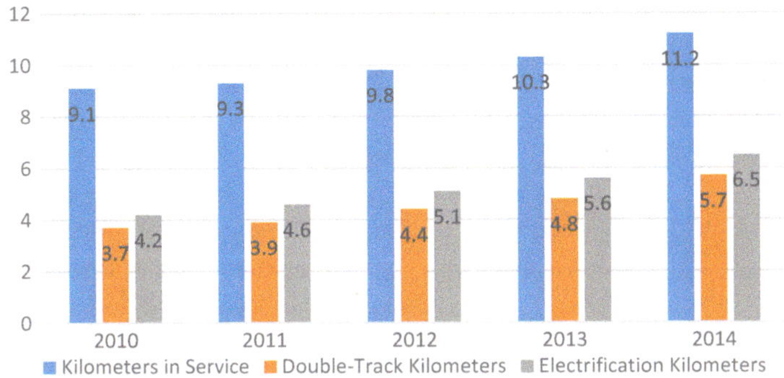

Fig. 1.1 km of railways in China (from National Railway Administration of the People's Republic of China)

Table 1.1 km of railways in China (from National Railway Administration of the People's Republic of China)

	2010	2011	2012	2013	2014
Kilometers in service	9.1	9.3	9.8	10.3	11.2
Double-Track kilometers	3.7	3.9	4.4	4.8	5.7
Electrification kilometers	4.2	4.6	5.1	5.6	6.5

1.2 High-Speed Railway Development in the World

For half a century, with the development of economy and technology in the world, high-speed railway, a major mobile equipment of railway passenger transport, got a rapid development. With Japan's Shinkansen, bullet train, France's TGV, Germany's ICE, Italy's ETR, and so on as the representative, high-speed railway gave birth to a new field of technologies with unique technique and complete system in modern industry. It also achieved remarkable achievements and gave a strong support for the development of the world economy and the progress of civilization.

(1) Development and progress of the Shinkansen, bullet train in Japan
Japan is the first country in the world to open up a high-speed train. Starting in 1964, Begin with 0 series high-speed trains, after 50 years of continuous improvement, 100 series, 200 series, 300 series, 400 series, 500 series, 700 series, N700 series, and E1–E7 series Shinkansen, bullet trains have been developed successively. At the same time, 300X, WIN350, STAR21, and other high-speed test trains are also developed. By the basic characteristics of the Shinkansen, bullet train can be summarized as follows: power-distributed type electric vehicle group, aluminum alloy train body, high light-weight level, and high aerodynamic performance. Light-weight non-bolster bogie, the semi-active suspension system is adopted in the 500 series and 700 series trains.

(2) Development and progress of the TGV (AGV) series high-speed trains in French
The development of TGV series high-speed train in France began at the end of 1960s. The plan is to design the diesel multiple unit (DMUs) at first and change to the development of high-speed electric multiple unit (EMUs) in 1975. Since the first TGV high-speed train vehicle has been successfully developed, the French have developed three generations of power concentrated TGV high-speed EMU. At this stage, the French is developing a new generation of high-speed trains—AGV power-distributed high-speed EMU.
(3) Development and progress of ICE/Velaro series high-speed trains in Germany
Germany ICE/Velaro series of high-speed trains is one of the world's most successful high-speed trains. The main representative models are ICE 1–3, ICT models, and E Velaro and RUS Velaro high-speed train. Its main features are that the power is centralized in the early stage and the latter is dominated by dynamic dispersion, the aluminum alloy train body, Bolsterless bogie and magnetic rail brake and eddy current brake braking system, moderate passenger number, complete function, high technical level, reasonable overall layout of structure, high-grade built-in, and high performance to operate and maintain.
(4) Development and progress of ETR series high-speed train in Italy
Italy ETR series of high-speed trains are all tilting train except ETR 500, so it is also called the Pendolino train. ETR series of high-speed trains include the first-generation ETR 401, the second-generation ETR 450, the third generation of ETR 460, ETR 470 and ETR 480, s220, and ETR 500 (power centralized high-speed tilting EMUs), and the fourth-generation ETR 600 models. At present, the Ansaldo Breda company is developing ETR 1000 high-speed train, called the type 1000. ETR 1000 high-speed train uses 4M4T marshaling structure, traction power is 9800 kilowatts, construction speed is 360 km/h, and the number passenger is 469. At present, the train is being tested and plans to put into use in 2016.
(5) Development of other typical high-speed trains in the world
The Swedish design a power centralized tilting EMU–X2000 type high-speed train. The trains can travel in the existing line and the working principle is using active train body swinging device to make the body swing an angle, to compensate for line under ultra high to ensure comfort while train is passing through the curve section, so as to improve the curving speed. Spanish designed and manufactured the Talgo series of high-speed trains, which is the representative models of high-speed railway; the tractive power is 8000 kilowatts, construction speed is 350 km/h, and the seating capacity is 300 people. British IC series high-speed trains are mainly used on the East and West Coast main line. The representative models are IC 225 intercity trains, construction speed is 225 km/h, which is the only single-ended dynamic Changbian power centralized high-speed train in the world.

(6) Technical characteristics and development trend of high-speed trains in foreign countries
At present, the main technical characteristics of foreign high-speed trains are described as follow: higher speed, many countries have been competing to develop high-speed train with the construction speed exceeds 360 km/h. The articulated high-speed train adopts the power dispersion technology, which has the better running quality, higher light-weight level, and the better utilization rate of the vehicle interior space. The train body tilting technique is adopted to effectively improve the speed of the curve; Focus on the overall aerodynamic shape of the train design, thereby reducing the running resistance of the train; Using permanent magnet motor traction technology, which has high power density, high power conversion efficiency, overload protection ability, and so on. By the design of the energy absorbing structure, the safety performance of the train is improved.

1.2.1 High-Speed Railway Development in China

Since 1990s, China began to carry out a large number of scientific research about the design and construction of high-speed railway construction technology, high-speed trains, operation and management of the basic theory and key technology organizations. At the same time, the Qinhuangdao–Shenyang dedicated passenger line was built and realizes the six time speed up for the existing railway. At December 2002, Qinhuangdao Shenyang passenger dedicated line was completed, which is the first railway passenger dedicated line designed and constructed by China own with the target speed of 200 km/h, infrastructure reserved 250 km/h. Independently developed the “China Star” EMUs in Qin Shen passenger dedicated line created “China Railway first speed”—321.5 K km/h.

In January 2004, the government approved the “medium- and long-term railway network planning,” to determine the railway network to expand the scale, improve the structure, improve quality, expanse the transport capacity rapidly, and improve the level of equipment quickly. By 2020, the national railway operating mileage reach to 10 million km, achieve separation of passenger and freight traffic in the busy main lines railway electrification rate and the double track rate reach to 50%, the transport capacity meets the needs of national economic and social development, and the main technical equipment reach or close to the international advanced level. In November 2007, the state issued “comprehensive transportation network in the long-term development plan,” which announced that, by 2020, the total size of the railway network will reach or exceed 120 thousand kilometers, and double track rate and electrification rate reach 50 and 60%. In October 2008, the Chinese government issued “Medium and long term railway network planning (2008 adjustment)," which announced that, by 2020, the national railway operating mileage will exceed 12 million kilometers and passenger dedicated line reach more than 1.6 million kilometers, double track railway electrification rate and the rate reach 50% and 60%, respectively.

What is more, a railway network with reasonable layout, clear structure, perfect function, and smooth convergence will be built and the transport capacity will meet the needs of the national economic and social development, and the main technical equipment reach or close to the international advanced level. The focus is to project the "four vertical and four horizontal" and the intercity passenger transport system in economically developed and densely populated areas. The "four vertical" passenger dedicated line includes Beijing Shanghai passenger dedicated line, including Bengbu to Hefei, Nanjing to Hangzhou passenger line, through the Beijing Tianjin to the Yangtze River Delta and the eastern coastal economically developed areas, Beijing–Wuhan–Shenzhen–Guangzhou passenger dedicated line, connecting the northern and the Southern China area, Beijing–Shenyang–Harbin (Dalian) railway passenger dedicated line, including Jinzhou to Yingkou passenger line and Hangzhou–Ningbo–Fuzhou–Shenzhen passenger dedicated line, connecting the Yangtze River, the Pearl River Delta, and the southeast coastal areas. The "Four horizontal" passenger dedicated line includes Zhengzhou Xuzhou to Lanzhou passenger dedicated line, connecting the northwest and East China, Nanchang–Changsha–Guiyang–Kunming passenger dedicated line, connecting the southwest, central China and East China, Shijiazhuang Qingdao to Taiyuan passenger dedicated line, connecting the north and East China and Nanjing Chongqing Wuhan to Chengdu passenger dedicated line, connecting the southwest and East China. At the same time, the construction of Jiujiang–Nanchang, Liuzhou–Nanning, Mianyang–Chengdu–Leshan, Harbin–Qigihar, Harbin–Mudanjiang, Changchun–Jilin, Shenyang–Dandong, and other passenger lines are carrying on with the purpose to expand the coverage of passenger dedicated line. In the ring Bohai Sea, Yangtze River Delta, Pearl River Delta, Zhuzhou and Xiangtan, as well as the Chengdu Chongqing and the urban agglomeration in the Central Plains, Wuhan city circle, Guanzhong Urban Group, on the west side of the Straits of urban group economically developed and densely populated areas in the construction of intercity passenger transport system, covering major cities and towns in the region. According to the medium- and long-term railway network plan, through the construction of Beijing–Shenyang, Shangqiu–Hangzhou–Beijing, Nanchang–Ganzhou railway passenger dedicated line and Beijing–Shanghai, Beijing–Guangzhou, Beijing–Haerbin, coastal, Longhai, Shangha–Kunming, Shanghai–Wuhan–Chengdu as the backbone, we built the "four vertical and four horizontal" high-speed rail network, while supporting the construction Guiguang, HeFu, and other high-speed rail line extension and further form the China's high-speed rail network with rich tentacles, high network access, and strong capacity.

On August 1, 2008, China's first full independent intellectual property rights, the world's first class level of high-speed railway Beijing Tianjin intercity railway traffic was put into operation.

On December 26, 2009, the longest railway built in one time and with the most complex engineering type in the world, Wuhan–Guangzhou high-speed railway, was put into operation with speed of 305 km/h.

On February 6, 2010, the world's first high-speed railway built in the wet Subsidence Loess Area, Zhengzhou–Xi'an high-speed railway connecting the central and Western of China with speed of 350 km/h, was put into use.

On June 30, 2011, the longest railway built in one time in the world, Beijing–Shanghai high-speed railway, was put into use. The length of this high-speed railway is 1318 km and connects the most two developed areas of eastern China, Beijing and Shanghai. The designed speed is 350 km/h and the initial operating speed is 300 km/h. In December 3, 2010, a new generation of "harmony" EMU CRH380AL in the Beijing–Shanghai high-speed railway from Bengbu to Zaozhuang creates a new record speed with 486.1 km/h in the test section.

On December 1, 2012, the world's first high-speed rail line in the cold region, Harbin–Dalian high-speed railway, was put into use. The length of this high-speed railway is 921 km and connects the main cities in the northeast of China. As a result, it will only cost 4 h and 40 min from Harbin to Dalian in the winter.

On December 26, 2012, the full line of Beijing–Guangzhou high-speed railway line was put into use. Beijing–Guangzhou high-speed railway passes through Beijing, Hebei, Henan, Hubei, Hunan, and Guangdong with length of 2298 km, which is the longest operating mileage of high-speed railway in the world. The designed speed is 350 km/h and the initial operating speed is 300 km/h.

By the end of 2014, Qinhuangdao–Shenyang railway, the Beijing–Tianjin intercity railway, Shijiazhuang–Taiyuan passenger dedicated line, Hangzhou–Shenzhen railway, Beijing–Guangzhou high-speed railway, Chengguan railway, Pipeng railway, Shanghai–Nanjing high-speed railway, Changjiu intercity railway, Hainan East Ring railway, Guangzhou–Zhuhai intercity lines, Changchun–Jilin intercity line, Beijing–Shanghai high-speed railway, Hefei–Bengbu high-speed railway, Shenyang–Dalian high-speed railway, Ningbo–Hnagzhou high-speed railway, the Tianjin–Qinhuangdao high-speed railway, Panying high-speed railway, Liunan dedicated passenger railway, Wuhan–Xianning intercity line and Shanghai–Hankou railway, Shanghai–Hangzhou section of Shanghai–Kunming high-speed railway, Guangzhou–Shenzhen section of Guangzhou–Shenzhen–Hong Kong Express railway, Shenyang–Harbin section of Beijing–Harbin high-speed railway, Zhengzhou–Xi'an section and Xi'an–Baoji section of Xuzhou–Lanzhou high-speed railway, HeNing section and HanYi section of NingRong high-speed railway, LiangYu section of HuRong high-speed railway, and GenWu section of NanGuang high-speed railway were all put into use.

By 2015, a number of new high-speed railways are put into operation, such as, Hefei–Fuzhou high-speed railway, Shenyang–Dandong high-speed railway, Jilin–Huichun high-speed railway, and Chengdu–Chongqing high-speed railway. The total operating mileage of China's high-speed railway comes to 1.9 million kilometers. After more than 10 years of unremitting efforts, with technical innovation, China's high-speed rail has made a series of significant breakthroughs in many technical fields, such as, high-speed trains, communication signals processing, traction power supply, operation management, security monitoring, and system integration. As a result, a high-speed railway technology system with Chinese characteristics and the overall level of technology reaching the world's advanced ranks is formed.

1.3 The Active Role of Mobile Communications for Railway

Along with the continuous development of modern railway transportation, higher and higher requirements for the wireless communication system are raised. Mobile communication system is crucial for the high-speed railway. Currently, application services such as train dispatching command, CTCS-3 level train operation control information, train dispatching order, radio train number check information, as well as dynamic monitoring information of signaling equipment are supported by the Global System of Mobile communication for Railways (GSM-R).

Due to the development of fourth-generation mobile communication technologies, high-speed railway broadband mobile communication system (LTE-R) should not be viewed merely as a GSM-R substitute, but also it can supply high-speed information transmitting channels for automatic train operation, train security video surveillance, train state monitoring and remote fault diagnostics, infrastructure wireless monitoring, emergency business disposal and passenger information service, and so on. LTE-R turns to be the information platform for the Internet of Things for railways, and also the basis of security assurance for high-speed railway operation.

(1) Dispatching Command and Safety Production

As the renewal and replacement of train dispatching radio communication system, railway mobile communication system is designed to support kinds of mobile voice communications, for instance, section business movement, emergency rescue, shunting marshaling operations, station yard wireless communication, and so forth.

Meanwhile, the requirements for mobile and fixed wireless data transmission, for example, radio train number transmission, train rear end air pressure, locomotive state information, train axle temperature detection, line bridge and tunnel monitoring, railway power supply status monitoring, crossing protection, and the like need to be addressed in LTE-R.

Safety information distribution and pre-alarm system takes mobile train as the main part, which ensures the construction wayside along rail tracks, the track maintenance, and the safety of both equipment and staffs in level crossing, train or station, thus reducing accidents.

(2) Train Operation Control Safety Protection

Railway mobile communication delivers the train-to-infrastructure security data transmission in CTCS-3 level train operation control system, providing a real-time transparent duplex transmission channel for the train control system, ensuring train secure operation at a high speed. Simultaneously, the railway mobile communication system is also capable of carrying safety data transmission of locomotive synchronization operation control, guaranteeing synchronization operation between heavy haul railway multi-locomotives, and improving operation efficiency.

(3) Railway Informatization

Passengers are viewed as the principal part of mobile information service system, which requires on-board ticketing services, mobile e-commerce and passenger mobile value-added services, and the like.

Railway network moving bodies such as locomotives, trains, and containers demand real-time dynamic tracking information transmission, to supply mobile transmission channels for real-time online information query and various management information systems. Obviously, railway informatization is an inevitable choice.

(4) Railway Mobile Internet

Railway Mobile Internet is considered as an integral part of "'the Internet plus 'railway'" strategy, whose development will contribute to accelerate the depth integration in the field of Internet and railway, to facilitate technological progress and efficiency promotion as well as organization reform of railway transportation, to promote the innovation and production on the railway sector, and to improve resource utilization efficiency and fine management level significantly.

In the complicated and rapidly changing railway environment, in order to achieve some advanced functions of large-scale high-speed railway network, such as train running status enquire, railway essential factors online level improvement, and train secure operation control, the next generation of railway mobile communication system with characteristics of large bandwidth, high real time, and high reliability comes to be an indispensable foundation.

1.4 GSM for Railway

Global System for Mobile Communications for Railway (GSM-R, GSM for Railway) is a communication system based on GSM technology, which strengthens the railway dispatching communication and is used in high-speed mobile environment. The China railway's overall goal of developing mobile communication network is to establish a comprehensive mobile communication platform for voice and data, and to build an integrated communication system with dispatching communication, train control, public mobile, and information transmission.

Railway digital mobile communication network construction is a systematic project, and closely related to railway dispatching communication, train control, and operation management. It should make full use of mobile communication technology, combine with the actual needs for railway transportation, and form a covering system-wide railway mobile communication network to provide a mobile integrated communication platform for railway transportation.

1.4.1 The Development of GSM-R

GSM-R is introduced from European railway special mobile communication system based on GSM technology, which adds the railway dispatching communication service and the high-speed railway mobility. GSM-R is an integrated economic and efficient railway wireless communication system. The development of GSM-R has experienced three stages, namely the standard formulation stage, the experimental verification stage, and the project implementation stage.

1.4.1.1 The Standard Formulation Stage

In 1992, the International Union of Railways (UIC) thought that GSM is gradually becoming the applicable standard of mobile communication, and found that the GSM technology can provide an ideal platform for new railway mobile digital communication system. Through the feasibility study, in 1993, the European railway decided to introduce the GSM technology as the foundation of the next-generation railway mobile communication system, which is GSM-R system. After that, UIC set the relevant standards and test, set up a standardized organization EIRENE, formulated a series of railway requirements specification, and designed indicators such as business functions, quality of service, and electromagnetic environment. At the same time, the continuous updating of GSM technology has laid a solid foundation for the GSM-R development.

1.4.1.2 The Experimental Verification Stage

To verify the reliability, mobility, and compatibility of GSM-R system, UIC set up another specialized organization MORANE, including railway operators, equipment manufacturers, and research institutions, which focused on the properties verification of GSM-R high-speed environment. From 1997 to 2000, the GSM-R system has been strictly tested and validated on high-speed railway in France, Italy, and Germany, respectively.

1.4.1.3 The Project Implementation Stage

Since 1999, some countries in Europe started operation test and commercial construction of GSM-R network.

Sweden is the first country to formally use the GSM-R network. In 1999, the first GSM-R network was built and put into use in Oresund Bridge from Sweden to Denmark.

From 2001 to 2004, Germany implemented the first-stage construction. From 2005 to 2007, the second stage was implemented. ETCS-2 system was tested on railway from Berlin to Leipzig. It completed debugging in 2005 and achieved commercial in 2006.

From 2002 to 2003, Italy took the test on the ETCS-2 system and the public GSM. From 2002 to 2005, the first stage was implemented.

From 2003 to 2008, France completed the basic construction.

Finland, Norway, Britain, Belgium, and Spain have successively carried out the nationwide GSM-R network construction.

1.4.2 GSM-R Key Technology and Engineering Measures

1.4.2.1 GSM-R Wireless Network

GSM-R is a mobile communication system for high-speed railway, which includes dispatching communication and train control information, and requires high quality of network service and maintenance. GSM-R wireless network optimization can draw lessons from mature GSM technology. However, GSM system is public communication services whose planning is limited in low speed and non-security. GSM-R system requires high-speed and reliable mobile communication business for railway communication.

The Main Differences Between GSM-R and GSM Wireless Networks

(1) Different frequency resources

GSM-R has only 4 MHz frequency bandwidth less than the public frequency resources, which face more limitation in the frequency planning. It should avoid the common frequency and adjacent frequency interference in the GSM-R network planning and optimization.

(2) Different covering ways

The public GSM mainly adopts planar network structure, while the GSM-R network is covered by the linear network structure.

GSM-R Wireless Network Reliability

GSM-R is related to safety and operational efficiency. Therefore, the reliability of GSM-R wireless network is higher than the public GSM network. In order to improve the reliability and maintainability of GSM-R wireless network, the corresponding measures have been taken in the railway above the speed of 300 km per h, plateau railway and heavy haul railway. The section with the speed above 300 km/h of Beijing–Shanghai high-speed railway adopt dual interleaving coverage as redundancy protection way. Qinghai–Tibet Railway and Daqin Railway employ co-site double base station.

1.4.2.2 GSM-R Core Network

GSM-R network is mainly used for special railway mobile communications business such as dispatching communication and the train control information. In the operation stage, the reliability and maintainability of GSM-R network have higher requirements than public GSM network. GSM-R core network should be provided 7×24 h of uninterrupted service, even if the planned downtime will also bring a lot of interference to the train operation and train operation organization. According to the maintenance requirements, GSM core network can interrupt service in less traffic period. In order to ensure the safe and reliable operation of the GSM-R core network, the following technical measures are usually adopted.

Equipment Protection Technology

Reliable and stable operation of the equipment is the basis of GSM-R core network business continuity. GSM-R core network devices include a circuit domain equipment (MSC, HLR, SCP, etc.) and packet domain equipment (SGSN, GGSN, etc.). It should prove functionality and performance of equipment panel, optimize implementation way, take into account the cost of project, and select the redundancy scheme (usually with 1:1, n: 1 or standby panel, etc.)

Network Protection Technology

(1) System-wide GSM-R core network public equipments in Beijing and Wuhan are host and backup for each other.

Beijing and Wuhan are respectively arranged system-wide railway GSM-R sharing equipment, such as GSM-R railway core switchboard, service control point (SCP), home location register (HLR), remote access authentication server (RADIUS), the domain name server (DNS), and GPRS home server (GROS). When the main device nodes have obstacles, standby nodes can quickly take control of all or part of the system-wide GSM-R sharing business.

(2) The railway administration has GSM-R switches. If there is breakdown, the impact will be controlled.
(3) GSM-R network employs multi-link protection. Each MSC interconnects at least with the two other MSCs. The voice relay between MSCs has at least two E1 circuits and utilizes different route transmission circuits. The packet domain also uses a reliable data link to duplex interconnect.
(4) The base station controller (BSC) arranged in the sub-line and the long line controls the influence of the BSC malfunction within coverage of the base station.
(5) GSM-R network adopts dual relay interconnection with Fixed users Access Switching (FAS).

Through the above measures, the risk of GSM-R network can be reduced. The business impact can also be controlled.

1.4.2.3 GSM-R Matching Equipment Technology

The GSM-R business is implemented by network and terminal matching equipment. The GSM-R system is introduced from European standard. Due to the different dispatching communication and related equipment standard configuration, the terminal supporting technical scheme becomes the main problem after the introduction of network equipment. At present, China's railway is mainly equipped with locomotive integrated wireless communication equipment (CIR), GSM-R handheld terminal, etc. CTCS-3 class train control equipment, DMS (train control equipment dynamic monitoring system) of signal specialty, is equipped with GSM-R SIM card. The terminal used by passenger transport train conductor is also equipped with a GSM-R SIM card. With the development of the business, the terminal will also be diversified and embedded.

1.4.2.4 GSM-R Project Implementation

GSM-R project implementation includes networking scheme, numbering plan, network parameter settings, joint debugging, project acceptance, etc. Clear and detailed design of the project various stages is the basis for the project smooth implementation. GSM-R system numbering scheme is derived from the European standard. The actual project numbering scheme is directly related to operation management and GSM-R business, which is the key to the GSM-R project implementation.

1.4.2.5 GSM-R Business and Implementation

GSM-R system is a railway mobile communication network, which is a platform of railway mobile communication service. The typical services are as follows:

(1) voice business: point-to-point voice call, voice group call, voice broadcast, multiparty communication, etc.

(2) data services: circuit and data service, packet domain data service.

railway-specific business: functional addressing, location-dependent addressing, railway emergency call, etc.

1.5 Next-Generation Mobile Communication System for Railway

Mobile communication system is one of the key infrastructures of high-speed railway carrying various services such as the railway dispatching command, trains run control, fault warning and danger notices, emergency rescue, etc. In order to further guarantee the safety and high efficiency of high-speed rail network as well as achieve convenient, comfortable and green transportation, new railway mobile communication services are constantly emerging, such as railway multimedia scheduling command communication, remote video monitoring, railway infrastructure monitoring, railway Internet of Things (LoT), station-yard wireless communication, mobile ticketing, tourist information services, etc. However, the service-carrying capacity of current railway mobile communication (GSM-R) is limited and cannot satisfy the demand of the new high-speed railway mobile communication services. At the same time, along with the rapid development of mobile communication technology and industry, the size of the GSM market shrinks, which forms a strong impact on the GSM-R industrial chain. Therefore, development of the next-generation mobile communication system for railway, realization of railway mobile communication system upgrading, and meeting the needs of new services development have become an irresistible trend.

Chinese railway mobile communication system has experienced from the first-generation simulation wireless railway dispatching system to the second-generation digital mobile communication system for railway (GSM-R) development. The GSM-R system is constructed in Qinghai–Tibet railway, heavy haul railway, high-speed railway, and passenger dedicated line. The railway communication equipment running status shows that simulation wireless railway dispatching system of 70,000 km common railway needs to be upgraded. According to the railway communications industry forecasts, GSM-R system life cycle will end up around 2020. Railway mobile communication system is faced with urgent industrial upgrading; moreover, evolution from narrowband to broadband has become the trend of the times. China Railway Corporation has made decisions of developing the third generation of broadband mobile communication systems (LTE-R), made the technology roadmap, and proposed the upgrade the transition scheme.

With continuous development of railway application, the railway application system gives more requirements of the next-generation communication system needs which are quite different from those in GSM-R network times, in terms of carrying service type and bandwidth requirements.

International Union of Railways (UIC) divides railway demands into two major categories of operating communication and support communication. According to the present status and development plan for Chinese railways, combined with the divided standards of UIC, we classify the next-generation mobile communication services of Chinese railway from multiple dimensions:

(1) Considering railway maintenance management system and the current professional business requirements in our country, based on UIC division of railway demands, it can be divided into driving-related service and passenger information service in accordance with the service attribute.
(2) According to the service types, it can be divided into voice service, data service, image service, and video service (Table 1.2).

Table 1.2 Next-generation mobile communication for railway service classification according to service attribute

Number	Service attribute	Service name
1	Driving-related service	Train control information
2		Locomotive synchronization control
3		Controllable train tail information
4		Dispatching command
5		Train wireless train number check
6		Train tail information
7		Dispatching communication
8		Operation and maintenance communication
9		Train safety warning
10		Train control equipment dynamic monitoring
11		Railway freight information system(transportation, equipment management)
12		Railway freight information system(railway freight car status information)
13		Chinese locomotive remote Monitoring and Diagnosis system(CMB)
14		Train coach running diagnosis system (TCDS)
15		Train working condition monitoring
16		High-speed railway power supply safety monitoring system (6c)
17		Maintenance and repair work card control system
18		Maintenance video monitoring system
19		Communications equipment monitoring system
20		Train security video monitoring
21		Infrastructure health management system
22		Disaster monitoring system
23		Rail gap video monitor system
24		Marshaling station wireless communication

(continued)

Table 1.2 (continued)

Number	Service attribute	Service name
1	Passenger information service	Passenger transportation management information system (voice, data, image, video)
2		Public security communication
3		Mobile ticketing
4		Passenger information system (PIS)

Next-generation mobile communication system LTE-R should have the following features: (1) support the mobility of up to 500 km/h, ensure reliable switching and service quality; (2) carry safety-related services such as train control and dispatching command, require network redundancy backup and overlapping field strength coverage; (3) limited frequency resource, need to use low-frequency carrier frequency aggregation; (4) due to carrying variety of services such as load remote video monitoring, railway infrastructure monitoring, uplink volume is greater than the downlink volume; (5) fast, reliable, traceability, and multi-priority voice group call, voice broadcast and emergency calls; (6) based on position addressing and functional addressing, call limitation based on the location, access matrix based on function, and other special services; (7) differentiated quality of service (QoS) requirements; (8) special networking methods: adopt the chain network along the railway, regional coverage, adopt a combination of chain and face shape in station-yard area; and (9) strict safety management system: LTE-R system should support the ease of use and traceability required by railway operation management system and safety responsibility cognizance.

In order to carry out the research in the next-generation mobile communication system for railway, China Railway Corporation established a specialized working group. Specific organizational structure is as follows:

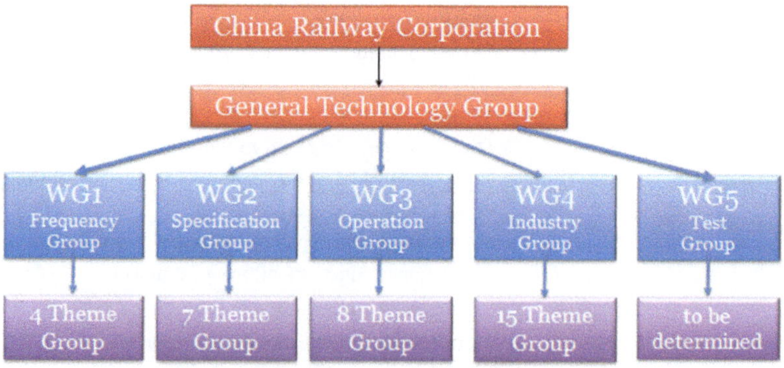

References

1. http://www.nra.gov.cn/fwyd/hyjc/gcjs2/952/201311/t20131126_3181.html
2. http://www.nra.gov.cn/fwyd/hyjc/gcjs2/952/201311/t20131126_3180.html
3. http://www.nra.gov.cn/fwyd/hyjc/gcjs2/952/201312/t20131214_3637.html
4. http://www.nra.gov.cn/fwyd/hyjc/gcjs2/952/201403/t20140307_5400.html
5. http://www.nra.gov.cn/fwyd/hyjc/gcjs2/952/201503/t20150306_12658.html
6. http://www.nra.gov.cn/fwyd/hyjc/gcjs2/952/201602/t20160203_20922.html
7. http://www.nra.gov.cn/fwyd/hyjc/gstl_/zggstL/201602/t20160215_21046.html
8. http://www.nra.gov.cn/fwyd/hyjc/gstl_/zggstL/gstllc/201602/t20160216_21067.html
9. Zhong Z, Li X, Jiang W (2007) Railway comprehensive digital mobile communication system. China Railway Publishing House, Aug-Sep
10. Hofestadt H (1995) GSM-R: global system for mobile radio communications for railways. In: International conference on electric railways in a United Europe. IET, 1995, pp 111–115
11. Sniady A, Soler J (2012) An overview of GSM-R technology and its shortcomings. In: 2012 12th international conference on ITS telecommunications (ITST). IEEE, 2012, pp 626–629
12. He R, Zhong Z, Ai B et al (2014) A standardized path loss model for the GSM-railway based high-speed railway communication systems. In: 2014 IEEE 79th vehicular technology conference (VTC Spring). IEEE, 2014, pp 1–5
13. Zhao L, Chen X, Ding J (2010) Interference clearance process of GSM-R network in China. In: 2010 2nd international conference on mechanical and electronics engineering (ICMEE). IEEE, 2010, pp V1-424–V1-428
14. Xun D, Xin C, Wenyi J (2010) The analysis of GSM-R redundant network and reliability models on high-speed railway. In: 2010 International conference on electronics and information engineering (ICEIE). IEEE, 2010, pp V2-154–V2-158
15. Jie S, Xiaojin Z, Tingting G (2010) Performance analysis of GSM-R network structure in China train control system. In: 2010 International conference on electronics and information engineering (ICEIE). IEEE, 2010, pp V2-214–V2-218
16. Briso C, Cortes C, Arques FJ et al (2002) Requirements of GSM technology for the control of high speed trains. In: The 13th IEEE international symposium on personal, indoor and mobile radio communications. IEEE, 2002, pp 792–793
17. Ning B, Tang T, Gao Z et al (2006) Intelligent railway systems in China. IEEE Intell Syst 21 (5):80–83
18. Matolak DW, Berbineau M, Michelson DG et al (2015) Future railway communications. IEEE Commun Mag 53(10):60–61
19. Bertout A, Bernard E (2012) Next generation of railways and metros wireless communication systems. ASPECT, Institution of Railways Signal Engineers
20. Ai B, Cheng X, Kürner T et al (2014) Challenges toward wireless communications for high-speed railway. IEEE Trans Intell Transp Syst 15(5):2143–2158
21. Ai B, Guan K, Rupp M et al (2015) Future railway services-oriented mobile communications network. IEEE Commun Mag 53(10):78–85
22. Tingting G, Bin S (2010) A high-speed railway mobile communication system based on LTE. In: 2010 International conference on electronics and information engineering (ICEIE), vol 1. IEEE, V1-414–V1-417
23. Dong H, Ning B, Cai B et al (2010) Automatic train control system development and simulation for high-speed railways. IEEE Circuits Syst Mag 10(2):6–18
24. Kim RY, Kwak JS, Hwang HC (2012) Technical challenges of railroad communications using long term evolution. In: 2012 International conference on ICT convergence (ICTC). IEEE, pp 563–564
25. Briso-Rodríguez C, Lopez CF, Fernandez JRO et al (2014) Broadband access in complex environments: LTE on railway. IEICE Trans Commun 97(8):1514–1527

Chapter 2
Key Issues for GSM-R and LTE-R

2.1 GSM-R Architecture

2.1.1 GSM-R Network Composition

GSM-R network consists of GSM-R digital mobile communication system (GSM-R system) and trunk transmission circuit.

GSM-R system contains four parts, which are network subsystem(NSS), base station subsystem(BSS), operation and support subsystem(OSS), and terminal device. Network subsystem includes mobile switching subsystem (SSS), mobile intelligent network (IN) subsystem, and general packet radio service subsystem. Fig. 2.1 shows the system structure of GSM-R and main interfaces.

2.1.2 Mobile Switching Subsystem

SSS mainly has several functions as follows: user service switching function, and user data and mobility management, and security management database functions as needed. SSS consists of a series of function entities, including MSC, HLR, and VLR. Mutual communication between the various functional entities by No.7 signaling protocol, each functional entity, is as follows:

a. Mobile Service Switching Center(MSC)
   MSC, as the core of the network, is in charge of mobility management and call control. Gateway MSC (GMSC) is a gateway office between GSM-R network and other communication networks.
b. Visitor Location Register(VLR)
   VLR, as a dynamic database, is responsible for storing the information of registered users, which have come into the control area, and to provide the necessary data call connection for mobile users. When the MS roams to a new

Z.-D. Zhong et al., *Dedicated Mobile Communications for High-speed Railway*,
Advances in High-speed Rail Technology, DOI 10.1007/978-3-662-54860-8_2

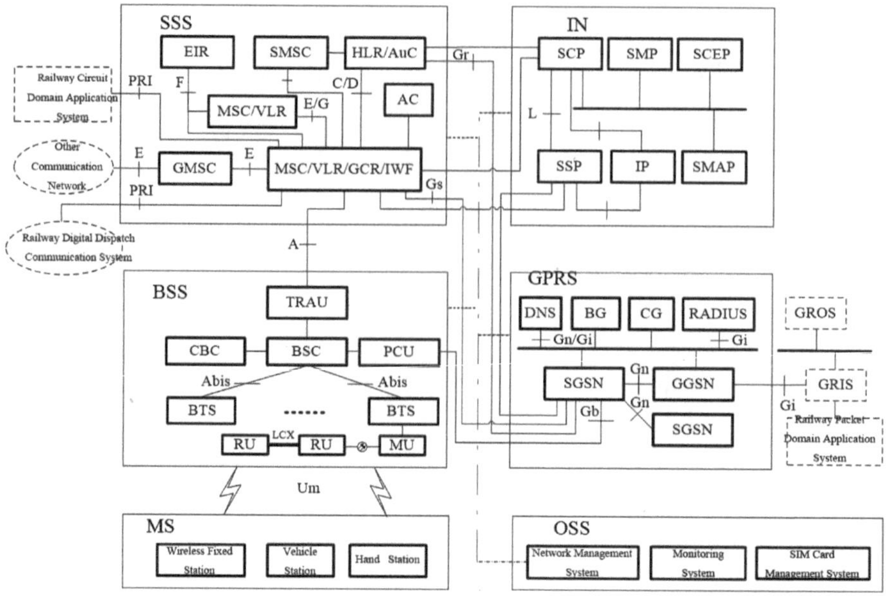

Fig. 2.1 GSM-R system structure

VLR area, the HLR initiates location registration to the VLR, and obtains the necessary user data; when the MS roams out of control, it needs to delete the user data. The VLR stores the ID list, which belongs to the user groups. When users roam, these information can be obtained by the home location register (HLR).

c. Home Location Register (HLR)
HLR is the mutual device for the CS domain and the PS domain, and it is also a database for mobile user management. HLR stores all the mobile user data in this area, such as the identification sign, the location information, the signing service, etc. When a user is roaming, HLR receives a new location information, and requires the former VLR to delete all user data. HLR provides routing information when the user is called.

d. Authentication Center (AuC)
AuC is the mutual device for the CS domain and the PS domain, which stores user authentication algorithm and encryption key entities. By HLR, AuC sends authentication and encryption data to the VLR, MSC, and SGSN, to ensure the legality and safety of communication. Each AuC and the corresponding HLR are matched, only passing the HLR and other network entities to communicate.

e. Interworking Functional Unit (IWF)
IWF is in charge of offering transformation of rates and protocols between GSM-R network and fixed-network data terminals. Its function depends on the interconnect services and the network structure.

f. Group Call Register (GCR)
GCR is used for storing the group ID of mobile users, and the mobile station makes use of voice group call service (VGCS), as well as voice broadcast service (VBS) calls the cell message. Besides, it should check whether the MSC, which starts to call, is charge of dealing with.
g. Short Message Service Center (SMSC)
SMSC is in charge of sending short message to MSC.
h. Acknowledge Center (AC)
AC is used for recording and storing relative information, which is about railway emergency call.
i. Equipment Identity Register (EIR)
EIR contains one or more databases, which can store IMEIs. These IMEIs can be classified as white list, black list, and gray list. According to the IMEI of the users, the network decides whether it will offer services for users.

2.1.3 Mobile Intelligent Network Subsystem

IN subsystem is the intelligent network functional entity, which is introduced into SSS. It separates the network switching function and the service control function, and realizes the intelligent control of the call.

GSM-R intelligent network consists of GSM service switching point (gsmSSP), GPRS service switching point (gprsSSP), intelligent peripheral (IP), service control point (SCP), service management point (SMP), service management access point (SMAP), and service context entering point (SCEP).

a. GSM Service Switching Point (gsmSSP)
As the interface between the MSc and SCP, gsmSSP has the function of service switching. gsmSSP can detect GSM-R intelligent services request, and it communicates with the SCP, requests the SCP response, also allows the service logic in the SCP affects call processing.
b. GPRS Service Switching Point (gprsSSP)
gprsSSP possesses the function of service switching. As the interface between the SGSN and SCP, it can detect GPRS intelligent service request. It communicates with the SCP, requests the SCP response, also allows the service logic in the SCP affects call processing.
c. Service Control Point (SCP)
SCP has the service control function, which contains the service logic of GSM-R intelligent network. Through the instructions issued by the SSP, it can complete the control of connecting and charging for intelligent network services, in order to achieve a certain part of the railway services. Meanwhile, it also has the function of service data, user data, and network data included, to provide the service control function in the implementation of GSM-R intelligent network service real-time extraction.

d. Intelligent Peripheral (IP)
   Under the control of SCP, IP offers a variety of specialized resources according to the corresponding service logical program. And these resources contain the receiver of DTMF, signal generator, record notice, etc.
e. Service Management Point (SMP)
   SMP can deploy and offer GSM-R intelligent network services. It has the management of the SCP service logic. Besides, the addition, deletion, and modification of the user service data are included. It can also manage and modify some related information of SSP.
f. Service Management Access Point (SMAP)
   SMAP has the function of service-managed access, and it can access to SMP for service manager. Besides, the SMAP can modify, add, and delete users' data and service performance by SMP.
g. Service Context Entering Point (SCEP).
   SCEP is used for developing and producing GSM network intelligent service first, after that, it tests and verifies these services. Third, service logic, management logic, and service data, which are verified and are of intelligent service, should be input into SMP.

2.1.4 General Packet Radio Service (GPRS) Subsystem

GPRS subsystem is responsible for providing packet-based traffic service for wireless users. It includes core layer and wireless access layer.

The core layer consists of several functional entities, such as SGSN、GGSN、DNS、RADIUS, etc.

The wireless access layer consists of PCU, base station, and terminal.

GPRS wireless access layer network should make full use of the equipment resources of GSM-R system, to protect investment; and it shares frequency resources of GSM-R system. Also, it uses the base station of GSM-R system to achieve wireless coverage, rather than increasing the GPRS system base station.

a. Service GPRS Support Node (SGSN)
   SGSN is the GPRS support node of MS service, and it can achieve mobility management and route searching.
b. Gateway GPRS Support Node (GGSN)
   GGSN is the gateway between GPRS network and extra data network. It achieves router selection, transformation into extra network protocol.
c. Domain Name Server (DNS)
   DNS is responsible for providing the domain name resolution functions for GPRS network internal SGSN, GGSN, and other network nodes.
d. RADIUS Service (RADIUS)
   RADIUS is responsible for the storage of user identity information, and completes the user's identification and authentication.

e. Packet Control Unit (PCU)
   PCU is charge of the data packet, wireless channel management, error detection, and automatic retransmission.
f. Border Gateway (BG)
   BG is used for interconnecting with different GPRS networks. It has basic security function; besides, it can add some related functions according to roaming protocol of different networks.
g. Charge Gateway (CG)
   CG can collect the ticket record of every GPRS support node, and also, it can store, back up, and merge the ticket, and then, transfers these ticket records to charge center.
h. GPRS Network Interface Equipment
   GPRS home server (GROS) is responsible for the implementation of GPRS GSM-R terminal (GPRS terminal for short) to check the current IP GRIS/M-GRIS address.
   GPRS interface server (GRIS) is responsible for the implementation of data forwarding, protocol conversion between GPRS terminal, and railway application system.

2.1.5 Base Station Subsystem

Through wireless interface, the BSS directly connects with the mobile station, which is responsible for the wireless signal receiving and transmitting and radio resource management. Besides, it connects with the MSC, to realize the communication between mobile users or mobile users along with fixed network, and to transmit system signal and user information.

BSS consists of the following part:the base station controller (BSC), the transcoder/rate adaptor unit (TRAU), the cell broadcast center (CBC), the base transceive station (BTS), the weak field devices, and other functional entities.

a. Base Station Controller (BSC)

BSC is the control part of BSS, which is responsible for the management of all kinds of interfaces, the management of the radio resources and wireless parameters, the signal processing of the call establishment, and the channel assignment in the cell.

b. Transcoder/Rate Adaptor Unit (TRAU)

TRAU is responsible for providing voice coding and rate adaptation functions between BSC and MSC, and it converts the 16kbit/s voice or data into 64kbit/s data.

c. Cell Broadcast Center (CBC)

CBC is responsible for managing the cell broadcasting message service.

d. Base Transceive Station (BTS)
BTS is the wireless receiving/transmitting device, which is controlled by the BSC and served for some cell, to complete the transformation between BSC and wireless channel. Also, it can realize the wireless transmission along with the related control functions between BSC and MS by the air interface. BTS has the rate matching, channel coding/decoding, modulation/demodulation, and other air interface physical layer functions.

e. Relay Transmission Equipment
Relay transmission device is used for the wireless coverage of GSM-R weak electric field area, including repeater station, trunk amplifier, leaky coaxial cable, antenna, and so on.

2.1.6 Operation and Support Subsystem (OSS)

OSS includes network equipment maintenance management system ("network management system" for short) and user management system.

a. Network Management System
   It provides the interface of the system devices for the operator, and collects and monitors the operation information and status of the whole network. Besides, it can produce system operation report according to the requirement of the operators, to provide the basis for network planning and adjustment. The basis contains operation and maintenance center (OMC) and network management center (NMC), and so on.
   According to the subject, the system can be divided into the exchange, grouping, intelligent network, wireless (including the base station and the weak field equipment), and other network management system.
b. Monitoring System
   - Interface monitoring system, including different interfaces: the Abis, A, PRI, Gb, Gn, and Gi, along with C, D, E, G, L, and No.7 signaling monitoring interface subsystem, comprehensive analysis, gateway, and network subsystems. It consists of collecting equipment, data processing equipment, a comprehensive analysis of the server, and the client; the main functions include acquisition, storage signaling and user data; parsing the raw signal and service data; real-time monitoring and display, information query, reporting capabilities and comprehensive analysis; and so on.
   - Other monitoring system can monitor the leaky cable, antenna, and tower, including on-site monitoring equipment and monitoring center equipment.

c. SIM Card Management System
   It is responsible for managing the related data of the network user, offering many operating functions shown as follows: opening account, deleting account,

and changing the authority of user service, to support the usual operation of service.

SIM card management system includes an application server, database server, data storage device, SIM card management terminal, SIM card maintenance terminal, gateway interface, SIM card-making terminal, SIM card reader, and so on.

2.1.7 Terminal

The terminal is a device which is used for direct operation and use of the GSM-R system, and is used for accessing the GSM-R network device, including a mobile station and a wireless fixed station.

a. Mobile Station

   Mobile station includes (locomotive, automobile) car station, hand station, train control data transmission equipment, train rail information transmission device, disaster prevention detection information transmission, and vehicle safety detection information transmission terminal.

   The terminal consists of mobile device and SIM card.

b. Wireless Fixed Station

   Wireless fixed station is the wireless terminal for nonmobile state, and it has the same service function with mobile station.

2.2 GSM-R Network Hierarchical Structure

2.2.1 Mobile Switching Network

GSM-R network structure should meet the needs of railway traffic control. In order to make the network structure simple, clear, and easy to operate, maintain and manage, the whole network is divided into mobile service sink network and mobile service local network.

a. Mobile Service Sink Network

The mobile service sink network is composed of a tandem mobile switching center (TMSC) and a trunk line that connects with the nodes.

The railway should be divided into three areas, setting TMSC in the center of the biggest area. Each TMSC is responsible for a number of mobile switching centers (MSC). TMSC is responsible for the transfer of long-distance traffic between different MSCs in this area. It also collects the long-distance traffic of the MSC in the area to other TMSC.

The network structure among TMSCs is mesh topology.

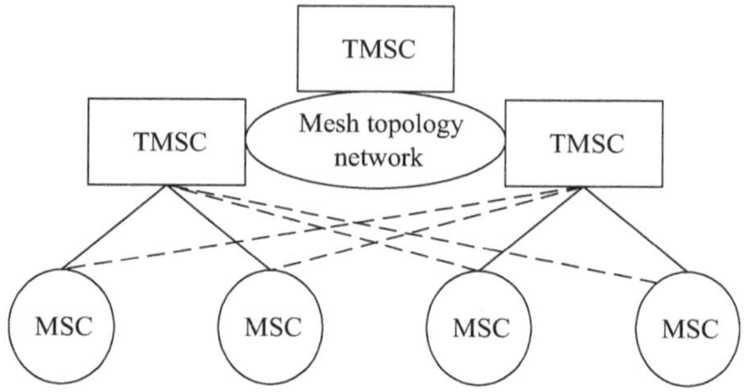

Fig. 2.2 The GSM-R network structure of the whole railway

b. Mobile Service Local Network
 The whole railway establishes several mobile service local networks.

Mobile service local network contains MSC, GMSC, HLR, and other equipments. A few mobile services share a HLR local network.

In a local network, it should be set one or several MSCs, or using one MSC to cover several mobile service local networks according to the amount of services. MSC is responsible for dredging or handling the traffic between different mobile users (or between mobile users and fixed users).

MSC should be connected to the adjacent TMSCs. According to the requirement, different MSCs can be directly connected.

The GSM-R network structure is shown in Fig. 2.2.

2.2.2 Intelligent Network

GSM-R intelligent network, based on the ITU-T/3GPP intelligent network, uses the CAMEL3 protocol standard in order to achieve some of the railway-specific services.

GSM-R intelligent network is composed of SSP, SCP, IP (intelligent peripherals), SMP, SMAP, and SCEP, as well as the link connecting these nodes. The network structure is shown in Fig. 2.3.

2.2.3 General Packet Radio Service Network

GPRS network should be divided into two levels: GPRS backbone network and GPRS local network.

a. GPRS backbone network
 The GPRS backbone network is composed of backbone routers and data links which connect with the nodes, and it is responsible for forwarding the data

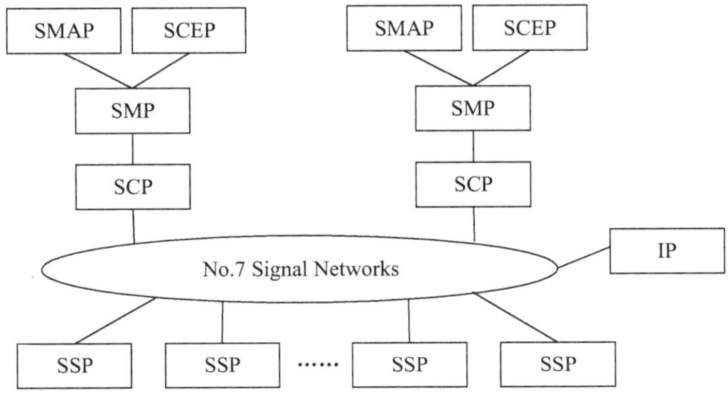

Fig. 2.3 The GSM-R intelligent network structure

services between different local networks. GPRS backbone network should be divided into three major areas, and establish the backbone routers in the large areas. Network structure between backbone routers is mesh topology.

b. GPRS local network

- The railway sets several GPRS local networks. Local network consists of SGSN, GGSN and DNS, RADIUS server and other devices, as well as the local area network connected to these devices.
- GPRS local network should be set up by the same access and the edge router accesses to the corresponding backbone router. In order to ensure the reliability of the network, the edge router should be set up in pairs, and access to different backbone routers, respectively.
- According to the requirements, different local networks can be directly linked.

The GPRS network structure is shown in Fig. 2.4.

2.3 LTE-R Architecture

The Evolved Packet Core (EPC), the subsystem of core network, is composed of Mobile Management Entity (MME), Home Subscriber Server (HSS), Serving Gateway (S-GW), Packet Gateway (P-GW), Multimedia Broadcast Multicast Service Gateway (MBMS-GW), Broadcast Multicast Service Center (BM-SC), Multicell/multicast Coordination Entity (MCE), routers and other equipment. The key technology of trunking communication system was proposed in 3GPP LTE R13 according to the special demand of railway operation. In addition, the Mission-Critical Push To Talk (MCPTT) server network element was identified based on 3GPP R13 standard. The LTE-R system block diagram is shown in the figure below.

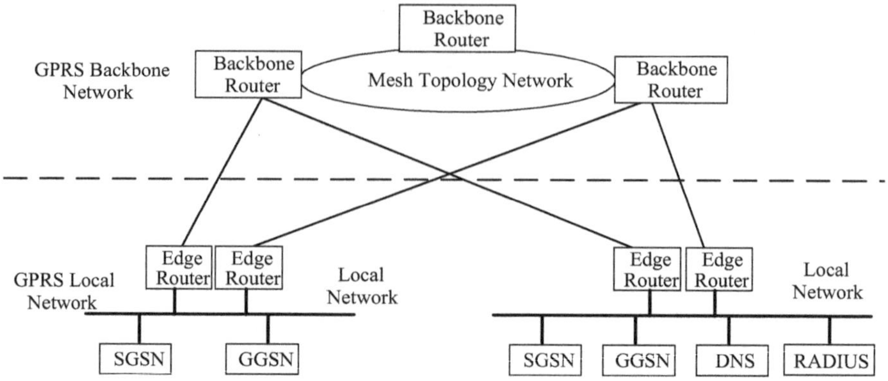

Fig. 2.4 The GSM-R network structure

MME: MME provides the necessary support for mobility management. Further functions include paging, security control, bearer control of core network, mobility control as idle mode UE. The MME is a key component of EPC,which is primarily responsible for control plane function related to user mobility and session management,and the main functions are as follows:

(1) Network Access Control (Fig. 2.5)

MME supports authentication and authorization for the UE. The authentication function manages whether to permit access request according to the usage of system resources. Safety management includes the following:

Authentication:MME realizes the mutual authentication and key agreement between network and UE through authentication function, in order to ensure the request of UE authorized in current network. Generally, the mobility management comes along with this function. International Mobile Subscriber Identification Number (IMSI), Globally Unique Temporary UE Identity (GUTI), and other identities are checked here。

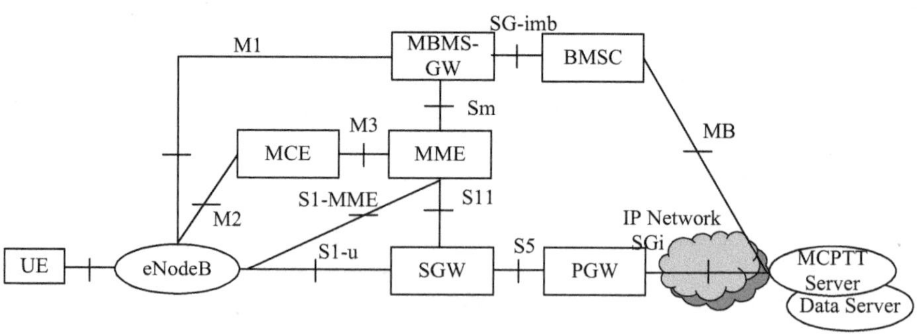

Fig. 2.5 LTE-R system block diagram

GUTI allocation:As a temporary user ID, GUTI protects the security of IMSI on the air interface. MME should assign the GUTI value after the first attachment to UE.

UE identification:User identification function is used to identify the effectivity. Equipment identification function is used to check the legality of the device.

Send function under the security context of AS:MME will contain AS security context in Initial Context Setup Request message for eNB. AS security context includes AS algorithm list and KSI, eNB will make choice of AS algorithm referring to the ability of the UE for realizing confidentiality and integrity protection of RRC signaling.

Confidentiality and integrity protection of NAS signaling:MME will add the list of NAS security algorithm, KSI and UE/MME selection security algorithm to UE NAS security mode command, and use the generated key to realize the NAS signaling confidentiality and integrity protection.

(2) Mobility management

Periodically update registered UE timer whose value is issued by the MME. Once the UE periodic timer is timeout, UE initiates periodic tracking area update. If the UE is not under UE E-UTRAN coverage, periodic tracking area update would perform when it gets back to the coverage area.

Attachment, detachment, tracking area list management, tracking area list update, handover, paging, and other general mobility processes.

After the MME cancels UE, the MME can notify the HSS by clearing UE message, i.e., subscription data and mobility management context of the UE.

Service request: UE-initiated and network-initiated service request. UE establish a security connection between the networks through service requests. Network initiates service request for the network downlink data transmit and UE signaling interaction scenarios.

Mobility restriction: According to area restriction information and access restriction information in user subscription, the mobility restriction is made for user.

Multiple PDN connectivity: MME supports multiple PDN connections for the same UE. If UE simultaneously initiated more than one PDN connection with the same APN, multiple PDN connections are to be connected to the same P-GW.

UE reachability: MME received the reachability request from HSS, and then stored UE reachability request. When the UE sent reachability request has been arrived, MME sends the reachability notification to HSS.

(3) Session management

It includes EPS bearer establishment, modification, and release;The access network side bearers establish and release;As interacting with the 2G/3G network, valid mapping between EPS bearer and the PDP context is performed.

(4) Network element Selection

P-GW selection: MME provides the channel using subscription information of users for assigning a P-GW for the 3GPP PDN connection.

S-GW selection: MME supports network topology to select an available S-GW for the UE.

MME selection: In the handover progress, based on the network topology, an available MME is selected for service UE.

HSS:HSS is a database for storing user subscription information; key functions include storing user subscription information, user authentication, and location information.

(1) User data storage and management

Storing its home user data, including user information, mainly IMSI, MSISDN, and IMEI/IMEISV, purged state identification, UE-AMBR, etc.; the ODB state identification, call blocking, roaming restriction; EPS APN contract information, the QOS contract data, PDN Type, etc.; location-related information, including MME identity, P-GW address, etc.; user charging related information; authentication information, including K; as well as user authentication algorithm ID.

For the non-3GPP access network, the HSS shall be able to store non-3GPP user data, the user data at least including a user information, mainly permanent user identity NAI, APN subscription information; roaming-related information, including the 3GPP AAA identification, P-GW identification, and so on; user subscription billing parameters; authentication information, including K, as well as user authentication algorithm identification; MIP contract information.

Corresponding operation management has been implemented based on the user subscription data, including accounts, sales households; user subscription data changes, including new and existing contracting business data; batch processing user data.

(2) User authentication

HSS provides a set or more sets of parameters to the MME according to the MME authentication request, and supports authentication-related processing. HSS and HLR equipment shared authentication center for the HSS supporting HLR function.

(3) Mobility Management

Store the MME addresses for current customer services, and storing the MME network capabilities related parameters.

With the location registration notification initiated by MME, HSS completes the user location registration and updates the current service MME address.

When the following conditions occur, HSS shall initiate the request of cancelation of the original MME and carry the associated write-off type: the user first attached network; the user moves to a new MME; the network enforces to change registration status or MME address of the user; the user is deleted; and so on.

As receipt of a request sent by the MME to clear the IE, HSS should be set to give the UE "UE clear" mark.

(4) Request notify processing from MME: According to request-specific information, HSS performs the appropriate action, such as update terminal information; set the current area to restrict access; update PDN GW address.

S-GW is responsible for connecting the eNBs and roam/switching between eNBs. S-GW is the gateway-oriented eNB end interface, the main functions are as follows:

(1) Session Management

EPS session management support functions include EPS bearer establishment, modification, and release. Storage and processing is under ECM-idle and ECM-connected state for terminal EPS bearer context.

(2) Mobility Management

S-GW helps accomplish the following mobility management program:
Based on the switch between interface X2 and S1;
Tracking area update;
The service request trigged by network side;
S1 connection release.

(3) Routing and data forwarding

S-GW has routing function as obtained the data form one node and forwarded to the next node. After the switch between eNBs or systems, S-GW users should send "end marker" packet to the source eNB, source SGSN, or source RNC, in order to help eNB rearrangement.

(4) QoS control

Support for the main Qos parameters bearded by EPS, including QCI, ARP, GBR, and MBR;

Support for terminals and network-initiated update bearer modification process based Qos;

Support for the bearer establishment/update access control: when resources are insufficient, the access is stetted with high ARP, and on the contrary, access is denied for low ARP;

Support for GBR and MBR bandwidth management for GBR bearer level.

Support for DSCP marking for bearer level of uplink and downlink data.

(5) Billing

EPS supports both offline and online charging functions. S-GW and P-GW support offline charging function, S-GW with the P-GW participated complete the online charging function. After the S-GW collecting billing information, generate CDR, it goes from interface Ga to the interface CG. In CG, the bill is post-merger processing, and then, passing through interface Bx to the billing system. It supports multiple charging modes: traffic, long, long time flow combinations, and so on.

(6) ISR (Optional)

Recognition: achieve appropriate treatment for EPS at ISR active and inactive state.

When active, S-GW updates only the new MME control plane address as update bearer process, as well as preserving the old SGSN information unchanged.

Under ISR activation status, S-GW changes TAU/RAU and receives MME/S4 SGSN delete session request.

P-GW: The gateway P-GW devices are responsible for connecting external packet-switched network and managing the connections between the external packets switched network and the user equipment (UE) devices. P-GW is a PDN-oriented SGi-terminated gateway which provides a stable IP access point for users as an anchor point for all the access techniques. The main functions of P-GW are shown below:

IP address assignment

For each PDN connection, UE must obtain at least one IP address (IPv4 or IPv6 prefix).

Session management

P-GW stores and processes the EPS of the terminals which are under the state of ECM idle and ECM connected to bear the context and addresses the corresponding external data network by APN. P-GW stores the mapping relationship borne by the downlink data SDF and S5/S8. For a PDN connection, P-GW supports the default bearer and the dedicated bearer.

Routing and data forwarding

P-GW has the function of utilizing GTP packet header and UDP/IP packet header to pack the PDU from the external data network, and regards corresponding address of the packet header as the identifier to utilize a point-to-point bidirectional channel to transmit the packed data to the terminal in the EPS network. For the GTP-U PDU in the external data network, P-GW removes its packed header and then forward to the external data network.

Related functions of external network access

P-GW can access the external IP network through transparent and non-transparent mode. In the non-transparent mode, P-GW should support the function of accessing RADIUS server and realizing the user identification. To guarantee the billing requirements of some data businesses, P-GW should generate the uniform RADIUS message according to the APN assignment on the P-GW side, and use the uniform RADIUS message to communicate with the external RADIUS server.

QoS control

① Support the main QoS parameters borne by EPS, such as QCI, ARP, GBR, MBR, and APN-AMBER;
② The initial bear level QoS parameters of the default bearer are assigned by the network according to the signature data (In the condition of

E-UTRAN, MME sets these initial parameters according to the signature data obtained from HSS). P-GW can change these parameters after the interaction with PCRF or based on local settings.

③ Support local settings PCC (strategy control and billing) rules.

④ Support the creation or modification of the dedicated bearer from UE and network sides, decide whether make the creation or modification or not, and assign QoS parameters for the bearer.

⑤ Support the realization of the bear level GBR, MBR bandwidth management function for the GBR bearer.

⑥ Support the APN-AMBR bandwidth management function of the uplink and downlink data streams for the non-GBR bearer.

DPI function

DPI can conduct the deep detection for the data message context in the application streams, and report the type and number of the data streams, cooperate with PCRF to accomplish the stream-based strategy control, and cooperate with the context billing to accomplish the stream-based billing function.

① Support the user-based packet filtering

② Support the user-level-based stream management control

③ Support the business-level-based stream management control

Billing function

Support online/offline billing and context billing functions. For offline billing system, when P-GW collects the billing information, it generates CDR to transmit to CG through Ga interface, and after the merged processing of call tickets, it transmits to the billing system through Bx interface. For online billing system, P-GW communicates with OCS through Gy reference point. CTF in P-GW generates the billing events, provides the billing information, assembles the billing information into billing events, and sends these billing events to OCF in OCS. It supports multiple billing modes: stream flow, time duration, stream flow, and time duration combinations and events.

PCRF selection

In the scenarios of attribution place and roaming place, it is possible to exist the condition where multiple PCRFs serve a single P-GW. P-GW conducts the selection for PCRF according to the defined process in TS 23.203, and in the meanwhile makes the PCC session in different terminals connect the correct PCRFs.

MBMS-GW

Gateway MBMS-GW devices are responsible for transmitting the data to eNB in the way of multicast.

BM-SC

BM-SC devices are responsible for the launch and authorization of the eMBMS business.

MCE

MCE devices are responsible for the session management and wireless assignment for the eMBMS business to accomplish the air interface dispatching.

Core Network Architecture

Based on the business requirements, LTE core network sets the network elements such as MME, HSS, S-GW, P-GW, MBMS-GW, BM-SC, MCE, router, and the 3GPP R13-based MCPTT server.

From the technical point of view, each core network element can apply either centralized or distributed type of architecture. But from the aspect of operation and the plan of the whole network and duration, the specific establishment type of network elements should be considered various factors such as reasonability, economy, characteristic of the network element, operation management system, and maintenance management. In the meantime, in considerations of the importance of the network element and the wide range of the failure's influence, redundancy and disaster robustness should be also considered in network element setting.

The networking mode of each element is analyzed as shown in Table 2.1.

Consider the interconnection between the core networks from two different manufacturers as the example; the network architecture and the interface explanations are shown in Fig. 2.6 and Table 2.2.

Interconnection network architecture of LTE-R and dispatching communication system

For the conventional dispatching voice communication business, it is required to interconnect the LTE-R core network to the conventional FAS dispatching communication system to realize the voice communication between the train driver and the FAS terminal of the dispatcher and the station attendant.

The feasible solution is regarding the FAS system as a multiple user terminal to customize the MCPTT Server connection between the gateway and the LTE-R system. The connection applies IP interface and the protocol is suggested to use SIP. The network architecture is shown in Fig. 2.7.

For the voice business such as trunking communication, the network applies the eMBMS bearer and adds the network elements such as 3GPP R13 defined MCPTT server.

For other data businesses, the network utilizes the application system server to interconnect the SGi interface with P-GW. The network architecture is shown in Fig. 2.8.

As the direction of railway mobile communication development, LTE-R will coexist with GSM-R in a long period of time. The LTE-R system and the GSM-R system will evolve from coexistence to interconnection and eventually realize the business taking over. In the evolution from GSM-R to LTE-R, the interconnection and the transition smoothness of both the networks should be fully considered.

The detailed interconnection strategy with GSM-R is shown below: In the GSM-R network, MSC is responsible for the calling of trunking voice business and the media processing. In the LTE-R network, MCPTT server is responsible for the calling of trunking voice business and the media processing. Thus for the interconnection between the GSM-R trunking voice business and the LTE-R trunking voice business, the only requirement for voice calling realization is the interconnection between MSC and MCPTT server.

Table 2.1 Networking mode analysis of each network element

Index	Network element name	Networking mode analysis	Networking mode	Redundancy setting
1	HSS	This type of network elements is for the management and strategy control of user data, business, and dispatching. It can apply the centralized and distributed networking. Centralized networking has a better economy, and can follow the existed GSM-R network to share the developed devices and management systems. In addition, if some conditions are satisfied, such network elements as HSS can be established together with HLR in the GSM-R network in order to promote the device utilization and protect the GSM-R network investment. Therefore, based on the above analysis, centralized networking has a significant advantage	Centralized setting, LTE network shared devices HSS, PCRF are set over the whole railway	Apply the remote disaster robustness system pairwise setting whose shared devices are synchronized by dedicated data link. The transmission channel is provided by two transmission systems with different physical routings
2	PCRF			
3	MME	The elements in this category are routers and gateways which can be applied by centralized and distributed networking. Distributed networking is better to distribute the network traffic to each core network node to avoid the network congestion caused by the heavy traffic of one node. Therefore, distributed networking has an obvious advantage	Distributed networking, each network element is set on each core network node	Each core network node and each network element is equipped with two devices, if condition permits, those devices can be remotely set. Those two devices apply the POOL mode to achieve the load balancing
4	S-GW			
5	P-GW			
6	MBMS-GW			
7	BM-SC			
8	MCE			

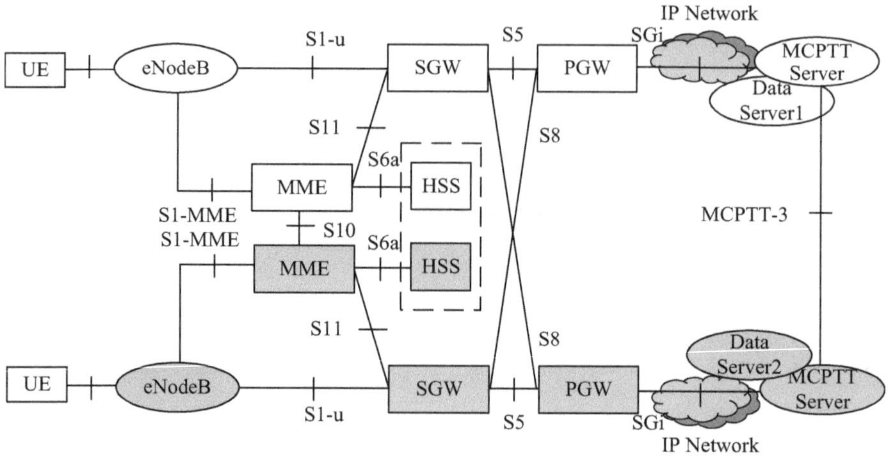

Fig. 2.6 Core network interconnection network architecture

Table 2.2 Core network interconnection interfaces

Interface	Description
S1-u	The user plane interface is between eNodeB and S-GW, which provides the user plane transmission of eNodeB and S-GW UDP/IP and GTP-U based protocol
S5	The interface which is responsible for the user plane data transmission and channel management between the Serving GW and PDN GW. It is used to process the Serving GW relocation and the PDN network required connection with the non-collocated PDN GW in order to support the UE mobility GTP-based or PMIPv6-based protocol
S1-MME	The control plane interface is between eNodeB and MME, which provides the S1-AP message transmission IP- and SCTP-based protocol
S10	The interface is between the MMEs, which is used for processing MME relocation and the transmission between MMEs
S8	The interface which is responsible for the user plane data transmission and channel management between the Serving GW and PDN GW. The difference with the interface S5 is that S5 is used for local Serving GW and local PDN GW, and S8 is used for Serving GW in the roaming place and PDN GW in the attribution place
S6a	The interface is between MME and HSS, which is used for transmission signature and data authentication
S11	The interface is between MME and Serving GW, which is mainly used for the transmission of the request message for creating, updating, and deleting the bearer
SGi	The interface is the connection interface between PDN GW and the external PDN
MCPTT-3	The interface is between MCPTTs from two different manufactures, which is proposed after R13 version

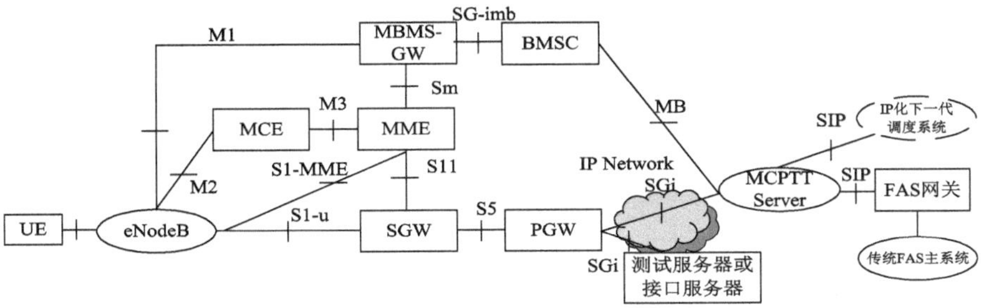

Fig. 2.7 Interconnection network architecture of LTE-R and dispatching communication system

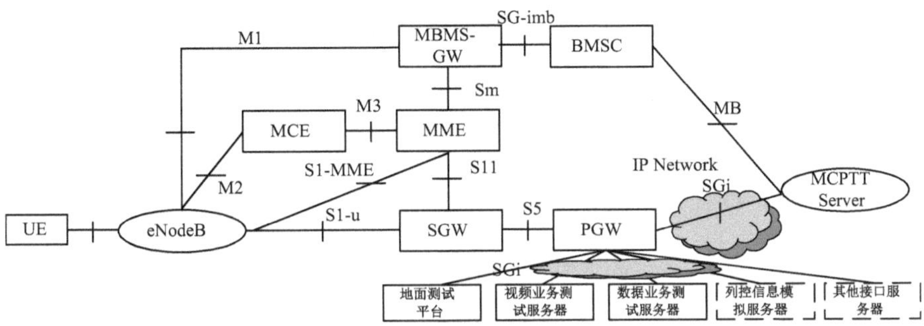

Fig. 2.8 Interconnection network architecture of LTE-R and GSM-R system

For the data business interconnection, the GSM-R has CSD and GPRS these two data communication types: CSD is from MSC to RBC through wired transmission network; and GPRS is from SGSN/GGSN to the interface server through the wired transmission network. The only requirement for realizing the grouped data interconnection with the GSM-R GPRS system is interconnecting P-GW with GGSN. The detailed interconnection network architecture is shown in Fig. 2.9.

2.4 Key Technologies for GSM-R

Railway has a close relationship with national economy and people's livelihood, reliability, availability, maintainability, security (RAMS) are always the key points of railway informatization construction, which are directly related to the property safety of the people. The wave propagation, wireless interference, wireless networking, encryption, and evaluation system of RAMS are the key factors which impact the transportation security and efficiency.

However, domestic research status indicates that the existing research results about the high-speed railway wave propagation channel cannot clarify the influence

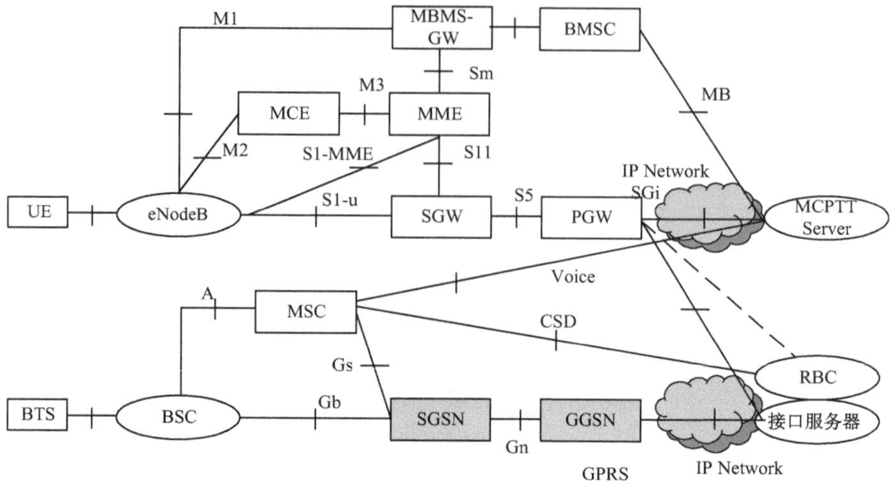

Fig. 2.9 Interconnection network architecture of LTE-R and GSM-R system

of the mobility. For the mathematical derivation of the multipath diameter distribution, although the theoretical derivation has a certain universality, different numbers of multipath propagation path environment of high-speed railway under targeted distribution modeling needs to rely on the measured data to construct a different path time-varying time-domain waveform. Using a variety of known probability distributions to fit the waveform, which is aiming to get the probability density function of the waveform, it is a good probability for the distribution of the number of multipath path constructed in accordance with the characteristics of high-speed railway in different propagation environments.

Although the current research has carried out aerial measurements and modeling studies, the antenna measurements methods of the multiple antennas, the multiple dimensions, the multiple positions in the scenes of high-speed railway and complicated situation and the antenna modeling method of angle domain, delay domain, and polarization domain are need to be studied. For deterministic modeling approach, there are some theoretical gaps in the structure composed of a complex space environment and radio wave propagation mechanism. Existing academic attempts are concentrated in the automotive field of communication; the field of rail transportation has not been touched. Therefore, it is an urgent need to make deterministic modeling methods to break the bottleneck, which is aiming to achieve modeling complex space environment and high-speed mobile broadband channel. For the half deterministic modeling approach, there is not a reliable solution to describe nonstationary characteristics of the fast time-varying channel, which is lack of the broadband channel modeling for the high-speed railway and complex space environment. Stochastic modeling is low complexity, which is essentially not be capable of characterization of nonstationary channel. Therefore, it is an urgent need to carry out research of modeling method for the high-speed railway complex environments.

Currently, there have been many researches on the reliability and availability model of the communication system. The reliability model for GSM-R network in ETCS based on Petri nets was described [1]. A channel transmission model and the factors which may lead to the wireless connection failure were proposed. The performance analysis of GSM-R network structure in China Train Control System was investigated [2]. A distributed antenna system was proposed [3], which is aiming to adapt to the tunnel. Radio coverage with antennas requires an accurate prediction of propagation loss inside and outside the tunnel, particularly when the communication system must maintain a high quality of service along the entire track [4]. The authors presented a comprehensive analysis and modeling of shadow fading in HSR environments [5]. In [6], it is observed that the handover rate and handover initiation delay increase and decrease with the standard deviation (STD) of shadow fading, respectively.

The existing GSM-R network has some definition for the QoS metrics of dedicated services and requirements specification, but not specifically for high-speed mobile scene. Moreover, the index set is not complete. An overview of shortcoming was presented in [7]. With the increasing speed of the railway, the reliable transportation of the train control information and other railway dedicated communication traffic has a more important influence on the security of the system. Railway mobile communication system evolves to the all packet-switched network, and the service types, flow characteristics, and user behaviors for high-speed railway are different with the public network service.

To guarantee the QoS of the dedicated service and the RAMS requirements for the high-speed railway, the system needs to measure the network traffic and combine the high-speed railway service with the characteristic of user behavior to build the high-speed railway mobile communication system business model, which is aiming to accurately model the various services and restudy the quantitative relationship between the applicability of the QoS metrics for the GSM-R system and the system Markov Model state statistical properties. The system would build the performance evaluation system which is suitable for the high-speed railway mobile communication network and meet the requirements of RAMS.

The key technologies of GSM-R are based on the hot and difficult issues which GSM-R system should solve immediately. The technologies build the theoretical foundation and the technological system to achieve secure and reliable transmission of information in the high-speed railway complex environment with the system thorough research. The key technologies could better guide GSM-R system construction and development. In general, the key technologies of GSM-R could be summarized as four aspects: (1) The radio wave propagation simulation modeling theory and method for high-speed railway, (2) The interference cancelation theory for GSM-R system; (3) The key technologies of safety data transmission for high-speed railway; and (4) The performance evaluation system for GSM-R system.

(1) The radio wave propagation simulation modeling theory and method for high-speed railway
The large-scale propagation model for high-speed railway includes the different propagation models which are suitable for different scenarios. The propagation model could optimize and modify the model according to the change of the railway surrounding environment. Moreover, the propagation model could also propose and verify the correction factor, which could build the accurate radio wave propagation model database for different high-speed railway scenarios (plains, mountains, cuttings, bridges, hills, stations, tunnels, viaducts, etc.).
The multipath fast fading distribution model mainly research the regularity of distribution for the multipath fast fading in high-speed railway and the field coverage margin due to multipath fast fading. Based on the small-scale fading and the distribution of multipath delay, the distribution of continuous burst error caused by multipath fading fast is studied in the model. Moreover, the model also studies antenna technology, channel estimation and equalization, multipath diversity, and error correction coding, which is aiming to overcome the small-scale fading impact on the security of data transmission.
Wireless coverage hole theory of high-speed railway mainly focusses on the formation mechanism about the wireless coverage hole of the high-speed railway and proposes the scientific definition, to build a model of wireless coverage hole which indicates the influence on the railway service.
On this basis, the model studies the influence of wireless coverage hole on the train control safety data transportation and analyzes on-board units of locomotive drifting in and out of wireless coverage hole, which is aiming to overcome the effects of the method of secure data transmission and eliminate wireless coverage hole technology.
(2) The interference cancelation theory for GSM-R system
The technology mainly studies the distribution of radio interference in high-speed railway scenario, which is taking GSM-R wireless networking and frequency planning into account. Moreover, the technology also discusses the mechanism of co-channel interference and adjacent channel interference in GSM-R system according to different radio wave propagation scenarios for high-speed railway (plains, mountains, cuttings, bridges, hills, stations, tunnels, viaducts, etc.). The technology combines the environment of high-speed railway with the interference propagation path and characteristics parameters outside of the statistical data analysis system.
The interference propagation path characteristic parameters, which is aiming to study the relationship between carrier to interference ratio and bit error rate (BER) and block error rate (BLER) and establish moderate computational complexity for train control system C/I analysis model.
The corresponding interference cancelation technologies and protective measures study interference cancelation methods which is based on reasonable engineering design and wireless network optimization. The technologies also take security data transmission needs of high-speed railway operation control system into consideration. The technologies also propose the interference

protective measures which is according to interference detection technology of Um, A and Abis interface monitoring, comprehensive monitoring system, and early warning technology of end-to-end security data transmission quality supervision

(3) The key technologies of safety data transmission for high-speed railway
Nonlinear distortion handle technology of high-speed railway radio channel mainly studies the radio channel nonlinear distortion mechanism under high-speed (250 km/h ~ 500 km/h), complex terrain, and varying environmental characteristics. On this basis, the technology studies the influence of nonlinear distortion on train control system security data transmission according to the sparsity of train control data, which is aiming to research the feasibility of the general packet radio service for train control data transportation. To reduce system latency, further research would study the fast synchronization of high-speed railway wireless receiver, channel estimation and equalization, and anti-Doppler frequency shift technology.
The redundancy technology in GSM-R system studies mechanism of redundancy technology, which is aiming to establish a variety of redundancy theoretical analysis model. The technology also researches the influence of different redundancy methods (interleaving sites coverage and two base station coverages with same site) on the network performance and service. In order to analyze the feasibility of mesh network in high-speed railway applications and establish a new train control security network model, the technology studies the application of mobile switching pool in high-speed railway.
According to the openness characteristic of wireless communication systems and the demand of high-speed railway train control security data transmission, safety protocol stack and information safety system mainly study transmission mechanism of security data in openness transmission system and establish dedicated railway safety data transmission protocol stack. The system proposes the new end-to-end information security system to remedy the existing security vulnerabilities of GSM-R system which is including end-to-end mutual authentication protocol, end-to-end encryption protocols, and processes. The system also investigates the integrity protection method of train control data and online key management method for train control system.

(4) The performance evaluation system for GSM-R system
The system mainly studies theoretical analysis model of the RAMS for GSM-R system, and deeply analyze the relationship between RAMS and network quality of service. According to the relationship between network quality of service and field coverage, radio interference, engineering design parameters (such as base station spacing, cell coverage radius, the overlapping area size, etc.), network operation, and maintenance, the system establishes the index evaluation system for GSM-R system.
The evaluation methods of GSM-R system indicators include establishment of the index ratings and tolerance for network quality of service, field strength coverage, network operation, and maintenance. The wireless communication

system RAMS indicators methods should meet the needs of reliable transmission for train control security data. Moreover, the methods also investigate the network quality service indicator of high-speed railway control data, wireless coverage level measurement, statistical analysis and measurement technology, and comprehensive evaluation method of network operation and maintenance.

2.5 Key Technologeis for LTE-R

The mobile communication system is an important part and the nerve center of the ground infrastructure for high-speed railway, which is aiming to provide accurate and timely information for train dispatching, train control, automatic train operation, train security video surveillance, train status monitoring, remote fault diagnosis, wireless monitoring for the infrastructures, emergency job processing, information dissemination and advertising, passenger information, and entertainment services. The system is the foundation and prerequisite to establish the operation safety and security system of high-speed railway. Currently, China's high-speed railway mobile communication systems are based on mature second generation and 2.5 generation narrowband mobile communication technology. However, the existing communication system is becoming the element limiting the number of running trains in areas with high train concentration, such as major train stations [8].

The main problems of sustainable development and the practical application are revealed in the following:

(1) The insufficient capacity and the difficulty networking limit the development of the applications. Due to the constraints of frequency resources, wireless networking is very difficult, and the co-channel interference and adjacent channel interference are serious in the major stations and hubs regions of high-speed railway. To ensure reliable transmission of train control (CTCS3) security data, the technology must use channel guarantee technology to improve the success rate of handover and many applications (e.g., wind warning, train status monitoring, real-time information transfer) are lack of development.
(2) The technology does not provide the real-time service to transmit security monitoring broadband data. The highest circuit domain data transmission rate is only 9.6kbps and the actual maximum packet domain data transmission rate is lower than 50kbps, limiting the development of the internet of things in railway. The technology also results in that the ground early warning monitoring data cannot upload to the train and the on-board monitoring data cannot transfer to the ground. The trains cannot communicate with each other directly, which would delay or hinder the timely failure processing.

The research about influence of high-speed mobile on handover and the corresponding measures are gradually increasing when the LTE system is applied to the railway environment. However, due to special requirements of the railway dedicated mobile communication on content, capacity, and quality, the research highly simplifies the radio wave propagation environment, wireless network of the train, and the communication requirements of passengers on the high-speed railway. Taking into account the communication requirements aggregation and group movement of the vast passengers, the performance indicators and implementations of the functions, such as handover control, should have some difference with each other. There is not particularly distinction among the currently LTE handover technology. Program selection, parameter optimization, resource allocation, automotive systems architecture, and other details of the design are closely related to the size and distribution of traffic.

Currently, some research focuses on Long-Term Evolution for Railway (LTE-R). A detailed evaluation of the BER and PSD for LTE-R suitably dimensioned for the high-speed railway channel was presented in [9]. In [10], the impact of mutual coupling on LTE-R MIMO capacity for antenna array configurations in high-speed railway scenario is investigated. The authors undertake stochastic delay analysis of train control services over a high-speed railway fading channel using stochastic network calculus [11].

Broadband mobile communication technology is an inevitable trend and selection of high-speed railway and urban rail transport development. The 7th world congress on high-speed rail was hold in Beijing in 2010. The Chinese Ministry of Railways and the International Union of Railways (UIC) clearly indicate that the evolution path of railway mobile communication system will be spanning third-generation (3G) mobile communication technology, and the GSM-R technology will directly develop to next-generation broadband mobile communication technology (LTE-R). However, there are some uncertainties when the existing research of the broadband mobile communication is directly applied to high-speed railway. The following three scientific questions need to address:

Scientific Question 1: The radio wave propagation mechanism of high-speed mobile and limited space environment. Due to the diverse surroundings of high-speed railway, complex electromagnetic environment, several limited spaces, and strong electrical interference, the wireless channel of high-speed mobile (more than 200 km/h) shows the fast time-varying characteristics of nonstationary. World wireless communications standards organization, including the Cooperation in the field of Scientific and Technical Research (COST) and business alliance partners (WINNER, Wireless World Initiative New Radio), are lack of the research about the combination of high-speed mobile, railway special application scenario, the radio wave propagation characteristics of broadband, and wireless channel model.

Scientific Question 2: The broadband, efficient and reliable data transmission mechanism under high-speed mobile. The fast time-varying channel, nonstationary characteristics, severe Doppler effect, frequent handover, and fast changing of application scenarios are caused by high-speed mobile. It is difficult to track the fast varying channel by the sparse pilot pattern design and channel estimation technology, which leads to the performance degradation of transmission rate, transmission efficiency, error rate, transceiver synchronization, channel estimation, demodulation, decoding, and other signal detection. The standards of LTE and LTE-A mainly focus on the guarantee of quality of service and data transmission rate for low-speed mobile scenarios. In the case of high-speed scenario (350 km/h), the corresponding indicators are almost blank, only maintaining basic communication. Therefore, it is an urgent need to study the key transportation technologies of high-data transmission rate to respond to the bad radio channel environment and maintain a high spectral efficiency in high-speed mobile scenario.

Scientific Question 3: The performance evaluation and optimization of the wireless resource management mechanism for high-speed railway. The high-speed railway should finish the performance evaluation and design requirements before the officially operation. Due to the complex radio wave propagation mechanisms and poor channel conditions, the specificity of high-speed railway is determined by the railway dedicated traffic, quality of service, and the severely limited spectrum. In order to establish the stochastic model of wireless resource management mechanism, the technology should take advantage of the stable railway speed, certain path, and predictive position and combine characteristics of wireless channel with physical properties to deeply describe the system features. Moreover, the technology proposes the analysis method to performance evaluation of wireless resource management mechanism and the optimization design, which takes queuing theory and stochastic network calculation into account.

2.5.1 The Application Requirements of the Next-Generation Railway Mobile Communication System

(1) Demand mining, definition, and classification of the traffic
The technology should mine the demand of each traffic department for the broadband mobile communication services and accurately classify the demand and the definition of traffic, which is aiming to predict the development trend of the railway mobile communication traffic.

(2) Traffic modeling
The technology need to establish the traffic model of air interface in wireless network side and core network side, which is aiming to confirm the parameters of air interface traffic model (including the RRC connection, uplink and downlink data rate, downstream traffic/upstream traffic, average connection duration, busy concentration factor, etc.) and the parameter of core network traffic model (including the number of users, uplink and downlink data traffic of network, types of traffic, tracking location area of recognizable traffic, user groups, etc.).

(3) The QoS of the traffic
The technology should analyze the QoS of various services and confirm the traffic metrics (including peak rate, average rate, the lowest rate, the access delay, end-to-end transmission delay, handover interrupt latency). Moreover, the technology also should confirm the QoS class identification of each service (QoS Class Identifier, QCI), including the resource type, priority, packet delay budget, packet loss rate, etc.

2.5.2 The Technology System and Network Architecture of the Next-Generation Railway Mobile Communication System

(1) The performance evaluation technology of TD-LTE and FDD-LTE in railway scenario.
In various railway typical scenarios (viaducts, cuttings, tunnels, stations, marshaling yard, etc.), the technology mainly studies the performance evaluation of TD-LTE and LTE FDD for high-speed adaptability, traffic bearing capacity, coverage capacity, interference, and analyze the mechanism of the differences between the two systems, which is aiming to propose the selection advice of the next-generation railway mobile communication system.
(2) The fusion networking technology of TD-LTE and FDD-LTE in railway scenario.
The technology should apply vertical handover, cell reselection, and load balancing to the fusion networking of TD-LTE and FDD-LTE in railway scenario. Moreover, the technology should take full advantage of TD-LTE and FDD-LTE and realize the data transportation of high transmission rate, reliable and low latency on the basis of low complexity, construction cost, and limited frequency.
(3) The network redundancy technology of the next-generation railway mobile communication
In order to propose the redundant network architecture, the technology makes full use of the access network of the next-generation railway mobile communication network (redundant wireless coverage, backups carrier frequency, BBU backup) and the redundancy backup technology of core network (MME/P-GW/S-GW/HSS), which could guarantee the reliability of the access network and core network.
(4) The network architecture of the next-generation railway mobile communication
In order to confirm the function of each network element and performance indicator, the technology should confirm the network architecture of the next-generation railway mobile communication system and define the interface function of each element.

2.5.3 Frequency and Bandwidth Requirements of the Next-Generation Railway Mobile Communication System

(1) Adaptability under different frequencies in railway scenario
Compare and make quantitative analysis of the performances within different frequencies under the various typical scenarios in railway, such as the viaduct, cutting, tunnel, railway stations, marshaling yards, and so on. And the performances are included wireless coverage, i.e., the cellular radius, channel capacity, interference, and so on.
(2) The spectrum requirements in the next-generation railway mobile communication system
According to the frequency distribution and national frequency planning in our country, as well as the 3 GPP LTE frequency planning, combined with the business requirements and system research, determine the next-generation railway mobile communication system spectrum requirements [12].
(3) The frequency planning technology in next-generation railway mobile communication system
Analyze the distribution characteristics of intercell interference and the influence to the performance of the system under different scenarios (yard, viaduct, and lines section) in railway. Then put the frequency planning and frequency reuse scheme into use.
(4) The analysis of the bandwidth demand in next-generation railway mobile communication system
Analyze the system bandwidth requirements in the next-generation railway mobile communication system, based on the business requirements in the next-generation railway mobile communication system by studying the volume calculation method.

2.5.4 The Key Technology in the Next-Generation Railway Mobile Communication System

In order to ensure the reliability and security of the next-generation railway mobile communication system, the key techniques are network technology, network planning technology, interconnection technology, high-speed adaptive technology, efficient transmission, and the information security technology [13].

(1) Network technology in the next-generation railway mobile communication system
The network technology mainly includes network hierarchy, mobility management entity (MME), home subscriber server (HSS), a service GateWay (S-GW), PDN GateWay (P-GW), the domain name server (DNS), the 3A

server (AAA), Policy and Charging Rules Function (PCRF), IP multimedia subsystem (IMS), base station equipment (eNodeB), the setting scheme of the data bearing network, routing scheme, numbering scheme, and the addressing principle.

(2) Network planning technology in the next-generation railway mobile communication system
Based on the cover characteristic which is linear cover in the railway scenario, analyze the relationship between the coverage level and the performances (transmission rate, system capacity, transmission reliability, and the success probability of the handover) of the system. To meet the different business QoS requirements, the cell edge coverage probability, the shadow fading margin allowance, the fast fading margin, the interference margin, and other technical indicators, we should improve the transmission reliability, the handover success probability, and system capacity [14].

(3) Interconnection technology in the next-generation railway mobile communication system
Interconnection technology mainly includes the interconnection between the next-generation railway mobile communication system and railway system application technology, such as scheduling, train running control, synchronous control of railway locomotives, and railway fixed telephone communication network connectivity.

(4) High-speed adaptive technology in the next-generation railway mobile communication system
Analyze the influence of wireless radio waves propagation and wireless channel caused by high-speed movement. Study the Doppler frequency shift estimation, adaptive antenna, and inter-carrier interference (ICI) cancelation technology. In order to improve the high-speed adaptability in the next-generation railway mobile communication system, and support the highest speed, mean while, guarantee the QoS requirements [15].

(5) Efficient transmission and the information security technology
In the high-speed movement scenario, the Doppler will seriously affect the communication quality and influence the transmission of train control signals and the passengers' signals. Therefore, the high-speed movement characteristics put forward the new challenge of the physical layer transmission techniques. It is of great significance to develop the high-speed railway multi-way relay network coding and the physical layer security, cognitive radio signal perception, recognition and carrier aggregation, the pilot signal design, and channel estimation in the high-speed railway multi-relay communications. The key problem is how to reduce the influence of the channel time-varying characteristics (i.e., serious Doppler principle caused by high-speed movement). Therefore, the cognitive radio signal perception, recognition, and carrier aggregation, as well as the pilot signal design and channel estimation in the multi-way relay networks, should be mainly studied. Take advantage of the multi-relay network coding to improve the communication reliability. Design

the effective transmission scheme from the perspective of physical layer security.

a. Network coding and physical layer security in high-speed railway multi-way relay networks
At present, the study of multi-way relay communication mainly focused on the low-speed movement scenario [16]. Therefore, it remains to be further studied in the high-speed scenario, especially the transmission design combined with the network coding, the combination and the joint design of other new techniques. In the typical physical layer security transmission, most assume that the transceivers and eavesdroppers are fixed, and the perfect CSI are known at the transceivers. However, the physical layer security in the high-speed railway should consider its own dynamic characteristic and the particularity of various scenarios. Thus, a new physical layer security which is suitable for the high-speed railway should be studied. By exploring the essence features of the multi-way relay communications in the high-speed railway scenarios, design a general model of multi-way relay network coding [17]. Based on the asymmetry of the communications within the carriage and train-ground communication, develop a joint optimization scheme between relay, base station, and users in the railway, as well as efficient and safe signal transmission methods.

 - The design of vehicle relay scheme against high-speed movement
   By exploring the relay algorithms combined with the physical layer network coding, develop the mechanism of joint design and optimization to against the serious Doppler principle caused by high-speed movement. Establish a general multi-way relay communication model for all kinds of communication requirements. Furthermore, design a vehicle relay scheme based on zero-forcing, minimum mean square error, user fairness, and maximum user throughput criterions.
 - The physical layer security algorithm design for high-speed railway
   By analyzing the security capacity in the high-speed railway scenario, develop exact broad band physical layer security transmission schemes with limited power, spectrum, and infrastructures, and analyze the secure transmit rate of these schemes.

b. Cognitive radio perception, recognition, and carrier aggregation in high-speed railway
Cognitive radio is an important way to solve the severely restriction of railway spectrum resources [18]. However, the existing research has not been involved the spectrum perception and modulation recognition algorithms in high-speed movement scenarios. The robustness algorithms of perception recognition by using a more effective mathematical tool need to be further studied. This algorithm will guarantee the control signal without interference and improve the spectrum efficiency of public networks. In addition, the study of carrier aggregation technology with high-speed

movement still remains on introduction. It is important to further study the carrier aggregation methods suitable for high-speed railway communications.

By effectively perceive the train control signal and the passengers' spectrum, to determine the duty criterion and carrier aggregation of train control signal and vehicle user signal. By the modulation recognition technology of unknown signal to explore the illegal signal evade method within the control signal spectrum and the restrain the interference to users.

- The study of spectrum perception in high-speed movement scenario
  The spectrum perception under high-speed movement includes a wide range of spectrum perception technology, as well as fast and efficient for the time-varying channel spectrum perception technology. Network node cognitive method can effectively improve the recognition efficiency. Spectrum perception technology, as a kind of signal means of security, can avoid train control signal being attacked by malicious signal. In addition, using of spectrum perception to seek the free spectrum, so as to enhance the passenger business information transmission.
- The study of modulation recognition in high-speed movement scenario
  Multiple signal and different modulation recognition technologies in high-speed movement scenarios are able to extract more effective signal characteristics, enhance the degree of differentiation of different signals. Effective mathematical tool can further improve the recognition performance. In view of the high-speed scenarios lead to Doppler extension, non-sensitive feature extraction frequency, it is important to design modulation recognition algorithms with low complexity, wide application range, and strong robustness.
- The design of carrier aggregation in high-speed movement scenario
  In view of the limited special railway spectrum and the discrete spectrum distribution, study the changes of aggregation caused by the position. Analyze impact on the wireless carrier aggregation caused by the surrounding environment, the running status, and the relative motion. The dynamic carrier aggregation methods under different mobile speeds and different spatial locations.

c. Pilot design and channel estimation in high-speed railway relay networks
   Aliasing time-varying channel estimation is an important component in wireless broadband high-speed railway relay networks, and it is also prerequisites to ensure reliable and efficient communication. There have not any effective and accurate time-varying channel estimation algorithms for high-speed railway relay networks by far. Therefore, the optimal pilot design remains to be studied.
   Explore the statistical feature of the aliasing time-varying channel in relay networks, by designing the optimal pilot pattern to realize the high precision aliasing time-varying channel estimation in relay networks.

- Optimal pilot design in relay networks
  In view of the high-speed railway wireless relay communication system, explore the aliasing time-varying channel modeling and analyze the statistical characteristics. Design effective base extension model to approximate the aliasing time-varying relay channel. In addition, explore the optimal pilot with the minimum channel error or maximum channel capacity. Study the features and methods of the parameters estimation and pilot design in relay networks time-varying channel.
- High precision aliasing equivalent time-varying channel estimation in relay networks
  Explore aliasing equivalent time-varying channel parameters in relay system. Design high precision aliasing equivalent time-varying channel estimation algorithms and derive the estimate error and CRLB lower bound. Study the channel capacity and the lower bound with imperfect channel state information [19].

d. The information security technology in the next-generation railway communication system
   Analyze the security threats faced by the next-generation railway mobile communication system. Study the authentication, encryption algorithm, key distribution and management in the next-generation railway mobile communication system.

2.5.5 *Hybrid Networking of GSM-R and the Next-Generation Mobile Communication System*

Wireless heterogeneous network can take fully reuse of the severely limited frequency resource by deploying low power stations. By different wireless accesses, this heterogeneous network can improve radio coverage, increase spectrum and energy efficiency, and enhance the fairness. But the related research about this area is still relatively weak, and some key problems are still in the blank. Therefore, we need to design the system framework based on the typical scenarios, inherent characteristics, and demands in high-speed railway. Furthermore, the effective management should be conducted by considering the features, such as linear coverage, intermittent service, and great difference between different business demands, so as to solve the limited resources, high mobility problems in next-generation high-speed railway communication systems. Then, we should design a heterogeneous mobile communication network suitable for the high-speed railway and a security system to meet the high reliable requirements.

(1) Bearing business in hybrid network

The next-generation railway mobile communication system can carry the existing GSM-R voice business, electricity expressway. Hybrid network can realize the business bearing from both networks and optimal allocation between these two networks.

(2) The connectivity and terminal blends in hybrid network
The database sharing mechanism between these two networks and the connectivity of the equipment from GSM-R and next-generation mobile communication system.

(3) Network selection, reside, and vertical handover technology in the hybrid network
Determine the choice of the network, network resides, and vertical switch during the switching protocol process. Establish a new network selection, network resides, and switching selection strategy, develop the features of these two networks. Reduce the ping-pang effect and improve the QoS.

(4) The use of GSM-R resources in hybrid network
Take full use of the existing GSM-R equipment, such as computer room, tower, power source, power amplifier, transmission, and so on. Reduce the costs of network construction and maintenance in the hybrid network.

The research on edge coverage provides reference and basis for the network planning design. And network planning scheme will also affect the coverage radius and fading margin settings, which affects the edge coverage level. Based on the network planning schemes in next-generation railway mobile communication, study the network information safety of and the interconnection with different railway applications such as scheduling communication interface.

2.5.6 The Evaluation and Optimization of High-Speed Railway Wireless Resource Management Mechanism

Study the high-speed railway wireless business modeling and performance evaluation system, the interference management mechanism of high-speed movement heterogeneous networks, and the high mobility wireless communication management mechanism. Model the typical business and evaluation system should consider the dedicated business and special QoS requirements. Under this business model and evaluation system, the spectrum resources serious limitations, high-speed mobility, and trajectories determination will cause a lot of challenges. Then study the interference and mobility management mechanism in the heterogeneous network.

(1) High-speed railway wireless business modeling and performance evaluation system

   a. Business modeling in high-speed wireless network
   Based on the business situation and demands of the high-speed railway mobile communication system, analyze the future business development trends and characteristics of high-speed wireless network. Further study the statistical features and user behavior characteristics of high-speed railway wireless networks. Study how to set up the business service system, flow model, and user behavior model based on the train control [20] and business requirements of high-speed railway mobile communication system.
   b. High-speed railway wireless network performance evaluation system
   Based on high-speed railway RAMS requirements, study high-speed railway wireless QoS indicators and evaluation methods. The research content includes the quantitative relationship between QoS index and the transient and steady-state solutions of the stochastic model, and how to build a high-speed railway broadband mobile communication system of performance evaluation system.

(2) Interference management mechanism in the high-speed mobile heterogeneous wireless network

   a. The high-speed railway heterogeneous wireless network architecture
   Study the high-speed railway heterogeneous wireless network system architecture, topology, the network planning, etc. Considering the infestation deployment between railway interval, small cell deployment in hot spots, relay deployment outside the train, and distribution antennas and Wi-Fi inside the train builds a radio coverage network architecture from a wide range of hot spots inside the train to improve the network capacity and ensure the safety of redundancy backup [21].
   b. The interference management mechanism of the high-speed railway heterogeneous wireless network
   By establishing the grouping markov channel model and the corresponding signal design method and cross-layer interference management mechanism, study effects of high-speed train mobile for signal transmission and the robustness interference management mechanism of heterogeneous network under imperfect channel state information. Take use of the fixed running speed, determined routine, and predictable position to conduct the interference management based on the interference prediction information. Based on the oriented [22] business resource management, develop real-time distributed resource management mechanism.

(3) High-speed mobile wireless mobility management mechanism

a. Mobility management mechanism in high-speed railway LTE-R wireless communication systems
Design LTE-R system wireless networking and vehicle-mounted system architecture with handover function in high-speed railway environment. To realize the location management in high-speed railway environment, the handover management mechanisms which used to guarantee the railway communication quality include handover trigger, handover algorithm, and pilot design [23].

b. Mobility management mechanism in high-speed railway large-capacity communication system
Build the large-capacity user mobility model in high-speed railway scenario. Study network mode, system architecture, handover management, and wireless resource management in multiple relays train-ground wireless communication network.

2.6 Summary

This chapter analyzes the network architecture and key technologies of GSM-R and LTE-R system. LTE-R as the next-generation communication system is the essential trend for the future HSR, and LTE-R will coexist with GSM-R in a long time. The two dedicated communication systems will be compatible, interoperability, and ultimately to achieve integration.

References

1. Wu H, Gu Y, Zhong Z (2009) Research on the fast algorithm for GSM-R switching for high-speed railway. J Railway Eng Soc 1:92–98
2. Jijing H, Jun M, Zhangdui ZH (2006) Research on handover of GSM-R network under high-speed scenarios. Railway Commun Signals 42:51–53
3. Ai B, Cheng X, Kürner T et al (2014) Challenges toward wireless communications for high-speed railway[J]. IEEE Trans Intell Transp Syst 15(5):2143–2158
4. Wang J, Zhu H, Gomes NJ (2012) Distributed antenna systems for mobile communications in high speed trains. IEEE J Sel Areas Commun 30(4):675–683
5. Guan K, Zhong Z, Ai B (2011) Assessment of LTE-R using high speed railway channel model. In: 2011 Third international conference on communications and mobile computing (CMC). IEEE, pp 461–464

6. TS 22.179 3rd Generation Partnership Project; Technical specification group services and system aspects; mission critical push to talk (MCPTT) over LTE; Stage 1. Tech. Specification. v13.1.0
7. Calle-Sánchez J, Molina-García M, Alonso JI, Fernández-Durán A (2013) Long term evolution in high speed railway environments: feasibility and challenges. Bell Labs Tech J 18 (2):237–253
8. Zimmermann A, Hommel G (2003) A train control system case study in model-based real time system design. In: Proceedings of IEEE IPDPS, pp 22–26
9. Shi J, Zhang X, Gao T (2010) Performance analysis of GSM-R network structure in China train control system. In: Proceedings of IEEE ICEIE 2010, pp 214–218
10. Briso C, Cruz J, Alonso J (2007) Measurements and modeling of distributed antenna systems in railway tunnels. IEEE Trans Veh Technol 56(5):2870–2879
11. Prasad MVSN, Singh R (2003) Terrestrial mobile communications train measurements in Western India. IEEE Trans Veh Technol 52(3):671–682
12. He R, Zhong Z, Ai B, Oestges C (2014) Shadow fading correlation in high-speed railway environments. IEEE Trans Veh Technol 64(7):2672–2772
13. Singh B, Aggarwal K, Kumar S (2005) Sensitivity analysis of handover performance to shadow fading in microcellular systems. In: Proceedings of the IEEE ICPWC, pp 446–450
14. Sniady A, Soler J (2012) An overview of GSM-R technology and its shortcomings. In: Proceedings of IEEE ITST, pp 626–629
15. Sniady A, Soler J (2014) LTE for railways: impact on performance of ETCS railway signaling. IEEE Veh Technol Mag 9(2):69–77
16. Guan K, Zhong Z, Ai B (2011) Assessment of LTE-R using high speed railway channel model. In: Proceedings of IEEE CMC, pp 461–464
17. Liu Y, Ai B, Chen B (2016) Impact of mutual coupling on LTE-R MIMO capacity for antenna array configurations in high speed railway scenario. In: Proceedings of IEEE VTC Spring, pp 1–5
18. Lei L, Hua J, Jiang Y, Shen X, Li Y, Zhang Z, Lin C (2016) Stochastic delay analysis for train control services in next-generation high-speed railway Communications System. IEEE Trans Intell Transp 17(1):48–64
19. Zhang Y, Xiong L, Jiang W, Ai B (2015) Analysis and research on spectrum requirements of LTE for railways. In: 6th International conference on wireless, mobile and multi-media (ICWMMN 2015), Beijing, pp 83–86
20. Banerjee S, Hempel N, Sharif H (2016) A survey of wireless communication technologies & their performance for high speed railways. J Transp Technol 06(1):15–29
21. Pan MS, Lin TM, Chen WT (2015) An enhanced handover scheme for mobile relays in LTE-A high-speed rail networks. IEEE Trans Veh Technol 64(2):743–756
22. Gao T, Sun B (2010) A high-speed railway mobile communication system based on LTE. In: 2010 International conference on electronics and information engineering (ICEIE), Kyoto, pp V1-414–V1-417
23. Wang F, Yuan X (2013) Zero-forcing multi-way relaying with sum rate maximization. Trans Emerg Telecommun Technol. doi:10.1002/ett.2707
24. Wang F (2014) Wireless MIMO switching: network-coded MMSE relaying and user group selection. Trans Emerg Telecommun Technol 25(5):507–514
25. Zhang J, Wang F, Zhong Z, Cabric D (2015) Local and cooperative spectrum sensing via Kuiper's test. In: Proceedings of IEEE ICC 2015, pp 597–584
26. Fan D, Zhong Z, Wang G, Gao F (2015) Doppler shift estimation for high-speed railway wireless communication systems with large-scale linear antennas. In: Proceedings of high mobility wireless communications (HMWC), Xian, pp 1–5

27. Lei L, Lu J, Jiang Y, Shen XS, Li Y, Zhong Z, Lin C (2016) Stochastic delay analysis for train control services in next-generation high-speed railway communications system. IEEE Trans Intell Transp Syst 17(1):48–64
28. Han J, Zhou K (2013) Interference research and analysis of LTE-R. In: 2013 IEEE 5th international symposium on microwave, antenna, propagation and EMC technologies for wireless communications (MAPE), Chengdu, pp 732–734
29. Baek JH, Kim GY, Yang DC, Choi HY, Kim YK, Yoon YK (2014) A LTE wireless communication interface test for on-board oriented train control system field test. In: 2014 International conference on information and communication technology convergence (ICTC), Busan, pp 690–694
30. Xu Q, Ji H, Li X, Zhang H (2016) Admission control scheme for service dropping performance improvement in high-speed railway communication systems. IEEE Trans Veh Technol 65(7):5251–5263

Chapter 3
Radio Propagation and Wireless Channel for Railway Communications

3.1 High-Speed Railway Propagation Scenarios

3.1.1 High-Speed Railway Propagation Scenarios Definition

As the scene partitioning for wireless channel modeling plays vital role in predicting radio wave propagation, it is necessary to partition radio wave propagation scene for HSR whose operating speeds are above 350 km/h. It is also important that the scene partitioning works as the basis for optimizing radio wave propagation prediction and the upper layer communication design.

Although special HSR scenarios such as cuttings, viaducts, and tunnels are different from propagation characteristics, there are still missing detailed and reasonable defined scenarios of wireless channel modeling for HSR [1]. The existing standard channel models related to HSR are deficient in the special HSR scenarios. For example, there are 17 dB errors of path loss between measurement data from Zhengzhou–Xian HSR passenger-dedicated line and prediction of Hata model which excludes diffraction loss [2]; the WINNER project only proposed WINNER D2 model whose working frequency is at 2–6 GHz, it is not suitable for GSM railway (GSM-R) wireless network operating at 930 MHz. The remained sections will descript the detail scene partitioning for HSR.

In this section, the HSR scenarios are classified into eleven scenarios based on references from several organizations and related standards mentioned in [1]. Furthermore, the practical investigations from several passenger-dedicated HSR lines and stations proved the reasonable partitioning for HSR scene [3–5].

- **Viaducts**: Viaduct is one of the most common scenarios along the rail, which is typically constructed with a series of arches for carrying the train across uneven ground. As there are strict requirements for the smoothness of rail to ensure the high speed of train (cruising speed up to 350 km/h), the appearance of viaduct is of great importance. There are two categories of viaducts which are defined according to line of sight (LOS) and NLOS conditions of radio wave

Z.-D. Zhong et al., *Dedicated Mobile Communications for High-speed Railway*,
Advances in High-speed Rail Technology, DOI 10.1007/978-3-662-54860-8_3

propagation, i.e., whether the direct ray between transmitter and receiver is blocked or not [1, 3].

- Viaduct-A: It related to the surface of viaducts that are lower than most scatters (e.g., trees and building). These scatters cause massive reflections and scatterings, which leads to severe shadowing in wireless propagation. Even the variation of fading distributions will follow the variation of these scatters (e.g., the swing of the trees), which causes the propagation channel under unexpected change.
- Viaduct-B: It is related to the surface of viaducts that are higher than most scatters, which is in the opposite of Viaduct-A [1]. In this category, most scatters that located around the Viaduct within a range of 50 m are lower than surface of viaduct, which brings about a dominant direct path (LOS ray) dominate the propagation channel. Generally, it is negligible for the effects of the scatters in this category.

• **Cuttings**: As a common scenario in HSR scenarios, cutting is designed to ensure the smoothness of rail when the train is passing through the uneven ground which constructed with large obstacle [1, 2]. The construction of cutting is with two steep walls that generally covered with vegetation and reinforced concrete, which can be either regular or irregular. The former is owing to the steep walls on two sides of the rail that are the same in slopes and depths; the latter is owing to the irregular of hills and mountains that cause the irregular up and down of steep walls along cutting.
There are generally three parameters to define the constructions of the cutting: crown width (mostly ranging from 48 to 63 m in China), bottom width (mostly ranging from 14 to 17 m in China), and the depth of cutting (mostly ranging from 3 to 10 m in China) [1, 4].
For the propagation channel, the cutting works as a container, the "wider" and "deeper" constructions make the container hold more multipath components. As illustrated in [4], the receive antenna located at the top of the HSR is sometimes lower or higher than the roof of the cutting. Under the former condition, the cutting will lead to more multipath components. Moreover, cross-bridges are always appearance along the cutting scenario for the purpose of bridging the gap between the two sides of the cutting. The existing cross-bridge will block the direct paths between transmitter and receiver in a short time, which will consequently lead to NLOS propagation and severe disruption of wireless communication [5].
• **Tunnels**: Under the purpose of durable high cruising speed of HSR train, the primary requirements for designing and building rail for HSR are keeping the smoothness and straight of the rail. For this purpose, tunnel construction is employed to ensure the high cruising speed of train in rolling terrain and mountainous region. Tunnel is an artificial underground passage and its cross section in HSR are generally vaulted or semicircle [1, 6–8]. As the smooth internal surface and limited structure of the tunnel, there are numerous

reflections and scatterings, which cause a waveguide effect of propagation inside tunnels. So there are many studies in terms of propagation channels related to waveguide theory inside tunnels.

- **Railway station**: Railway station helps trains for regularly stopping and loading/unloading passengers. Generally, the speed of the train will change dramatically near the railway station. When the trains are near to the station, the speed of the trains will drop less than 80 km/h. Further, it will drop to 3–5 km/h while entering the station
  - Station A: This kind of railway station indicates the medium- or small-sized one [1]. The passengers and the load/unload platform are generally near to the rails. This construction can satisfy the medium traffic requirements. Moreover, the station mostly does not have awning on the top of the rails. Similar to the radio wave propagation considered in suburban environment, both LOS and NLOS conditions should be considered in this scenario.
  - Station B: This kind of railway station indicates the large-sized station with busy traffic requirements. For each day's operation, there are usually more than 60 thousand people and 6000 train appearance (these number will more easily to boom during the vacation). The Guangzhou South railway station, Beijing South Railway station are representative for this kind of railway station. On the top of the rail, there is usually a big awning exists, which makes the radio wave propagation similar to indoor scenario [1]. Furthermore, the base stations are generally located outside/inside the awnings. This structure leads to significant characteristics in radio wave propagation when the train moves into or out of the station.
  - Station C: This kind of railway station indicates the marshaling stations and container depots where the freights are loaded into or unloaded from the carriages and the carriages are marshaled [1]. The rails and carriages are dense, which expects higher level requirements for train operating signal system. Meantime, the dense structure consisted of dense metallic carriages leads to richness reflection and scattering components in propagation channel.

- **Combination scenarios**: The actual HSR scenarios are far more complex, especially in China where the railway tracks are continuous more than thousands of kilometers with constant variation of the topography. This existing circumstance brings about the combination scenario that several propagation scenarios may exist in one communication cell. The challenging task for prediction of radio wave propagation is urgent need of research. Therefore, there are two categories of combination in HSR: tunnel group (Combination scenario A) and cutting group (Combination scenario B) [1].
  - Combination scenario A: When the train passes through the terrain of multi-mountain environment, the train will experience a number of tunnels. These tunnels are not continual, which causes the train frequently moves in and moves out of the tunnel. In this process, the radio wave propagation will greatly under serious fading at the beginning or the end of the tunnel [1].

  - Combination scenario B: In the cutting group, the structure of the cutting changes frequently with the changes of terrain, e.g., the steep wall may transitorily disappear. The constantly changes among different cutting structures make the wireless communication undergo a disruptive change.

- **In-carriage**: To provide a high-quality service for passengers, the corresponding wireless communication intra the carriage is very important. The relay transmission (In-carriage-A) and direct transmission (In-carriage-B) are defined as the two categories of in-carriage scenario for wireless communication.

  - In-carriage-A: In the relay transmissions case, a so-called moving relay station will be mounted at the ceiling of the carriage [1, 9]. The base station communicates with the relay, then, the relay provides the communication service to the passengers. The channel characteristics of link between base station and relay will experience various fading and loss depending on the different scenario partitions talked above. Furthermore, the radio wave propagation intra the carriage is less likely to be influenced by the changes of the outside environments. Therefore, several typical indoor channel models can be employed to guide the channel propagation intra the carriage [1].
  - In carriage B: The direct transmission indicates the link that is developed directly from base station to the passenger inside the carriage. In this scenario, the penetration loss of the carriage has an additional effect on the radio wave propagation.

The proposed scene partition scheme is based on comprehensive consideration of three categories attributes in HSR: the physical attribute, user attribute, and coverage of wireless network.

The first attribute is related to radio wave propagation mechanisms between base station and passenger (e.g., LOS, NLOS, reflection, etc.). The second attribute is related to the requirements of user who need the high quality of service. The third attribute is related to the various wireless networks covering approach [1]. The partitioning for different scene is of great importance, especially for developing the accurate path loss prediction models and propagation channel characterization in HSR scenarios. The following sections are mainly concentrated on the channel modeling in different HSR scenarios.

3.1.2 Propagation Scenarios of Wide-Sense Vehicle-to-X Communications

Vehicle-to-X (V2X) communication and train-to-X (T2X) communication have been more and more important over the past few years, since road safety and railway safety are increasingly noticed by the public. As shown in Fig. 3.1, V2X and T2X are collected together to constitute the complete concept—Wide-Sense Vehicle-to-X (WSV2X), in order to form the comprehensive understanding. The

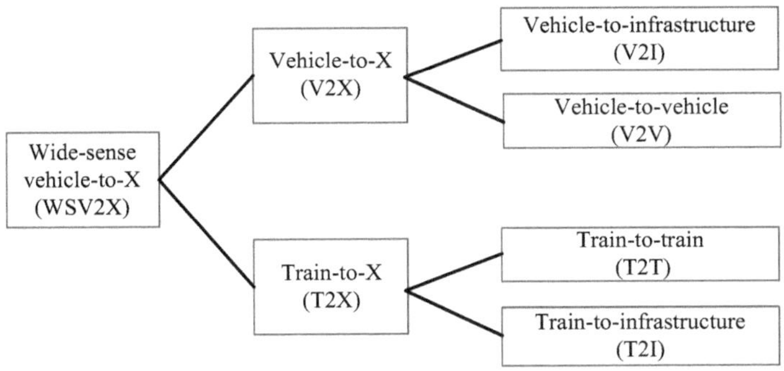

Fig. 3.1 Block diagram of the constitution of the concept of wide-sense vehicle-to-X (WSV2X)

deployment of wireless communication technology in vehicular and railway networks is the basic idea behind WSV2X communications. In this way, vehicles, trains and infrastructure build a wireless network so that they can exchange control and traffic information, such as road barriers, traffic accidents, etc., through the wireless communication link.

The standardization and the application related toWSV2X are as follows.

(i) Intelligent transport systems (ITSs) [10] for V2X communications: ITSs are advanced applications which aim to provide innovative services relating to different modes of transport and traffic management. They help various users get a better understanding of transport networks, and make safer, more coordinated, and "smarter" use of them. Some significant efforts have been made to realize the ITSs, such as IEEE 802.11p [11] and ETSI ITS G5 [12]. Part of the Wireless Access in Vehicular Environments (WAVE) initiative [13] is developed to operate at 5.9 GHz band, with 75 MHz bandwidth and seven 10 MHz channels. Systems based on IEEE 802.11p as well as alternative systems are also being developed in the USA, European Union, and some Asian countries.

(ii) Communication-Based Train Control (CBTC) system [14], Global System for Mobile Communications Railway (GSM-R) [15], and Terrestrial Trunked Radio (TETRA) [16] are the three main standard systems for T2X communications. (1) CBTC works at 2.4 GHz, which connects the train and track equipment for the traffic management and infrastructure control by telecommunications. (2) As an international wireless communications standard for railway communication, GSM-R is used for communication between train and railway regulation control centers and applications. GSM-R works at 900 MHz band and can guarantee performance at speeds up to 500 km/h without any communication loss. (3) TETRA works at the 400 MHz band and it mainly serve for government agencies and emergency departments such as police forces, fire departments, and ambulance. Besides, it is used by transport services, and the military.

(iii) Although there is no official standard for T2T communications so far, many academic efforts have been made to achieve the direct T2T communications. The most representing system is called Railway Collision Avoidance System (RCAS) [17].

(iv) Long-Term Evolution for Railway (LTE-R) [18, 19] and LTE for V2X communications are researched and under industrial consideration. In the near future, the time of the wideband WSV2X communications can be expected to come. For the preparation, the wideband channel characteristics of WSV2X communications are under research by a large number of scholars.

In the next few sections, we will focus on the scenarios, characters, and modeling philosophies of the propagation and wireless channel of WSV2X, since the propagation and channel characterization are always a fundamental topic with high research interest.

The properties of the scenarios in which the cars, trains, and infrastructures communicate with each other dominate the channel characteristics of WSV2X channels. Four main scenarios for the V2X channels (highways, suburban streets, rural streets, and urban streets, defined by [20]) and nine main scenarios for the T2X channels (viaduct, water, suburban, cutting, mountain, rural, tunnel, urban, and station) are shown in Fig. 3.2. There have been many studies on the channel characteristics of these scenarios. As seen in Fig. 3.2, the scenarios in the blocks with the same color have similarities on the characters of the environments, and are expected to show similar properties of channels. Although each single scenario has already been well researched independently, no study on these channels in the similar scenarios has been found. In order to give some rough inspirations of the joint analysis of the V2X and T2X scenarios, some common senses of the propagation mechanisms in every group of comparable V2X and T2X scenarios are summarized in Fig. 3.2.

The differences of characteristics between WSV2X channels and traditional cellular communications channels are summarized in accordance with the following dimensions.

(i) Heights of Tx and Rx: the Tx and the Rx in V2X channel and T2T channel are generally at the same height and in similar environments. (1) LOS: the LOS is relatively harder to be kept compared to CC channel and T2I channel. (2) Diffraction: in CC channel and T2I channel, the wave mainly propagates in the vertical plane, so the obstructs are the roof tops of buildings, top of the cutting walls, and terrain changes. However, the propagation in V2X and T2T channels mainly happens in the horizontal plane.

(ii) Frequency of communication: the carrier frequency of V2X channel is 5.9 GHz, which is higher than that of CC channel (700–2100 MHz) and T2X channel (400 MHz, 900 MHz, and 2.4 GHz).

(iii) Distance between Tx and Rx: the distance of communication is normally dozens of meters or hundreds of meters, which is shorter than that of typical CC and T2I channels (1–3 km).

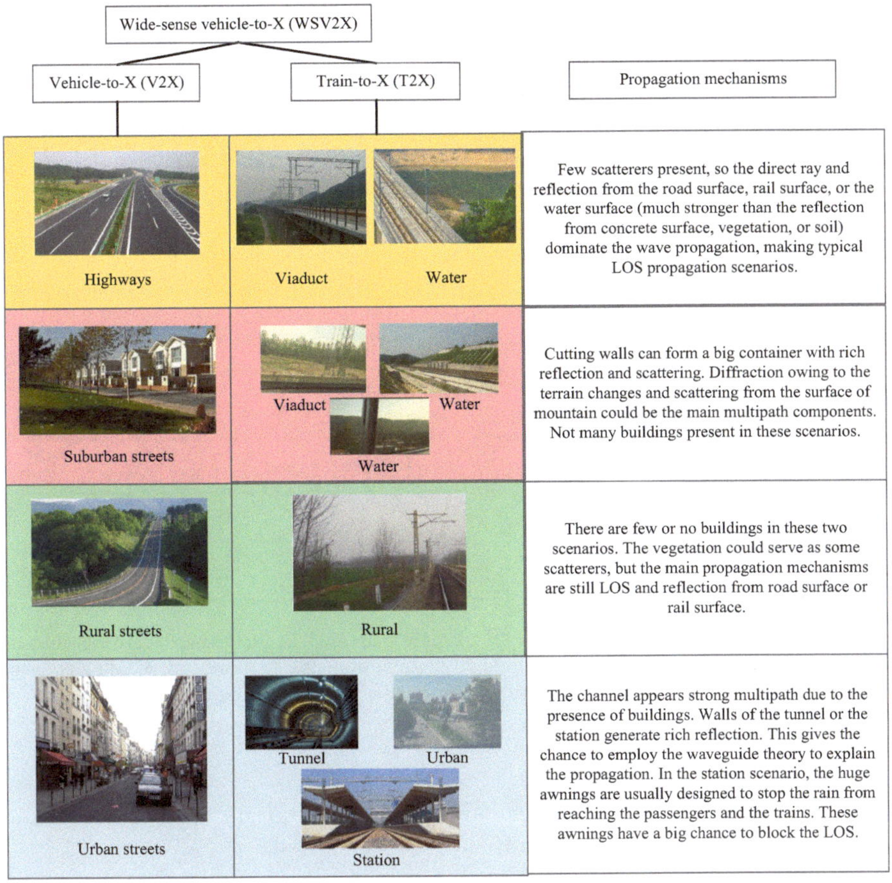

Fig. 3.2 Typical scenarios of WSV2X channels. The scenarios in the same *light colorful blocks* are similar or comparable

(iv) Nonstationarity: The WSV2X channel is the typical nonstationary channel; that is, the channel statistics change within a rather short period of time [20].

Since the WSV2X channel changes as Tx, Rx, and scatterers are moving around, the channel impulse response (CIR) which contains all information about the channel is time-variant. Figure 3.3 offers a panorama of metrics for the WSV2X channel classified in accordance with the antenna configuration, bandwidth of the system, and different domains. All the parameters can be assigned to totally six domains:

(i) loss and fading: path loss, shadow fading, and small-scale fading,
(ii) time domain: coherence time and stationary time,

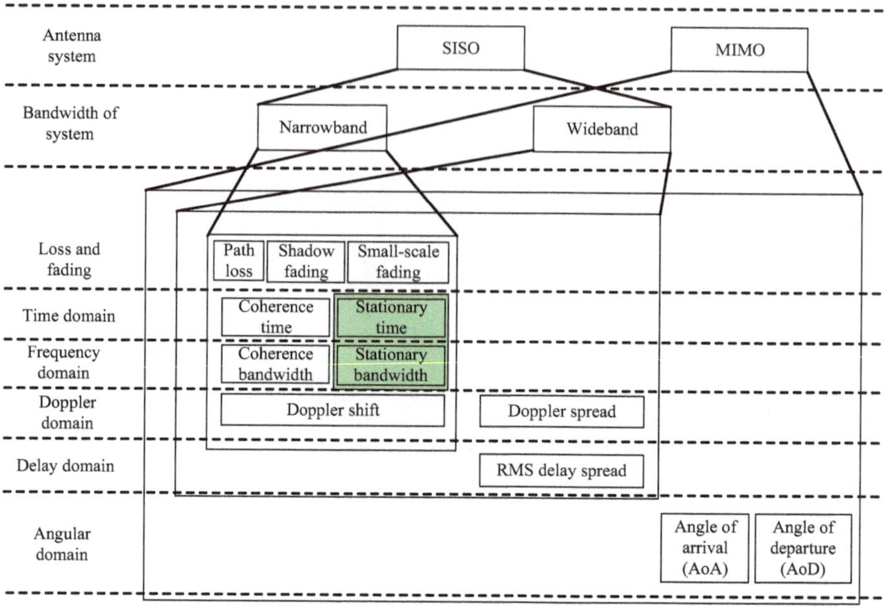

Fig. 3.3 Panorama of metrics for WSV2X channel classified in accordance with the antenna configuration and bandwidth of the system and different domains. Stationary time and stationary bandwidth (in the block with *green red color*) are the two specific parameters in the nonstationary channel

(iii) frequency domain: coherence bandwidth and stationary bandwidth,
(iv) doppler domain: Doppler shift and Doppler spread,
(v) delay domain: RMS delay spread,
(vi) angular domain: angle of arrival (AoA) and angle of departure (AoD).

Figure 3.4 (which is the measured results in [21, 22]) gives an example of the time-varying power delay profile (average squared magnitude of impulse response) in the WSV2X channel. Up till now, most of the metrics given by Fig. 3.3 in the V2X channel have been well studied by many researchers, but there is no deep insight or comprehensive understanding of these characters for wideband MIMO T2X communication systems. There are still quite limited ergodicity and the reliability of the experimental data and corresponding parameter extractions, and more works in different scenarios at various frequencies are needed. According to the research results of [20, 23–28], similar or comparable scenarios of T2X, V2X, and standard scenarios are collected together, and three preliminary categories are obtained as follows.

(i) "Relatively open scenarios": this category is defined by the scenarios with a relative wide open space but few scatterers or strong reflectors.

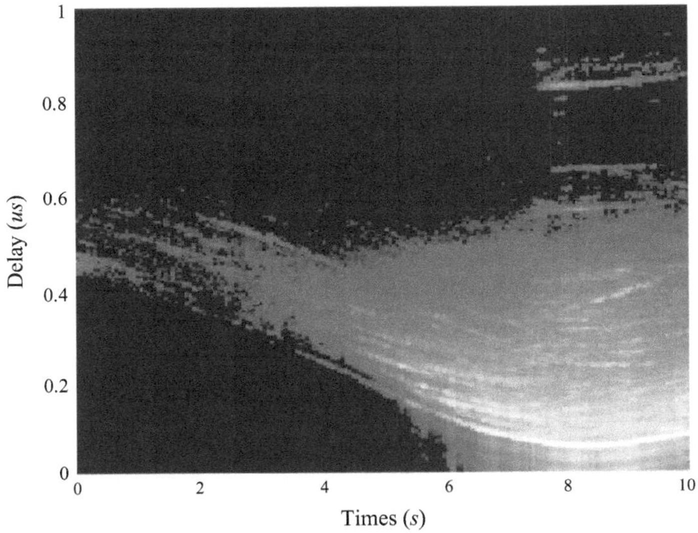

Fig. 3.4 Example of the time-varying power delay profile (average squared magnitude of impulse response) in theWSV2X channel. It can be found that the delay of the first taps varies because the *Tx* and *Rx* approach each other firstly and move away from each other later. The LOS tap experiences fading, the gap in the delay domain between strong components (clusters), and the LOS tap changes with time, and there is splitting of clusters over time as well

(ii) "Semi-closed scenarios": this category indicates that the scenario is surrounded by some walls, buildings, or terrain but still has some free or open space.
(iii) "Relatively closed scenarios": this category mainly includes the limited space and closed scenarios, such as the T2X tunnel, T2X station, V2X urban, standard indoor office/residential, standard typical urban microcell, and standard large indoor/hall.

There are primarily three types of channel modeling approaches forWSV2X channel: deterministic channel models, stochastic channel models, and geometry-based stochastic models.

3.1.2.1 Deterministic Channel Model for WSV2X Channels

Deterministic channel model was pioneered by Wiesbeck [29–31]. Based on ray-tracing techniques, deterministic channel models use the geographical and morphological information from a database to model the propagation channel in a specific location. Generally, the direct path, specular reflections, and diffuse scattering are included in 3D ray-optical approach. In order to determine single scattering processes, all surfaces of the structures in the scenario, such as buildings, trees, vehicles, and traffic signs are divided into tiles. The corresponding time-variant CIR $h(\tau, t)$, which well characterizes the frequency-selective channel and can be expressed as

$$h(\tau, t) = \sum_{k=1}^{N(t)} a_k(t) \cdot e^{j(2\pi f \tau_k(t) + \varphi_k(t))} \cdot \delta(\tau - \tau_k(t)), \tag{1}$$

where the K-th multipath component at time t is formulated by an amplitude $a_k(t)$, a delay $\tau_k(t)$, and an additional phase shift $\varphi_k(t)$. More detailed information about the deterministic modeling approach can be found in [32, 33].

3.1.2.2 Stochastic Channel Models

Basically, the stochastic channel models can be divided into two types: narrowband stochastic channel models and wideband stochastic channel models. Unlike deterministic channel models, stochastic channel models do not require to determine the impulse response in a specific location. Narrowband stochastic channel models mainly focus on characterization of the fading statistics as well as the Doppler spectrum. Some researches and interesting results about narrowband stochastic channel models can be found in [34–37]. Wideband stochastic channel models contain relative more information about the investigative channels, which usually create the statistics of the received power with a certain delay, Doppler shift, and angle of arrival, etc.

3.1.2.3 Geometry-Based Stochastic Models

Geometry-based stochastic modeling (GSCM) is similar to the stochastic modeling, but it uses a simplified ray tracing along with random scatterers. The GSCM is widely used in MIMO channel modeling [38]. Generally, the GSCM models can be divided into the regular-shaped GSCMs (RS-GSCMs) and irregular shaped GSCMs (IS-GSCMs) [39], and both of them have been researched in the V2X channel and the T2I channel. Figure 3.5 gives sketches of the geometry-based stochastic model of T2I channel [40] in the cutting scenario and V2 V channel [41] with scatterers in realistic positions.

3.1.2.4 Summary and Selecting a Suitable Modeling Approach

There are some specific advantages and disadvantages of the modeling methods offered above.

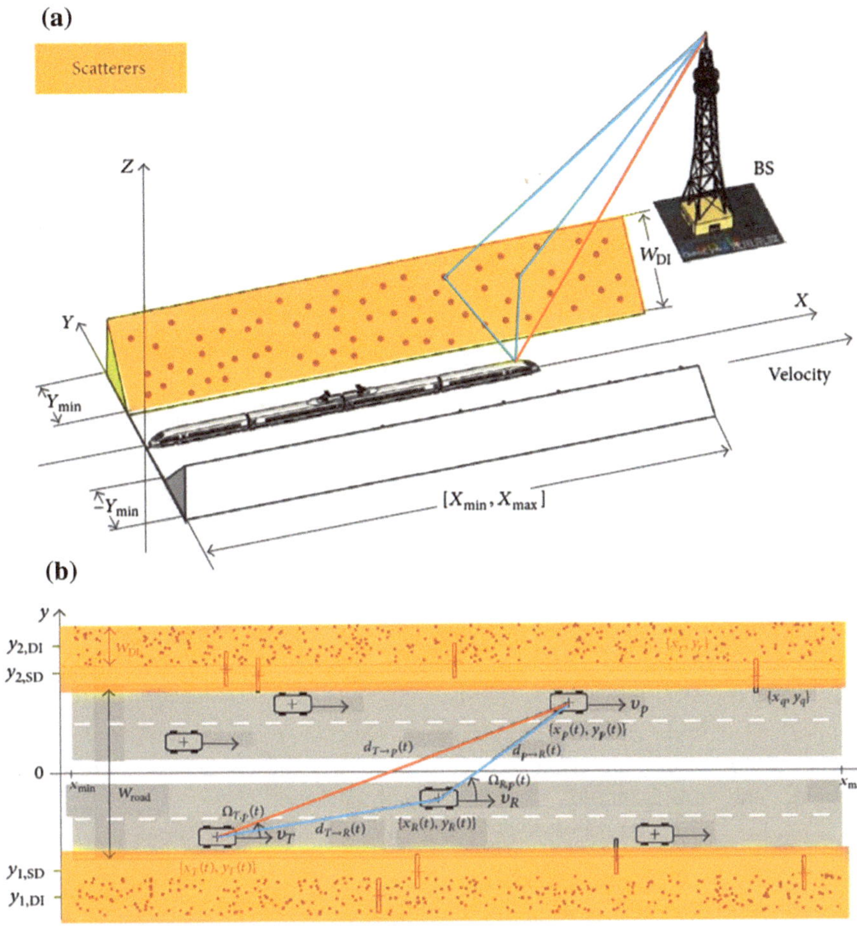

Fig. 3.5 **a** Sketch of the geometry-based stochastic model of *T2I* channels in the cutting scenario with scatterers in realistic positions [40]. **b** Sketch of the geometry-based stochastic model of V2V channels [41]

(i) The biggest advantage of deterministic channel model is that it offers the most accurate simulation of the realistic channel model including the nonstationarity of the channel naturally. But, it requires highly accurate topographical databases.

(ii) Stochastic channel model is a relatively flexible way to describe channels in different scenarios. However, the main drawback of the application of such kind of modeling method to the WSV2X channel is that it does not consider the nonstationarity of the channel.

(iii) Geometry-based stochastic model can reflect the realistic behavior of theWSV2X channel and implicitly describes the nonstationarity of the channel. But it still needs too much computation compared with stochastic channel model.

Traditional concepts of V2X and T2X are summarized together to constitute a more general concept—wide-sense V2X (WSV2X). This chapter offers an overview of the development of the WSV2X channels, such as the standardization, scenarios, characters, and modeling approaches. This chapter integrates the common senses of V2X, T2X, and standard scenarios and forms a new panorama of the scenarios classified by the similar physical characters and propagation mechanisms.

3.2 High-Speed Railway Channel Measurements

3.2.1 Measurement Methods and System

This section presents various methods and related systems to measure the radio channels in high-speed railway (HSR) networks.

Measuring the wireless radio channel is better known as channel sounding: A transmitter sends out a known signal that excites the channel, the receiver stores the received signal and can thus estimate the radio channel from the known transmit signal [42]. Generally, there are three types of channel sounding methods from the perspective of system addressing the key feature of radio channel in narrowband, wideband, and multi-antenna wireless communication systems, respectively. Narrowband sounding only provides fading characteristics of wireless channel but does not provide information regarding the multipath components (MPC). While dispersion characteristics of wireless channel can be observed via wideband sounding method. A multi-antenna sounding can add spatial information and can be used for MPC extraction. Different measurement approaches can also be divided into time-domain and frequency-domain approaches. One can obtain the impulse response of wireless channel directly via a time-domain channel measurement. Alternatively, a frequency-domain channel measurement can estimate the transfer function. The choice of channel sounding method will usually depend upon the application foreseen for the measurement data.

In HSR communication systems, radio channel is rapidly time-varying due to the high mobility of the train and the fast changing propagation environments. As a result, the characteristics of HSR channels, e.g., nonstationarity and large Doppler shift, significantly differ from those of low mobility mobile cellular communication channels [43, 44]. Furthermore, HSR channel differs significantly from those available for other mobile cellular systems [45, 46]: (1) Diverse scenarios and propagation conditions: propagation channel environment is highly variable (e.g., tunnels, viaducts, terrain cuts, and so on). (2) Line of Sight (LoS) conditions dominance: current HSR routes and network plans ensure LoS in most of the cases, thus multipath components contribute less between the mobile terminal and base station (BS).

In a time-variant system, the repetition period T_{rep} of the transmit signal is of fundamental importance [47]. Transmit signal should be designed properly to meet the requirement of channel measurements in HSR environments. For the channel sounding of time-variant channels it has to be ensured that the channel is *doubly underspread*, i.e., that the channel fulfills a two-dimensional Nyquist criterion [48]

$$2v_{\max}\tau_{\max} \ll 1.$$

In other words, T_{rep} should be at least twice the maximum expected Doppler shift, depends on the carrier frequency and vehicular speed. A doubly underspread channel is both dispersion underspread and correlation underspread, and is characterized by having its spreading function and correlation function concentrated about origin of the delay τ and Doppler v plane. This implies that the product of the maximum delay, $\tau_{\max}$, and Doppler, $v_{\max}$, should be small, such that their product is much smaller than 1. A small $\tau_{\max}$ and $v_{\max}$ result in a large coherence bandwidth and coherence time, respectively.

First, the sampling rate or snapshot repetition time $T_{\text{rep}} = Kt_s$ must be shorter than the coherence time of the channel, i.e., the channel must not change during the sounding of one snapshot. This implies

$$t_s \leq \frac{1}{2v_{\max}K},$$

where t_s is the length of sounding sequence and $K = T_{\text{rep}}/t_s$ is the repetition period. Second, the length of the sounding signal must always be longer than excess delay $\tau_{\max}$ of the channel in order to avoid overlap between consecutive sounding signals. This inequality takes into account the delay dispersion of the channel by

$$t_s \geq \tau_{\max}.$$

Thus, the values of snapshot repetition time should be fulfilled by [49]

$$\tau_{\max} \leq t_s \leq \frac{1}{2v_{\max}K}.$$

For example, when the multipath components are assumed to extend over 1 ms and the distance of two base stations (BSs) in HSR communication systems is 1.2 km, which corresponds to 4 ms, the length of the sounding signal has to be in excess of 5 ms to ensure that the movement of the multipath components is still within the observable window [50]. Besides, the frequency band of future HSR communication systems is recommended to 800 MHz frequency band for high-priority service and support low-priority service in the 1.8 GHz frequency band [51]. At 1.8 GHz, the maximum expected Doppler shift can be accommodated within 600 Hz for a maximum train speed of 360 km/h, then the snapshot repetition time T_{rep} should be at most smaller than 833 μs. As for multiple-input multiple-output (MIMO) measurements in HSR environments, these requirements are essentially the same with the added requirement of multiple transmissions and multiple receptions. In order to capture Doppler spread completely, the switch time of all antennas should be completed within the coherent time of the channel.

Channel measurement systems for HSR have to be able to measure and store fast fluctuations of the time-varying, wideband, double-directional propagation channel. Architecture of a typical channel sounding system for mobile communication networks is shown in Fig. 3.6. There exist different types of channel sounders [52]. One can distinguish between channel measurement systems used to collect narrowband channel response data, wideband channel response data, and channel response data collected using multiple transmitting and/or receiving antennas [53].

While current channel sounding techniques are quite mature, there are still some challenges in HSR channel measurements that should be addressed including [44]:

(1) How to transmit a particular waveform without interference to other existing wireless networks in HSR channel measurements.
(2) How to perform MIMO measurement in HSR environments.
(3) How to increase the measurement efficiency in HSR channels.

In fact, measurement using the dedicated channel sounding equipment can only be performed in a certain fixed scenario. It is hard to reflect the overall channel properties due to diverse scenarios on HSR. Moreover, since the train can travel with a maximum speed of 360 km/h and usually the wireless coverage of a channel sounder is 1 km, the recording time in the coverage is approximately 20 s, which may not be adequate to obtain the statistical property of the time-varying channels by the collected data [45]. An efficient channel identification method should be proposed for HSR environments.

A good example of HSR channel sounding system based on existing cellular network with high measurement efficiency and low measurement restriction is shown in Fig. 3.7. This system consists of the LTE railway network and an LTE

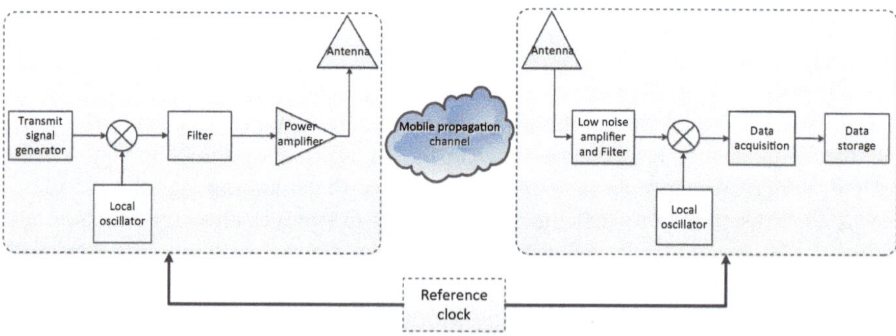

Fig. 3.6 Block diagram architecture of a typical channel sounder in mobile networks [52]

sounder. The LTE sounder is used to collect the channel data in the whole coverage area of the network, which makes continuous measurements feasible. To sum up, the channel measurement system for HSR environments should adopt specific measurement method considering measurement coverage, interference against existing wireless network, MIMO design as well as measurement efficiency to get correct and accurate measurement data to investigate the radio channel propagation characteristics.

3.2.2 *Measurement Campaign*

High-speed railway channel measurement campaigns pose additional challenges compared to measurements in cellular environments. Many measurement campaigns [54–107] for different HSR environments were presented in the literature. We briefly review some recent typical HSR radio channel measurement campaigns according to the scenarios, measurement equipment, measurements' setup parameters (i.e., carrier frequency, bandwidth, and antenna configuration), and estimated channel statistics, as shown in Table 3.1. Most of measurements for HSR can be categorized as follows:

3.2.2.1 Channel-Sounder-Based Measurements

There are several HSR measurement campaigns using standard commercial multidimensional channel sounders so far. In 2006, RUSK measurement campaign was conducted at 5 GHz in Germany between Siegburg and Frankfurt by Medav, TU Karlsruhe, and TU Ilmenau as rural moving networks which is also known as the D2a model in the Winner II model [54]. It is the first time that a standard channel sounding system used in HSR scenarios. This sort of measurement is perfect for one site channel investigation with high bandwidth and MIMO configuration, but measurement efficiency is relatively low if someone wants to conduct a long

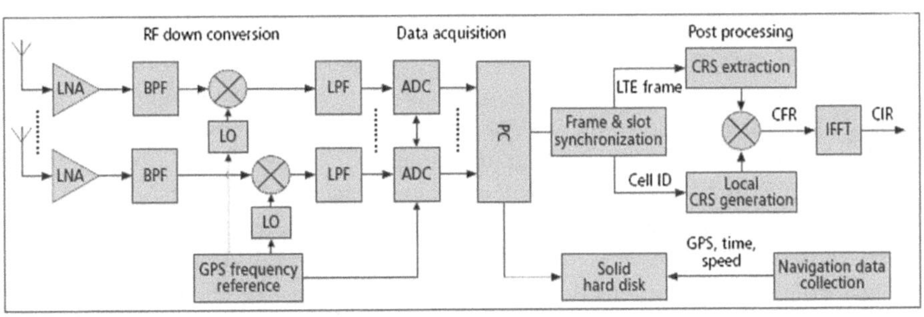

Fig. 3.7 LTE-based HSR channel sounding scheme in [44]

Table 3.1 Standardized path loss model (in dB) for hsr environments. Reprinted from Ref. [1], Copyright 2014, with permission from IEEE

Urban	$\Delta_1 = -20.47$ $\Delta_2 = -1.82$
Suburban	$\Delta_1 = 5.74\log_{10}(h_b) - 30.42$ $\Delta_2 = -6.72$
Rural	$\Delta_1 = 6.43\log_{10}(h_b) - 30.44$ $\Delta_2 = -6.71$
Viaduct	$\Delta_1 = -21.42$ $\Delta_2 = -9.62$
Cutting	$\Delta_1 = -18.78$ $\Delta_2 = 51.34\log_{10}(h_b) - 78.99$
Station	$\Delta_{1\text{-}3}4.29\log_{10}(h_b) - 70.75$ $\Delta_2 = -8.86$
River	$\Delta_1 = 8.79\log_{10}(h_b) - 33.99$ $\Delta_2 = -2.93$

distance measurement under different scenarios and also the data process for MIMO measurements will be a huge challenge.

3.2.2.2 Railway-Network-Based Measurements

Due to practical limitations on HSR, some researchers have attempted to perform the channel measurement campaigns based on railway network instead of traditional channel sounders. The basic idea is to exploit the signal transmitted from the railway network to enable continuous measurements along the rail track. One of the biggest advantages of these measurement campaigns is that a wider coverage can be achieved for a group of terrains such as plain, foothill, urban area, and tunnel. A series of GSM-R channel measurements were conducted in viaduct scenarios on HSR in China [57–69, 85–95]. For channel characterization purposes, the GSM-R signal is regarded as a narrowband CW signal. To enable wideband channel characterization, the common pilot channel signal in the dedicated WCDMA network was collected and analyzed to extract the multipath properties [93, 94]. However, the measurement setup will be imposed restrictions by the existing railway networks.

As shown in Table 3.1, most of the existing HSR measurements are conducted in the following scenarios: open space, viaduct, cutting, hilly terrain, tunnels, and stations. Regarding carrier frequency, most of the measurement campaigns in the literature were conducted at the carrier frequency of 930 MHz in GSM-R systems [57–69, 75–78, 85–92]. It is worth mentioning that all of the aforementioned measurements were for narrowband channels with bandwidth of 200 kHz. Wideband channel measurements with higher bandwidths, i.e., 10–100 MHz, and higher carrier frequencies, i.e., 2.1–5.2 GHz, were reported in rest of them.

On antenna configuration side, the majority of HSR measurements campaigns so far have concentrated on single-input single-output (SISO) systems. In MIMO systems, where multiple antennas are equipped at both ends, are essential for providing higher capacity to meet the requirements of future high-speed data transmissions [99]. The channel measurement, particularly the MIMO channel measurement at high moving speeds, remains to be a challenging task. As to far, only very few measurement campaigns were conducted using multiple antennas at either the Tx, i.e., single-input multiple-output (SIMO) systems [54, 108], or Rx, i.e., multiple-input single-output (MISO) systems [54]. Hence, MIMO wideband channel measurement campaigns with carrier frequency and bandwidth larger than GSM-R ones are needed for future HSR communication system developments. The first measurements that combine MIMO at speeds of 300 km/h are reported in [96]. Moreover, the MIMO measurement system does not use a switched array, but records channels in parallel.

Several statistical channel metrics, which provide a more condensed characterization, have been derived and widely adopted from HSR measurement campaigns: path loss, fading statistics, Doppler spread, and delay spread. One can conclude that large-scale fading statistics, i.e., path loss (PL) and shadowing are the most concerned parameters of channel investigation in HSR environments. The Ricean K-factor is another important parameter in link budget and channel modeling. Therefore, many papers presented the estimation of K-factors in different scenarios, e.g., open space [54], viaduct [57, 58, 66, 67, 70, 71], cutting [75–79], and hilly terrain [81]. In [61, 67, 68, 78], the spatial/temporal variations, e.g., fade depth (FD), level crossing rate (LCR), and average fade duration (AFD), were investigated. FD is a measure of variation in the channel energy about its local mean due to small scale fading and it is calculated from the difference in signal levels between 1% and 50%. Measurements in viaduct scenarios have shown that FD is independent of the viaduct's height but is affected by the number and closeness of surrounding scatterers that are higher than the viaduct [61, 68]. LCR is defined as the expected rate at which the received signal crosses a specified level in a positive-going or negative-going direction, while AFD is defined as the average period of time for which the received signal is below this specified level, i.e., threshold. LCR and AFD were statistically modeled as functions of the structural parameters of the viaduct and cutting scenarios in [68, 78]. The results showed that the severity of fading in viaduct scenarios is greatly reduced compared with that in open space scenarios, since fewer reflected and scattered paths in viaduct scenarios are expected at the receiver which leads to smaller values of LCR. Obstacles around the viaduct can cause minor variations of the LCR values but have no significant impact on the AFD. Cutting's dimensions have also very minor impact on the AFD of the received signal, while surrounding obstacles and crossing bridges over the cutting have no influence on the LCR and AFD. The stationarity interval, defined as the maximum time duration over which the channel satisfies the wide sense stationary (WSS) condition, of HSR channels was investigated in [88, 90] based on measurements. It showed that conventional channel models offered stationary intervals much larger than the actual measured ones. Doppler behavior and angular information of HSR channels in open space scenarios were analyzed in [55], while power delay profiles (PDPs) were investigated in [54, 56, 88, 93, 94].

Ref	Scenario	Measurement type	Carrier frequency	Bandwidth	Antenna configuration	Train velocity	Channel statistics
[54]	Open space	Channel sounder based (MEDAV RUSK sounder)	5.2 GHz	120 MHz	SIMO	350 km/h	PL, SF, K, DS, PDP, AS
[55]	Open space	Channel sounder based (PropSound)	2.5 GHz	50 MHz	MISO/SIMO	290 km/h	DS, AoA, AoD, PAS, DF
[56]	Open space	Channel sounder based (VSG + VSA)	2.6 GHz	20 MHz	SISO	370 km/h	PL, DS, DF, PDP
[57–69]	Viaduct	Railway network based (Willtek 8300 Griffin)	930 MHz	200 kHz	SISO	350 km/h	PL, SF, PDF, LCR, AFD, CDF, FM
[70]	Viaduct	Channel sounder based (PropSound)	2.35 GHz	10 MHz	SISO	240 km/h	PL, DS, K
[71]	Viaduct	Channel sounder based (PropSound)	2.35 GHz	50 MHz	SISO	196 km/h	DS, K, SF
[72]	Viaduct	Channel sounder based (VSG + VSA)	2.6 GHz	20 MHz	SISO	370 km/h	PL, SF, DS, K
[73, 74]	Viaduct	Channel sounder based (PropSound)	2.35 GHz	50 MHz	SISO	200 km/h	PSD, DF, AoA, K
[75–78]	Cutting	Railway network based (Willtek 8300 Griffin)	930 MHz	200 kHz	SISO	350 km/h	PL, SF,ψ K, FD, LCR, AFD
[79, 80]	Cutting	Channel sounder based (PropSound)	2.35 GHz	50 MHz	SISO	200 km/h	PL, K, SF, DF, DS
[81, 82]	Hilly terrain	Channel sounder based (Tsinghua university, THU)	2.4 GHz	40 MHz	SISO	295 km/h	PL, SF, K
[83]	Tunnel	Channel sounder based (Helsinki University of Technology, TKK)	2.154 GHz	30 MHz	SISO	N/A	PL, DS
[84]	Tunnel	Channel sounder based equipment	2.154 GHz	30 MHz	SISO	N/A	PL
[85, 86]	Station	Railway network based (Willtek 8300 Griffin)	930 MHz	200 kHz	SISO	N/A	PL, K, SF, FD, LCR, AFD
[87]	Crossing bridge	Railway network based (Willtek 8300 Griffin)	930 MHz	200 kHz	SISO	N/A	PL, K, SF, FD
[88–92]	Various	Railway network based (Willtek 8300 Griffin)	930 MHz	200 kHz	SISO	350 km/h	PL, PDF, DS, PDP, SI,SF,ψ K, LCR, AFD
[93, 94]	Various	Railway network based (R&S TSMQ Radio Network Analyzer)	2.1 GHz	3.84 MHz	SISO	240 km/h	PL, K, DS, PDP
[95]	Various	Railway network based (Universal Software-defined Radio Peripheral, USRP)	2.1 GHz	3.84 MHz	SISO	300 km/h	PL, PDP

(continued)

(continued)

Ref	Scenario	Measurement type	Carrier frequency	Bandwidth	Antenna configuration	Train velocity	Channel statistics
[96]	Various	Channel sounder based (Eurecom Express software-defined radio Card)	800 MHz/2.6 GHz	5/10/20 MHz	MIMO (2*2)/ MISO/SISO	300 km/h	PDP, DF
[97]	N/A	Channel sounder based	2.2 GHz/5.2 GHz	20 MHz	SISO	270 km/h	PL
[98]	N/A	Channel sounder based (PropSound)	2.35 GHz	100 MHz	SISO	N/A	PL, DS, K

SISO single-input single-output, *MISO* multiple-input single-output, *SIMO* single-input multiple-output, *PL* path loss, *DS* RMS delay spread, *K* Ricean K-factor, *PDP* power delay profile, *AS* angular spread, *AoA* angles of arrival, *AoD* angles of departure, *PAS* power azimuth spectrum, *DF* Doppler frequency, *SF* shadow fading, *FD* fade depth, *LCR* level crossing rate, *AFD* average fade duration, *PDF* probability density function, *CDF* cumulative distribution function, *FM* fading margin, *PSD* power spectrum density, *SI* stationarity interval, *VSG* vector signal generator, *VSA* vector signal analyzer

3.3 Narrowband Channel Characterization of High-Speed Railways

3.3.1 Path Loss

3.3.1.1 Path Loss Model

Since currently the Hata model is widely used in the HSR engineering implementations in China, we develop our standard based on the Hata's formula. The classical Hata model includes three scenarios: urban, suburban, and open area. The path loss model in urban is considered as the basic formula and the correction factors are added to lead to the models in suburban and open area. The standard Hata model in urban (with the large city correction factor) is as follows

$$\begin{aligned} PL_{\text{Hata}} = {} & 74.52 + 26.16\log_{10}(f) \\ & - 13.82\log_{10}(h_b) - 3.21(\log_{10}(11.75h_m))^2 \\ & + [44.9 - 6.55\log_{10}(h_b)]\log_{10}(d), \end{aligned} \tag{3.3.1}$$

where f is the carrier frequency in MHz. hb and hm are the BS effective antenna height and the vehicular station antenna height (against the surface of rail track in the HSR) in meters. d is the T-R separation distance in kilometers.

The modified path loss model in HSR based on the Hata's formula is expressed as [100]

$$\begin{aligned} PL_{\text{Proposed}} = & \Delta_1 + 74.52 + 26.16\log_{10}(f = 930) \\ & - 13.82\log_{10}(h_b) - 3.21(\log_{10}(11.75h_m))^2 \\ & + [44.9 - 6.55\log_{10}(h_b) + \Delta_2]\log_{10}(d), \end{aligned} \tag{3.3.2}$$

where Delta1 and Delta2 are the correction factors for the proposed model. The reasons of using these two factors are: i) Delta1 is used to normalize the constant in the model to ensure a sufficient fit; ii) it has been found that the special railway environments and constructions usually affect the path loss exponent, so we add Delta2 in the model. We also remain some basic formula of Hata, e.g., the frequency term, so that it can still be easily extended to some other frequency bands in the future.

The correction factor Delta is derived from the difference between the optimal path loss curve, i.e., the Least Square (LS) regression fit curve. According to a visual inspection, we find that Delta can be modeled as a function of the logarithmical hb, expressed as [100]

$$\Delta_i = p \cdot \log_{10}(h_b) + q, \tag{3.3.3}$$

where p and q are obtained by the LS fit. This expression is also consistent with the formula in Hata model. We enforce $p = 0$ if no distinct linearly decreasing or increasing is observed, and use the averaged value of the measured correction factors as q. Since the empirical formula should be as simple as possible for usability, in the regression fit, we do not introduce new parameters of the railway constructions and topographical features into the model. This is because those parameters will significantly increase the model complexity, and more importantly, those parameters of the environments are usually not available to the engineers when they design the system. It is expected that Delta follows a similar expression against hm, however, no measurement is available to verify this expectation and develop a model.

The above table summarizes the estimated correlation factors for each scenario based on the LS regression fit. To remove the effect of the BS antenna pattern on the measured path loss, a calibration was conducted. As shown in the table, the derived correlation factors have the similar terms to the Hata formula, and can be easily extended into the Hata model.

3.3.1.2 Validation

To evaluate the goodness of fit (GoF) of the path loss model, the coefficient of determination R-Square and the root mean squared error (RMSE) are employed. *R*-Square is a measure of how successful the fit is in explaining the variation of the data, and RMSE is a measure of the differences between values predicted by a model and the values actually observed. *R*-Square ranges from -infty to 1, with a value closer to 1 indicating that the regression model fits the data better; and an RMSE value closer to 0 indicates a fit that is more useful for prediction.

To validate the proposed path loss model, we use the measurements from two different HSR lines [102]: i) "Zhengzhou–Xian" (ZX) line, which is used to propose the model; and ii) "Beijing–Shanghai" (JH) line (with 104 HSR cells), which is only used to validate the proposed model. The R-Square and RMSE of the proposed model, and four standard models: Hata, ITU-R, 3GPP, and WINNER, are estimated in each cell (Fig. 3.8).

The above four figures show the GoF comparisons of using the measurements in the ZX and JH HSR lines. The results of the LS regression fit in each cell, which is the optimal curve, are plotted for comparison. We can see that in both HSR lines, the proposed model outperforms other four standard path loss models, and the performance is very close to the optimal result (the green curves). We also note that the values of R-Square for the other four standard models are even less than 0 in a significant percentage of cases, which means the fit is actually worse than just fitting a horizontal line.

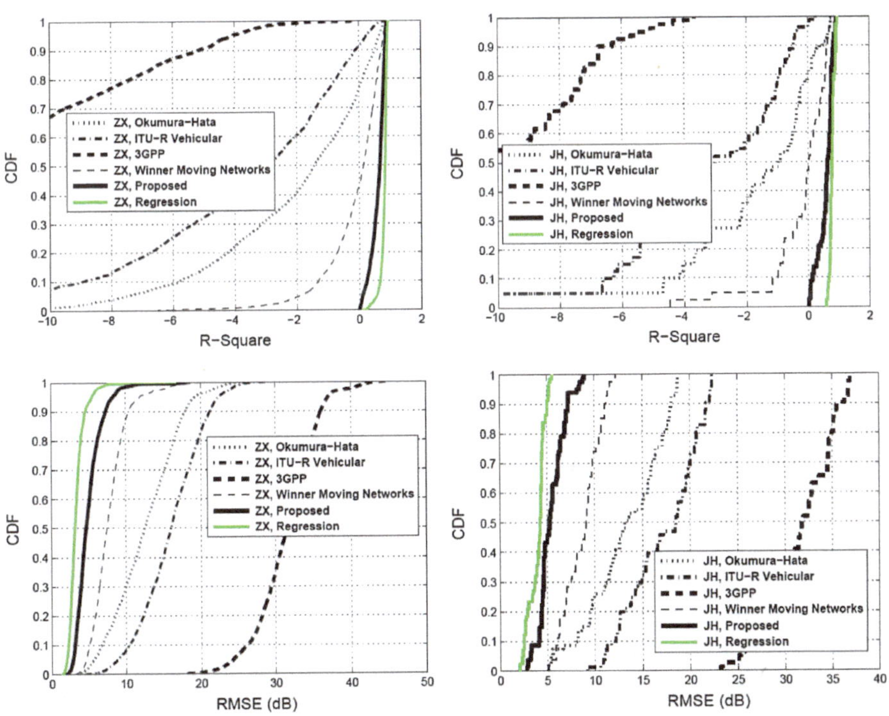

Fig. 3.8 Model validation with R-Square and RMSE. Reprinted from Ref. [44], Copyright 2014, with permission from IEEE

3.3.2 *Shadow Fading*

3.3.2.1 Standard Deviation

Our measurements suggest that the zero-mean Gaussian distribution fits the data (in dB) well in each environment [2]. We use the Kolmogorov–Smirnov (KS) test, with a CI of 95%, to validate the zero-mean Gaussian distribution. The statistic of the KS test is defined as the maximum value of the absolute difference between the cumulative distribution function (CDF) of the measured shadow fading components Y1 and the CDF of the estimated distribution Y2, which can be expressed as

$$D_{\mathrm{KS}} = \max(|F(Y_1) - F(Y_2)|). \tag{3.3.4}$$

The following table summarizes the KS passing rate of the zero-mean Gaussian distribution in each environment. It is found that the KS passing rate is larger than 84% in all the environments (Table 3.2).

The mean value of the standard deviation of shadowing in each environment is presented in the above table. We can see that mean value of sigma in the railway environments ranges from 2.7 to 3.7 dB. We also note that sigma in HSR scenarios is significantly less than in classical cellular systems. This is because in HSR, the high BS leads to clear LOS propagation (i.e., the signals are usually less shadowed), and thus reduces the shadowing effects.

3.3.2.2 Autocorrelation Characteristics

For the measurements from each BS, we estimate the autocorrelation coefficient of shadow fading. In our measurements the autocorrelation coefficient has been found to follow an exponential decay function, expressed as

$$\rho(\Delta d) = \exp\left(-\frac{\Delta d}{d_{\mathrm{cor}}}\right) \tag{3.3.5}$$

d_{cor} represents the decorrelation distance, which depends on the scenario and is usually defined to be the distance at which the correlation drops to 1/e. The

Table 3.2 Passing rate of the Gaussian distribution and standard deviation of shadow fading in railway environments. Reprinted from Ref. [2], Copyright 2014, with permission from IEEE

Environment	KS passing rate (%)	Mean, sigma (dB)
Urban	96.97	3.19
Suburban	85.48	3.33
Rural	93.61	2.85
Viaduct	91.92	2.73
Cutting	91.60	3.63
Station	84.59	2.77
River	91.09	3.09

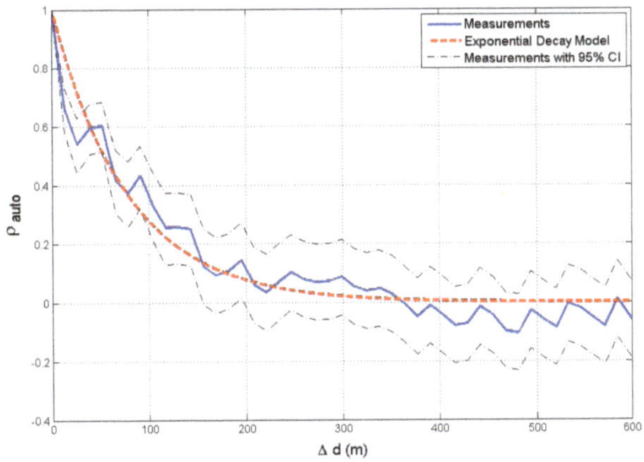

Fig. 3.9 Example plot of the measured autocorrelation coefficient in one cell of the viaduct environments, together with an exponential decay model. Reprinted from Ref. [43], Copyright 2014, with permission from IEEE

decorrelation distance reflects how fast the large-scale parameters are changing over the route (Fig. 3.9).

Example plot of the measured autocorrelation coefficient in the rural environments is shown in the above figure, where we can see that the exponential decay function offers a good fit to the measurements. The curves of the measurements with 95% CI are plotted, which are estimated. We can see that the CI is reasonably narrow before the autocorrelation coefficient drops to 0.2, which shows that the estimation of dcor with 1/e threshold has sufficient accuracy. Note that the tightness of CI depends on the number of samples. As we use the 40-wavelength sliding/nonoverlapped window to remove small-scale fading, in each cell the number of shadow fading samples is mostly less than 300. This size of data set limits the tightness of CI. (Table 3.3)

For the estimated d_{cor} in each environment, we examine the dependency of d_{cor} on the parameters of each scenario: *h*, theta, *D*, and h/theta. Those parameters affect the received power in the HSR cell for a particular BS antenna, and are also examined in the modeling of cross-correlation in the following subsection. However, we find that d_{cor} is independent of those parameters. Therefore, we summarize the mean value of d_{cor} for each railway environment in the above table.

Table 3.3 Shadowing autocorrelation analysis. Reprinted from Ref. [2], Copyright 2014, with permission from IEEE

Environment	Mean, dcor (m)	Correlation coefficient [95% CI]
Urban	57.12	0.28, [0.04 0.49]
Suburban	112.48	0.38, [0.34 0.42]
Rural	114.79	0.25, [0.18 0.32]
Viaduct	115.44	0.23, [0.19 0.27]
Cutting	88.78	0.34, [0.28 0.39]
Station	101.22	0.42, [0.36 0.48]
River	114.58	0.19, [0.07 0.31]

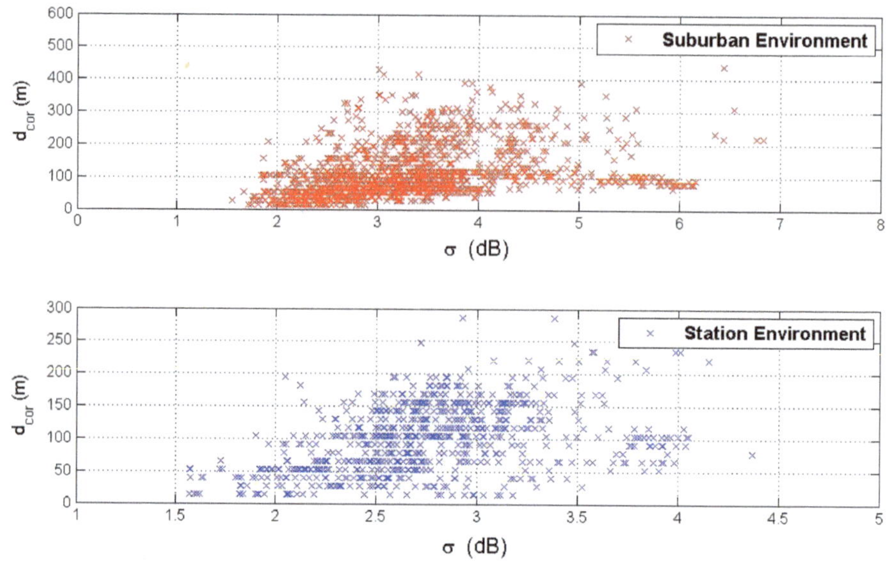

Fig. 3.10 Example plots of decorrelation distance in suburban and station environments. Reprinted from Ref. [43], Copyright 2014, with permission from IEEE

We can see that in most of the railway environments, d_{cor} is larger than 100 m; only in the urban and cutting environments, a d_{cor} less than 90 m is observed. We conjecture that it is caused by the rich reflection/scattering components in urban and cutting environments (Fig. 3.10).

Furthermore, we note that in each cell a higher standard deviation sigma of the shadow fading usually corresponds to a larger d_{cor}, as shown in the above figure. It is conjectured that the scatterers that increase the decorrelation distance usually lead to a larger variation of shadow fading. We therefore derive a correlation coefficient [101], and the coefficient is summarized in the above table, together with the 95% CIs. The values of CIs indicate a reasonable accuracy of estimation, because of the large data set. Generally, Gamma is larger than 0.2, and its positiveness implies that larger sigma values are associated with larger d_{cor}.

3.3.2.3 Cross-Correlation Characteristics

For each NC (in a particular scenario), we estimate the cross-correlation coefficient. 95% CI of the estimated is plotted in the above figure. The estimated CIs are limited by the number of shadowing samples in each cell, as discussed before. It is found that the 95% CI is reasonably narrow. A relatively wide CI is only observed for a small value of coefficient. This ensures a sufficient accuracy. In the following, we discuss the variation of coefficient and propose a statistical model to characterize the cross-correlation property (Fig. 3.11).

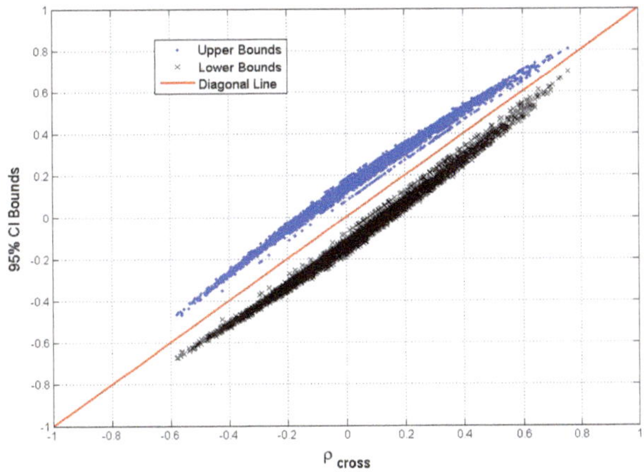

Fig. 3.11 Example plots of the 95% CI bounds of cross-correlation coefficient in all environments. Diagonal line is plotted for reference. Reprinted from Ref. [43], Copyright 2014, with permission from IEEE

3.3.2.4 Discussions

It is found that the cross-correlation coefficient exhibits a large fluctuation from cell to cell. It is conjectured that cross-correlation coefficient is affected by factors such as antennas, BSs, environments, etc. We thus examine some of these factors in the following and then propose a heuristic model.

Several factors of the scenario are considered: environment where the NC is located; separation distance D between the two BSs; heights h of the two BSs; and tilt angles theta of the antennas against the BS towers [101]:

- Environment: It has been found that the specific environments in railways significantly affect the large- and small-scale characteristics. We also find that cross-correlation coefficient changes in different environments. Therefore, we distinguish different environments when we model cross-correlation coefficient. We have sufficient measurements in each environment to ensure an accurate analysis.
- Separation distance *D*: A dependency of cross-correlation coefficient on *D* is not observed in our measurements. One possible reason is that *D* in our measurements mostly is around 3–4 km, which does not cover a large range so that we have few "realizations" to examine the dependency of cross-correlation coefficient on *D*. We also note that, as reported in some other measurements, a clear dependency of cross-correlation can only be observed when TX/RX separation distance varies within 1000 m, which is not a realistic case for HSR deployment. Since we use the operative GSM-R network in the measurements, changing *D* to have more realizations is not feasible in our current work. We therefore do not consider the impact of *D* on cross-correlation coefficient in the following modeling.
- h and theta: For a particular antenna gain pattern, the BS antenna height h and angle theta determine the gain and the received power at different locations. In another word, the channel characteristics can be considered as h- and

theta-dependent. A large h, or a small theta usually lead to better coverage in most of the railway cell. We therefore introduce a heuristic term of h/theta to represent the impact of BS on the cross-correlation characteristics. The difference of h/theta between the two BSs can be expressed as

$$\xi = \left| \frac{h_1}{\theta_1} - \frac{h_2}{\theta_2} \right| \tag{3.3.6}$$

A small xi means the two BSs in the NC generally have similar impacts on the shadow fading of the two TX-RX links.

In the following, we present the measured cross-correlation coefficient and the proposed model based on above analysis. Note that there is only one realization of xi in urban scenario, therefore, we do not develop the xi-based model of cross-correlation coefficient in urban to avoid misleading conclusions.

3.3.2.5 Model

Example plots of cross-correlation coefficient measured in the viaduct and cutting environments are shown in the above figure. Our first observation is that the cross-correlation coefficient generally has a large variation, and a distribution ranging between −1 and 1 should be used to describe the variation. Meanwhile, the mean value of cross-correlation coefficient is found to follow a linear function [101]. It is observed that in the suburban, rural, viaduct, station, and river scenarios, cross-correlation coefficient decreases with increasing xi; while in the cutting scenario, cross-correlation coefficient is found to increase with xi. Our measurements show that the cutting structure, leads to a negative cross-correlation coefficient at small values of xi. Note that in all the six scenarios, a small value of cross-correlation coefficient is generally observed at large xi, which follows the physical insight that a small xi implies that the scenario difference between the two links is small and therefore a large cross-correlation is observed. Finally, the standard deviation of cross-correlation coefficient is found to be independent of xi [101]. Note that when calculating the standard deviation, we drop the sets with less than 20 samples, because the size of those sets cannot guarantee sufficient accuracy of standard deviation estimation. Measurements for other environments were verified, though relevant plots are not shown here due to space limitations (Fig. 3.12)

Summarizing, our model of cross-correlation coefficient is as follows [101]:

- We use the truncated Gaussian distribution bounded between −1 and 1 to describe the variation of cross-correlation coefficient.
- Mean value of cross-correlation coefficient is modeled as a linear function of xi, expressed as

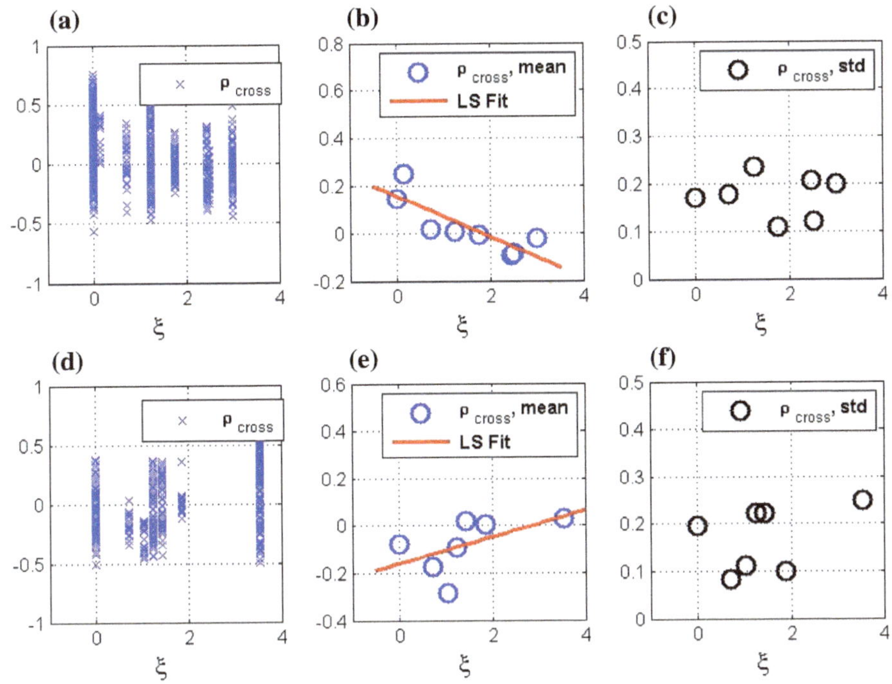

Fig. 3.12 Example plots of cross-correlation coefficient. **a–c** Viaduct environment. **d–f** Cutting environment. Reprinted from Ref. [43], Copyright 2014, with permission from IEEE

$$\overline{\rho} = a \cdot \xi + b, \tag{3.3.7}$$

where a and b are the tunable parameters and are obtained by using an LS regression fit.

- Instead of modeling standard deviation as a function of xi, we simply average it. This is because (i) no distinct dependency of the standard deviation on xi is observed; and (ii) it reduces the estimation error and avoids misleading conclusions (Fig. 3.13).

In the above figure, we show the example CDF plots of the estimated cross-correlation coefficient case in the viaduct and cutting environments. It is found that the truncated Gaussian distribution indicates a reasonable fit, which has a KS passing rate larger than 87% in all the six environments. The goodness of fit of the Uniform distribution is examined, and it generally has a KS passing rate lower than 50% (Table 3.4)

In the above table, we summarize the obtained parameters of the cross-correlation model. Note that our model is derived from the measurements conducted with a particular GSM-R system, and it is therefore limited to these conditions, e.g., the range of xi. The root mean squared error (RMSE) is calculated

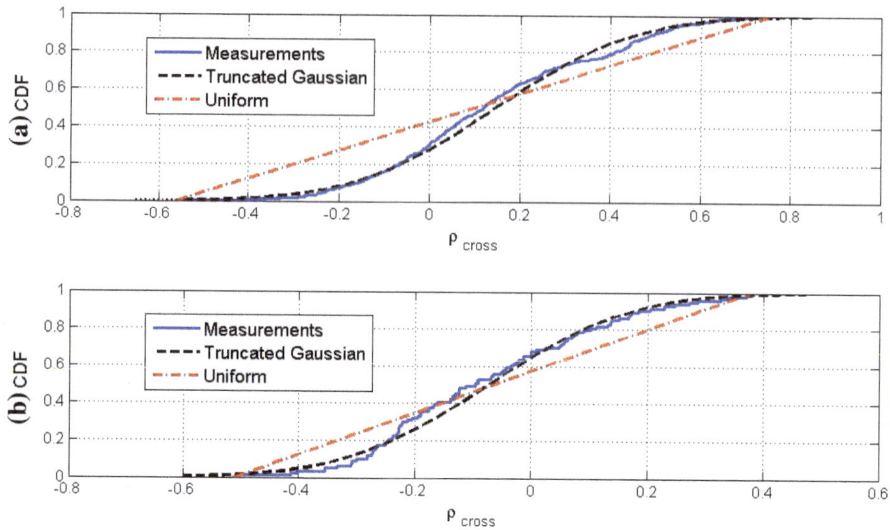

Fig. 3.13 Example plots of cross-correlation coefficient, together with the CDFs of truncated Gaussian and uniform distributions. **a** Viaduct. **b** Cutting. Reprinted from Ref. [43], Copyright 2014, with permission from IEEE

Table 3.4 Cross-correlation model parameters in railway environments. Reprinted from Ref. [2], Copyright 2014, with permission from IEEE

Environment	a	b	Mean Value of Sigma_cross	RMSE
Urban	−0.055	0.25	0.16	0.08
Suburban	−0.016	0.066	0.18	0.07
Rural	−0.086	0.16	0.17	0.06
Viaduct	0.056	−0.16	0.17	0.06
Cutting	0.056	−0.16	0.17	0.09
Station	−0.053	0.23	0.14	0.09
River	−0.016	0.22	0.21	0.03

and summarized in the above. We can see that the RMSE is generally less than 0.1, which means the model has a reasonable fit.

3.3.2.6 Model Implementation

Since it is very difficult to conduct the practical channel measurements in HSR, a recipe of the generation of the large-scale fading channel is very useful for system design. Based on the results in this paper, it is possible to model the shadowing of the radio channel by generating a sequence of values (in dB) that have desired normal distribution, and possess the necessary correlation properties. We describe below a list of steps to follow in order to generate the shadow fading channels in HSR with the desired properties [101]:

- Choosing a particular environment of HSR.
- Generating a covariance matrix K. Perform the following factorization

$$K = P \Lambda P^{\mathrm{T}}, \tag{3.3.8}$$

where P is the matrix whose columns are the eigenvectors of K and Lambda is the diagonal matrix of eigenvalues. Generating two independent identically distributed zero-mean, unit-variance Gaussian random variables ×1 and ×2. The sequences s1 and s2, which both have the desired covariance matrix K, can be obtained by

$$[s_1 \quad s_2] = \left(P\sqrt{\Lambda}\right)[x_1 \quad x_2] \tag{3.3.9}$$

- Generating two Gaussian random variables S1 and S2 by

$$[S_1 \quad S_2] = \sigma \cdot [s_1 \quad s_2] \cdot R, \tag{3.3.10}$$

where sigma is the standard deviation of the shadow fading. Matrix R is an upper triangular matrix that satisfies the equation

$$R^{\mathrm{H}} R = \begin{bmatrix} 1 & \rho \\ \rho & 1 \end{bmatrix} \tag{3.3.11}$$

*S*1 and *S*2 can thus be considered as the shadow fading components of two neighboring BSs in one NC, with the desired correlation properties.

3.3.2.7 Model Validation

To validate the proposed correlation model, we use the measurements from 10 cells in another HSR lines: "Beijing–Shanghai" line, whose measurements were not used in the development of the above models. The measurement system is the same as reported in Section III. We generate the shadow fading components with the same number of samples to the measurements. The generated sequences of shadow fading, with full auto- and cross-correlation properties, are compared with the measurements in "Beijing–Shanghai" line, and both first- and second-order statistics are validated as follows [101] (Fig. 3.14).

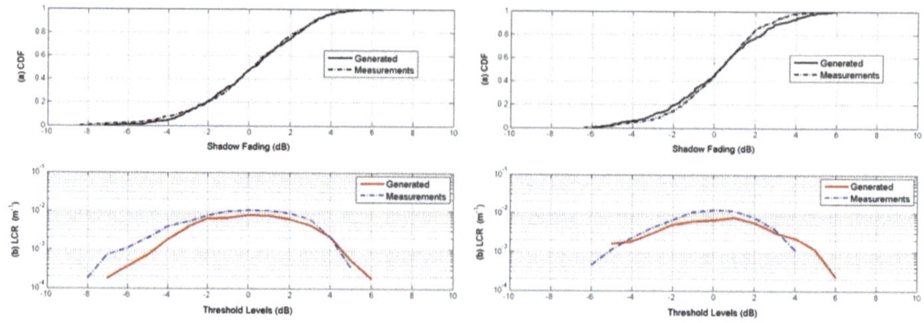

Fig. 3.14 Validations using the measurements of "Beijing–Shanghai" HSR. Reprinted from Ref. [43], Copyright 2014, with permission from IEEE

- First-Order Statistics Validation

We compare the CDF of the generated shadow fading components with measurements. Examples of CDF comparisons in suburban and rural environments are presented in the above figures, where we can see that the generated sequences offer a reasonable fit to the measurements.

- Second-Order Statistics Validation

We use the level crossing rate (LCR) of the shadow fading to validate the second-order statistics of our model. LCR is defined as the times that the signal crosses a given threshold level from up to down within a unit of length (1 m). Examples of LCR comparisons in suburban and rural environments are presented in the above figures. We find the LCRs of the generated shadow fading to be fairly close to the measurements in both cases. Measurements for other environments were verified, though relevant plots are not shown here due to space limitations.

3.3.3 Small-Scale Fading

Power fluctuates around a (local) mean value on a very-short-distance scale and these fluctuations happen on a scale that is comparable with one wavelength. Therefore, these fluctuations called small-scale fading which is caused by interference between different Multi-Path Components (MPCs) [103]. Small-scale characteristics and parameters of the narrowband channel for High-Speed Railways are described here.

930-MHz narrowband measurement campaigns for HSR propagation channels have been carried out along the 458-km-long "Zhengzhou–Xian" HSR line in China [104–107, 109–111]. In [112], measurements have been carried out in the Line 10 tunnels of Madrid's subway, between Tribunal and Príncipe Pio stations.

3.3.3.1 Fade Depth

Fade Depth (FD) measures the variation in the signal energy about its local mean due to small scale fading. It is an important channel parameter from the perspective of system design as it determines the required fade margin and link budget for an acceptably low system outage probability. FD is defined as the difference in power levels (in decibels) between the 50 and 1% level values for each case [113].

- **Cutting scenario**

Five deep cuttings along the HSR track are chosen and numbered 1 to 5 to investigate the small-scale fading behavior in [105]. We obtain the 50 and 1% values from the empirical cumulative distribution function (CDF), as shown in Fig. 3.15. The results of FD are summarized in Table 3.5.

It is found that the FD for the cutting scenario is around 17 dB, which is close to the 18.5 dB obtained for Rayleigh fading. This is a result of the steep walls on both sides of the cutting. They retain the reflection and scattering components and lead to severe small-scale fading with crown width ω_{down} and bottom width ω_{down}. Finally, we carry out the regression fit using a linear combination, and the results lead us to the formulation

$$\text{FD(dB)} = 25.26e^{-0.013(\omega_{\text{up}} + \omega_{\text{down}})} + 11.49e^{-0.00045(\omega_{\text{up}}\omega_{\text{down}})}. \qquad (3.3.12)$$

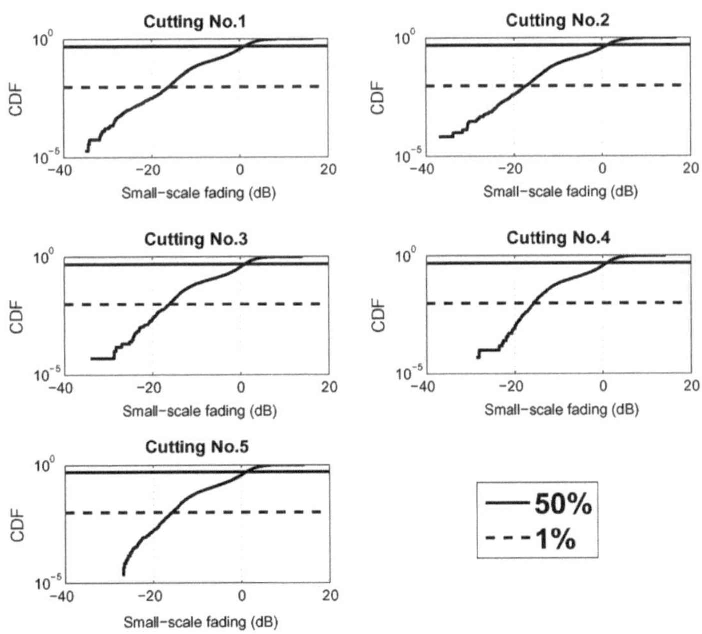

Fig. 3.15 CDF of the measured small-scale fading for each cutting based on multiple measurements. Reprinted from Ref. [105], Copyright 2013, with permission from IEEE

Table 3.5 Analysis results of each cutting. Reprinted from Ref. [105], Copyright 2013, with permission from IEEE

	Cutting Number	1	2	3	4	5
Fading depth	FD (dB)	17.43	18.53	17.37	16.79	16.94
LCR(crossings) per wavelength	10 dB	0.011	0.0199	0.0123	0.0089	0.0084
	0 dB	0.721	0.751	0.629	0.718	0.751
	−10 dB	0.220	0.250	0.189	0.206	0.206
	−20 dB	0.011	0.012	0.0079	0.0025	0.0055
AFD (wavelengths)	10 dB	95.15	50.06	81.25	112.1	118.73
	0 dB	0.517	0.493	0.573	0.505	0.496
	−10 dB	0.319	0.326	0.382	0.331	0.326
	−20 dB	0.306	0.312	0.315	0.311	0.307
Results *of v's*	a_{LCR}	0.91	0.95	0.76	0.90	0.95
	b_{LCR}	0.16	0.15	0.17	0.18	0.18
	cL_{CR}	−0.23	−0.21	−0.23	−0.24	−0.24
	a_{AFD}	0.42	0.43	0.54	0.44	0.42
	b_{AFD}	0.021	0.02	0.032	0.022	0.021
	c_{AFD}	0.54	0.50	0.43	0.49	0.63
	k_1	0.0245	0.033	0.0316	0.0143	0.032
	k_2	3.851	1.643	4.519	3.078	2.934
	k_3	−0.00382	−0.001	−0.00811	−0.00331	−0.00164
	σ(dB), $d \leq 200$ m	4.88	4.41	4.89	4.14	3.92
	σ (dB), $d \geq 200$ m	4.24	4.55	4.38	4.51	4.57

Obviously, (3.3.12) presents a small prediction error and explains the variation of the data successfully. It shows that even in a small variation range of FD, the structural parameters of cutting still significantly affect the fading behavior. The formulation also implies that the "wide" cutting (with great $\omega_{up} + \omega_{down}$ and $\omega_{up}\omega_{down}$) results in few received reflected and scattered waves from the steep walls, and considerably reduces the severity of fading.

- **Viaduct scenario**

In [106] the measurements cover different viaduct four cases: Four BSs with a fixed 20 m relative antenna height and different viaduct heights of 10, 15, 20, and 25 m are utilized. The results are summarized in Table 3.6. According to the measurements, the FD for the viaduct scenario is around 15 dB, which is smaller compared to the 18.5 dB4 for the Rayleigh channel. This is because the train is 204 m long, 3.8 m high, and 3.3 m wide, and the receiver antenna is slightly higher than the train roof. Therefore, few reflected and scattered paths from the surface of the rail and the ground below the viaduct can get to the receiver antenna. Even though the

Table 3.6 Parameters and analysis results of each viaduct case. Reprinted from Ref. [106], Copyright 2013, with permission from IEEE

	Viaduct	Case 1	Case 2	Case 3	Case 4
Structure parameters (m)	H	10	15	20	25
	h	20	20	20	20
Measurement	Cellular radius (m)	2000	2665	2502	1336
Parameters	Sampling interval (cm)	14	10	10	10
	Measurement times	3	2	3	3
	average speed (km/h)	260	295	180	65
Results of fade depth (dB)	1%	−14.89	−13.44	−14.28	−14.84
	50%	1.07	0.94	1.14	1.05
	FD	15.96	14.38	15.42	15.89
Rate of the small scale best fit distribution	Ricean	81.02%	92.29%	95.69%	76.01%
	Nakagami	13.88%	6.42%	3.64%	18.33%
	Rayleigh	4.94%	1.28%	0.67%	5.12%
	Lognormal	0.16%	0%	0%	0.27%
	Suzuki	0%	0%	0%	0.27%
LCR(crossings) per wavelength	0 dB	0.533	0.656	0.649	0.655
	−5 dB	0.293	0.319	0.334	0.348
	−10 dB	0.145	0.151	0.191	0.160
	−15 dB	0.025	0.012	0.022	0.026
AFD(wavelengths) per wavelength	0 dB	0.152	0.156	0.130	0.164
	−5 dB	0.016	0.013	0.015	0.026
	−10 dB	0.002	0.002	0.004	0.005
	−15 dB	0.00011	0.00020	0.00007	0.00056
Shadowing standard	$d \leq 400$ m	4.91	6.19	4.44	4.19
Deviation σs (dB)	$d > 400$ m	2.44	3.58	1.8	2.07

scatterers are close to the track (as in case 4), the high viaduct (H = 25 m in case 4) can still result in clear LOS and few scatterers, and reduces the severity of fading. Note, though, that even scatterers that are mostly lower than the viaduct have some impact on FD. We can observe that the FD in case 2 is the smallest because the scatterers (e.g., buildings in the small town) in that case are fewer than in other cases. In case 4, the scatterers (e.g., buildings and vehicles) in the town adjacent to the track lead to a 1.5 dB higher value of FD than in case 2. This phenomenon shows that a reduction in the severity of fading caused by the viaduct is moderately affected by the number and closeness of surrounding scatterers.

- **Station scenario**

As for train station scenarios [110], no matter whether the Tx is far or near, the semi-closed station leads to larger extra loss and more complicated fading behavior than the open station. When the Tx is far, FD is 14.0–17.2 dB in semi-closed station and 10.6–14.4 dB in open station. While, when the Tx is far, FD is 21.8–27.6 dB in semi-closed station and 18.0–22.1 dB in open station.

In the crossing bridge-related regions [111], the FD is between 14.88 and 27.96 dB, whereas the Max. FD is between 19.80 and 45.55 dB.

3.3.3.2 Level Crossing Rate and Average Fade Duration

In railway communications, the received power often undergoes heavy statistical fluctuations [114], which leads to a drastic increase of the bit error rate. The Level Crossing Rate (LCR) and Average Duration of Fades (ADF) help to know how often the received signal crosses a given threshold per time unit, and for how long on average the signal is below a certain threshold. They can help in the selection of transmission bit rates, word lengths, and interleaving algorithm [115].

Providing a mathematical formulation, the LCR is defined as the expected value of the rate at which the received field strength crosses a certain level r in the positive direction. This can also be written as

$$N_R(r) = \int_0^{\infty} \dot{r} \cdot pdf_{r,\dot{r}}(r, \dot{r}) d\dot{r} \quad \text{for} \ r \geq 0, \tag{3.3.13}$$

where $\dot{r} = dr/dt$ is the temporal derivative, and $pdf_{r,\dot{r}}$ is the joint pdf of r and $\dot{r}$.

The ADFs can be simply computed as the quotient of the cdf of the field strength and LCR.

$$ADF(r) = \frac{cdf_r(r)}{N_R(r)}. \tag{3.3.14}$$

- **Viaduct scenario**

The results of the four cases with different viaduct heights H in high-speed railway viaducts in [106] are illustrated in Fig. 3.16.

The threshold levels range from −20 dB to +10 dB. The LCR and AFD values for four typical threshold levels are tabulated in Table 3.6. For the LCR, case 1 presents the least crossings due to the presence of the viaduct which reduces the severity of fading and decrease the crossings, whereas a given threshold level is crossed more frequently for case 4 in general which is a result of the large number of scatterers around the viaduct. For the AFD, we find results to be fairly similar in all cases. The statistical results show that, in general, the severities of the fading in the viaduct scenario are greatly reduced for all four cases compared to other environments. This is because the viaduct with high H creates “clear” LOS, leads to fewer reflected and scattered paths to get to the receiver, and results in small values of LCR and AFD. Moreover, we also find that the influence of H on the variations

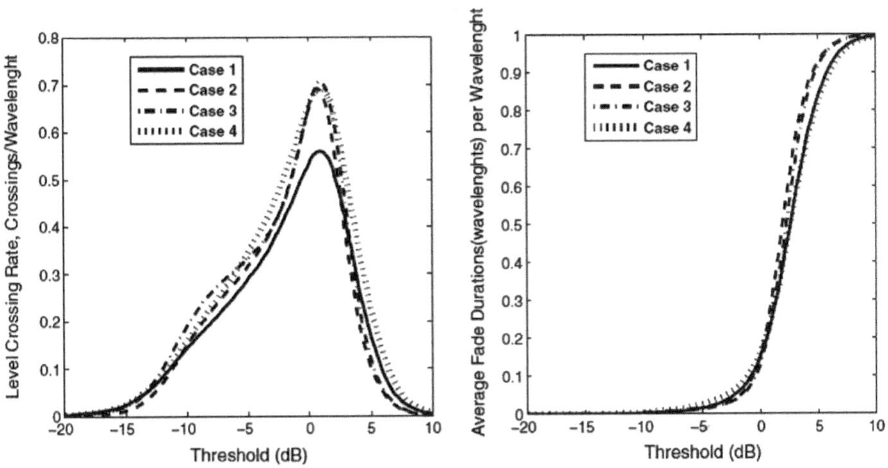

Fig. 3.16 LCR's and AFD's for the four cases. Reprinted from Ref. [106], Copyright 2013, with permission from IEEE

of LCR and AFD for the four cases is not significant. This is because the viaduct reduces the severities of the fading and leads to small values of LCR and AFD for all four cases. Therefore, tiny variations of LCR and AFD occur when H changes.

- **Cutting scenario**

Besides, the results of LCR and AFD in cutting scenario in [4] for four typical threshold levels (R) are tabulated in Table 3.5. It is found that for $R = 0$ dB, there are nearly 7 crossings in an observation window of 10 wavelengths, whereas for $R = -20$ dB, there are just a few crossings in an observation window of 1000 wavelengths. As to AFD, fade duration on average lasts for 0.52 wavelengths at $R = 0$ dB in an observation window of 1 wavelength, and for $R = -20$ dB, the values of AFD are close to 0.31 wavelengths.

- **Station scenario**

The results of LCR and AFD in different station scenarios are studied in [110]. The semi-closed station has larger standard deviations of the shadow fading, and larger values of FD, LCR, and AFD than the open station in every corresponding case, because more structures exist both on the front and the broad side of the semi-closed station than on that of the open station. Furthermore, whether the Tx is far from or near the station, the zones inside the station always have larger FD, Max. FD, LCR, and AFD than the corresponding zones outside the station. This reveals that the multiple reflection and scattering can be effectively retained by the limited space inside the station; therefore, the FD is stronger than in the wider space after the station.

3.3.3.3 Amplitude Distribution of Small-Scale Fading

In [104], high-speed railway propagation scenarios are divided into two regions: Region 1—inside the bottom area of the antenna; and Region 2—outside the bottom area of the antenna. After removing the path loss and large-scale fading from the raw data, we investigate the small-scale fading behavior in Regions 1 and 2. We first examine the empirical distribution of the fading amplitudes. Four distributions, namely, Ricean, Nakagami, and Rayleigh, which are widely used in modeling small-scale fading, and lognormal are tested using the Akaike's Information Criteria (AIC). AIC is a measure of the relative goodness of fit of a statistical model and the Kolmogorov–Smirnov (KS) test is used to verify the model selected by the AIC-based method to ensure that a satisfactory fit is obtained. We use the 20 wavelengths sliding window described above to estimate the parameters of each distribution and conduct the AIC tests.

Figure 3.17a, b shows the relative frequency of AIC selecting each of the candidate distributions as best fit. It is found that the Ricean distribution provides the best fit in a majority of the cases. We also see that the best-fit rate of each distribution in Region 1 is close to Region 2, which means that the directional

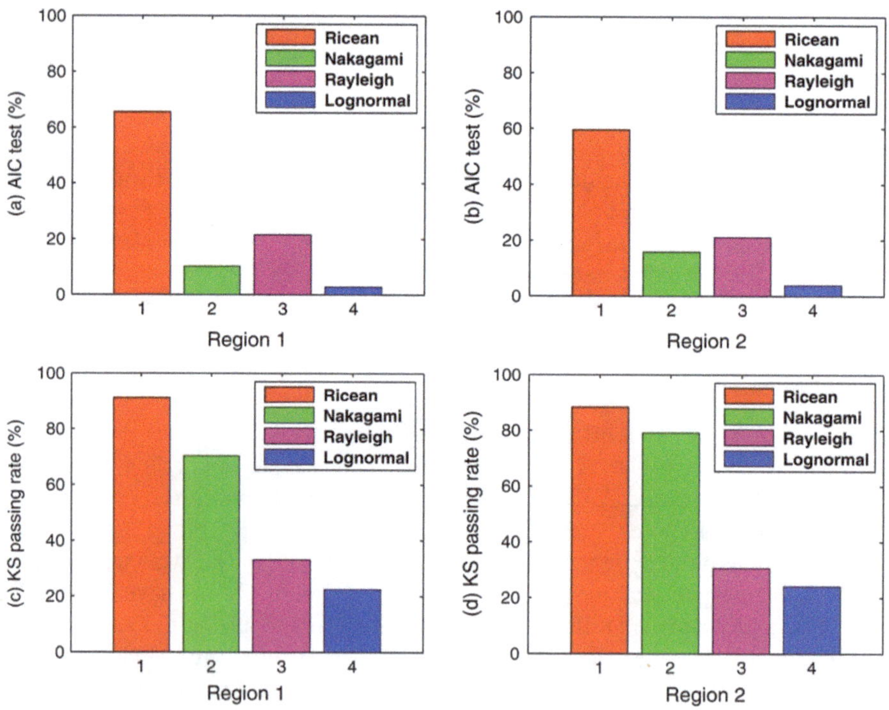

Fig. 3.17 Relative frequencies of AIC and KS tests selecting a candidate distribution as best fit to small-scale fading amplitudes. Reprinted from Ref. [104], Copyright 2015, with permission from IEEE

transmitting antenna does not significantly affect the small-scale fading distribution. Then, the Kolmogorov–Smirnov (KS) passing rate of each distribution is recorded as a measure of the goodness of fit, as shown in Fig. 3.17c, d. It is found that the passing rate of the Ricean distribution is generally larger than 80%, which verifies that it offers a satisfactory fit. We henceforth suggest the Ricean distribution in HSR environments.

- **Cutting scenario**

The model with the highest Akaike weights is the best distribution to describe the data set. As for 5 cutting scenarios in [105], Fig. 3.18 shows the plots of the Akaike weights for different candidate distributions based on multiple measurements in five cuttings. It can be observed that the Ricean distribution has the best fit and Nakagami is the second best. Rayleigh and Lognormal fits are not suitable for cutting scenarios. This is as expected since there is a clear LOS path due to the high transmitting antennas in HSR cutting scenarios, and the Ricean distribution is commonly used to describe propagation channels with a dominant signal.

- **Viaduct scenario**

As for viaduct scenario in [106], Fig. 3.19 shows the Akaike weights for different candidate distributions based on multiple/repeated measurements. Our analysis

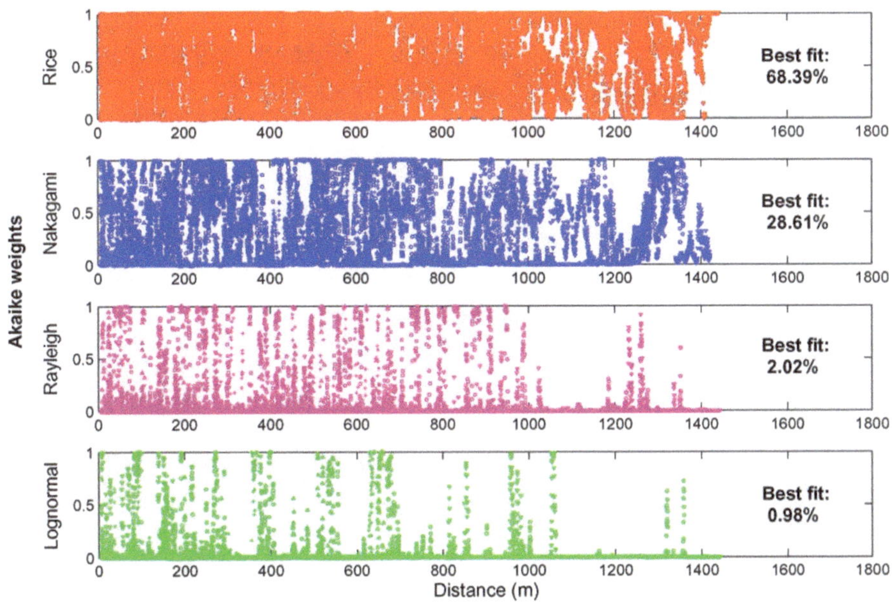

Fig. 3.18 Akaike weights and the percentage of the best fit for four candidate distributions based onmultiple measurements. Themeasurements for cuttings Nos. 1, 2, 3, 4, and 5 are plotted with circles, squares, +'s, Δ's, and ∇'s, respectively Reprinted from Ref. [105], Copyright 2013, with permission from IEEE

shows that: (i) overall the Ricean, Rayleigh, Nakagami, Lognormal, and Suzuki distributions show the best fit for 87.66, 9.50, 2.71, 0.09, and 0.04% of all cases. This is not surprising since the Ricean distribution is widely used for an LOS propagation scenario; (ii) the performance of Lognormal and Suzuki distributions are poor for almost all cases6. The rates of the best fit for each case are tabulated in Table 3.6. It is found that the rate of best fit for Ricean distribution in case 1 is 81.02%, while it is 92.29% and 95.69% in cases 2 and 3, respectively. This is because a higher value of H leads to fewer reflection and scattering components at the receiver, which results in the Ricean distribution fitting better for a higher viaduct. A large number of scatterers around the track in case 4 leads to a drop of best-fit rate of Ricean distribution to 76.01%.

- **Station scenario**

Train station can generate multiple scattered waves and different clusters of reflected waves, making the small-scale fading even worse than the Rayleigh fading [110]. The AIC test indicates that the Rician distribution offers the best fit for the small-scale fading in most cases, except in zones C_near and D_near, whereas Nakagami distribution is the best fitting result in these two zones with the parameter m smaller than 1. This reveals that the train station can lead to strong fluctuations of the signal strength, which is even worse than the Rayleigh fading (corresponding to $m = 1$ in Nakagami distribution) that is thought to be typical in the NLOS environment. The huge steel awning, steel frames, metal pylons, and indicators of the station generate multiple scattered waves and different clusters of reflected waves. Moreover, this is the typical environment of the Nakagami fading.

- **Crossing bridges scenario**

In crossing bridges scenario, the following zones are defined [111]. Zone A: When the Rx is in the narrow space under the crossing bridge but the LOS is still kept, the directed wave and multiple reflected waves from the track and the bridge are the main components of the received power; Zone B: When the Rx has passed the bridge but the LOS is still kept, the directed wave and reflected waves are the main components; Zone C: When the Rx has passed but the LOS is blocked by the bridge, there is no directed wave from the Tx; hence, the reflected and diffracted waves dominate; Zone D: When the bridge is near the Tx, the LOS is blocked before the Rx passes the bridge. Hence, Zone D is the case when the Rx is under the bridge but under a non- LOS (NLOS) condition. No LOS exists, and the multipath propagation dominates. Figure 3.19 shows the relative frequency of AIC selecting each of the candidate distributions as best fit in all the propagation zones based on multiple measurements (Fig. 3.20).

- **Tunnel scenario**

In [112], Nakagami-m distribution has been found to be a very good fitting for mobile radio channel fast fading. Parameter m, which indicates the severity of the amplitude fading, is estimated by using the moment-based method in all the testing

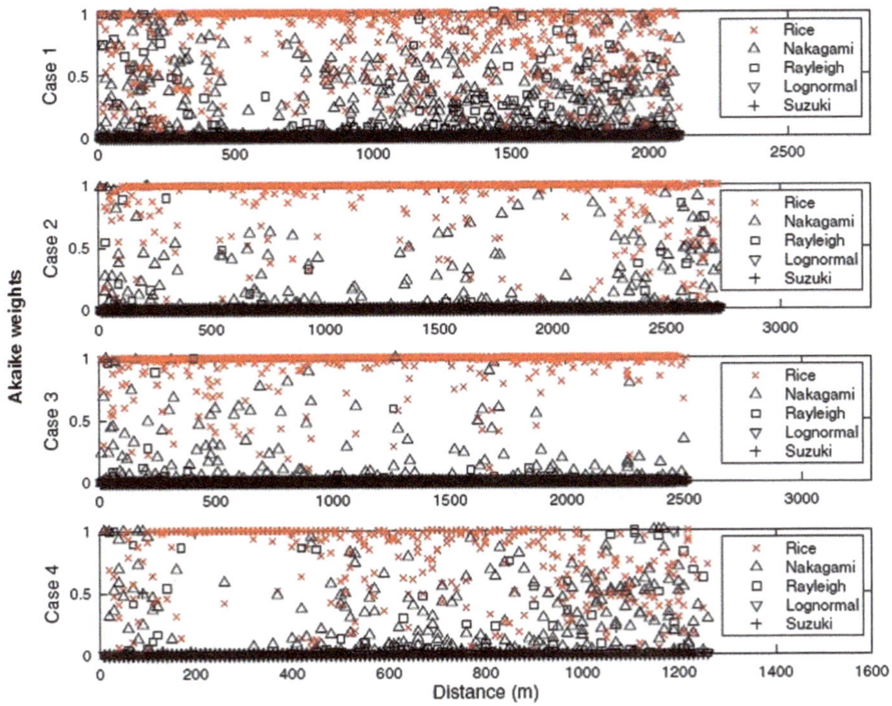

Fig. 3.19 Akaike weights for five candidate distributions based on multiple measurements. Reprinted from Ref. [106], Copyright 2013, with permission from IEEE

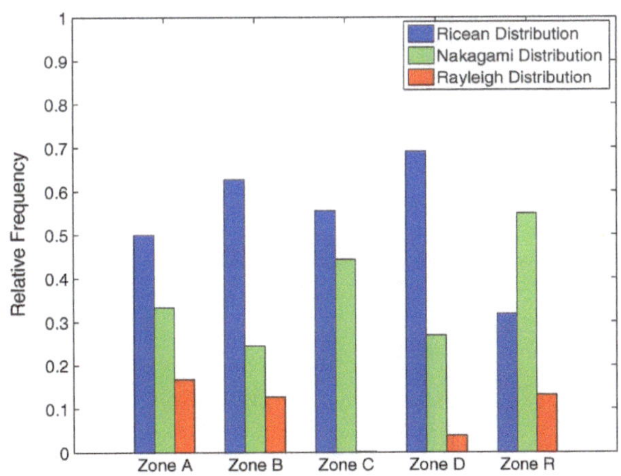

Fig. 3.20 Relative frequency of AIC selecting each of the candidate distributions as best fit in all the propagation zones based on multiple measurements. Reprinted from Ref. [111], Copyright 2014, with permission from IEEE

and comparison cases. According to Lee's Criteria, we choose the window with a length of 91 samples to separate the local mean from the fast fading data. Details are shown in [112].

3.3.3.4 Envelope Autocovariance of Small-Scale Fading

Due to the multipath propagation, motion of the terminals, and variations of the environments in the wireless channels, Doppler (frequency) dispersion occurs, which is described by the important parameter of coherence time (or equivalently, coherence distance) [116]. It can be calculated from an important second-order statistic: the envelope autocovariance function $\rho(\Delta d)$, where Δd indicates distance difference.

For the railway viaduct scenario], $\rho(\Delta d)$ determines the correlation of received envelope as a function of change in receiver position and is useful for studies in correlation properties, written as

$$p(\Delta d) = \frac{E|\{[r(d) - E(r(d))][r(d+\Delta d) - E(r(d+\Delta d))]\}|}{\sqrt{\mathrm{Var}[r(d)]\mathrm{Var}[r(d+\Delta d)]}}, \tag{3.3.15}$$

where *r*(*d*) denotes the received envelope at *d*, *E*[·] and Var[·] denote the expected value and the variance of [·] respectively. Note that since we assume stationary statistics, $E(r(d)) = E(r(d + \Delta d))$.

Figure 3.21 shows the absolute value of envelope autocovariance function $\rho(\Delta d)$ with a Δd up to 3 m. We present the coherence distance Dc, which is important in the design of wireless receivers that employ spatial diversity to combat

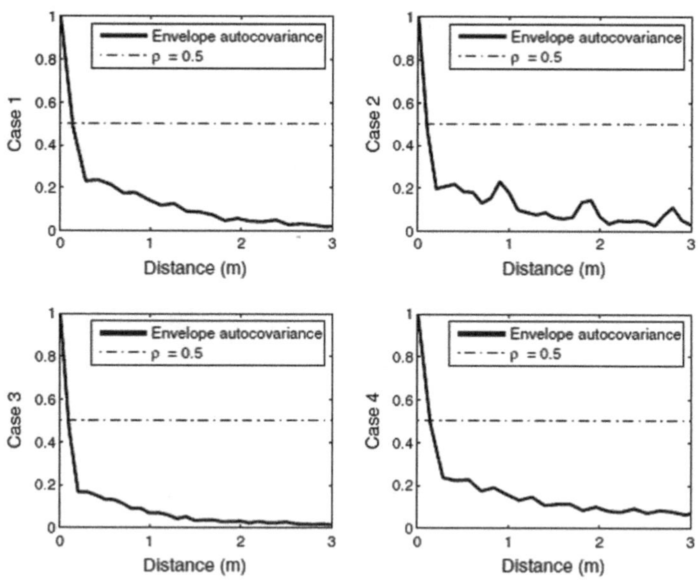

Fig. 3.21 Absolute value of the envelope autocovariance function $\rho(\Delta d)$. Each one is the average result based on multiple measurements. $\rho(\Delta d) = 0.5$ is marked.. Reprinted from Ref. [106], Copyright 2013, with permission from IEEE

spatial selectivity. A convenient definition for the Dc is the value that satisfies the equation $\rho(\Delta d) = 0.5$. According to Fig. 3.21, Dc's are 14, 10, 10, and 10 cm for cases 1 to 4.

Note that the calculated Dc's equal to the sampling interval, therefore, the real coherence distance should be smaller than 10 cm, which is smaller than λ/2 at 930 MHz. It shows that sufficient decorrelation can be achieved using a spatial separation of less than λ/2, and effective diversity systems for GSM-Railway can be implemented using antennas less than λ/2 apart.

3.3.3.5 Ricean *K*-Factor

To model the fading characteristics in a unified manner, the Ricean distribution is utilized based on the above findings. The key parameter of this distribution is the Ricean *K*-factor. The Ricean *K*-factor measures the severity of fading [117]. The knowledge of the values of the *K*-factor can thus be used in the design of different wireless communication techniques [118]. A realistic model of the *K*-factor is important in link budget calculations and system simulation.

As mentioned before, high-speed railway propagation scenarios are divided into two regions (Region 1—inside the bottom area of the antenna and Region 2—outside the bottom area of the antenna) in [104].

(1) Overall Estimation: For a fast simulation, we can model the linear scale *K*-factor as a random variable that has mean μ_K and standard deviation σ_K that changes between Region 1 and Region 2, but is otherwise independent of the location within the cell.

Figure 3.22 shows that in both regions, the distribution of μ_K over the cells follows the lognormal distribution, and σ_K follows the Gaussian distribution. Note that the cdf plots for Region 2 are not presented due to space limitations. The parameters of the distributions are summarized in [104], and the *K*-factors in both regions are fairly close to each other.

(2) Distance Dependence: A more detailed model describes the decibel-valued *K*-factor as a function of distance [103] in each cell; this model is also used for some special HSR environments (see [105–107, 109]). Here, we examine the distance dependence of the *K*-factor in the practical HSR cells based on our measurements. The model is expressed as

$$K_i(\mathrm{dB}) = a_i d + b_i, \quad \begin{array}{c} i = 1, \quad 0 \le d \le D \\ i = 2, \quad D \le d \le 4\,\mathrm{km} \end{array}, \tag{3.3.16}$$

where d is the separation distance in meters, and a and b are the coefficients that can be estimated by linear regression using a minimum mean-square error criterion. $i = 1, 2$ is the index of Regions 1 and 2. It is observed that the variations of the

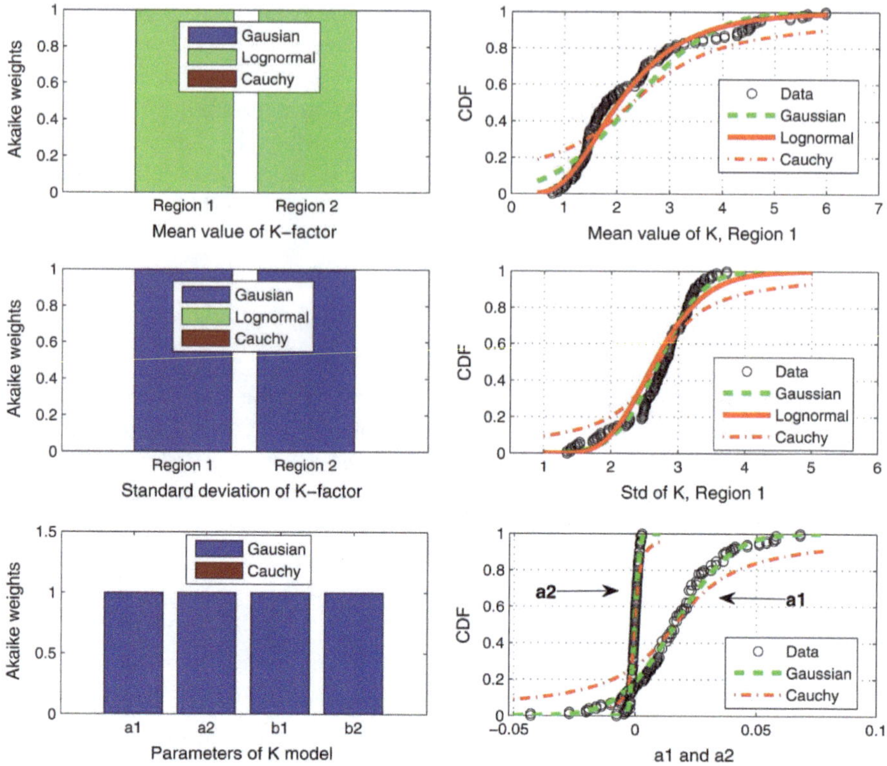

Fig. 3.22 Akaike weights of cell-to-cell distributions and the cdf plot for small-scale fading: mean value, standard deviation, and model of Ricean *K*-factor. Reprinted from Ref. [104], Copyright 2015, with permission from IEEE

measured a and b, from cell to cell, can be modeled as Gaussian variables, as shown in Fig. 3.22.

The values of a and b can be found in Table 3.7 of [104]. Several observations are worth noting: In Region 1, the *K*-factor increases with distance. In contrast, we do not observe distance-dependent variations of *K*-factors in Region 2, where the mean value and the standard deviation of a_2 are both very small, compared with other distance-dependent *K*-factor models. This is also different from the results in [119], which develops a distance-dependent *K*-factor model in a special HSR scenario—a 1200-m-long viaduct. We note that the variation of *K*-factors against distance is very sensitive to the environments. The practical HSR cells usually consist of many different and randomly distributed scenarios. Therefore, the *K*-factors in the practical HSR cell (with a large size) indicate a distance-independent behavior.

Table 3.7 Parameters of time dispersion at two frequencies

Region index		I	II	III	IV
Corresponding scenario		Tunnel	Cutting	Viaduct	Cutting
950 MHz	Mean excess delay (ns)	374.66	193.11	60.61	31.40
950 MHz	Average RMS delay spread (ns)	218.67	125.40	50.01	40.20
950 MHz	Average number of multipatli	6	5	1–2	1–2
2150 MHz	Mean excess delay (ns)	270.68	195.68	66.31	43.62
2150 MHz	Average RMS delay spread (ns)	120.44	135.18	52.03	38.81
2150 MHz	Average number of multipath	6	5	1–2	1–2

(3) Variations in K-Factor: For each cell, the deviation of the measured decibel scale K-factor from the previously mentioned distance-dependent K-factor model can be written as

$$\triangle K = K_{\text{measure}} - K_{\text{linear}}, \tag{3.3.17}$$

where K_{linear} is the distance-dependent K-factor model of (3.3.16). The mean value of ΔK in each cell is close to 0.

For the standard deviation of ΔK, after changing it to the linear scale, we find that it can be modeled by the Gaussian distribution, as shown in Fig. 3.23. Note that the cdf plots for Region 2 are not presented due to space limitations. Then, we calculate the autocovariance function $\rho(\triangle d)$ of ΔK for each cell and record the

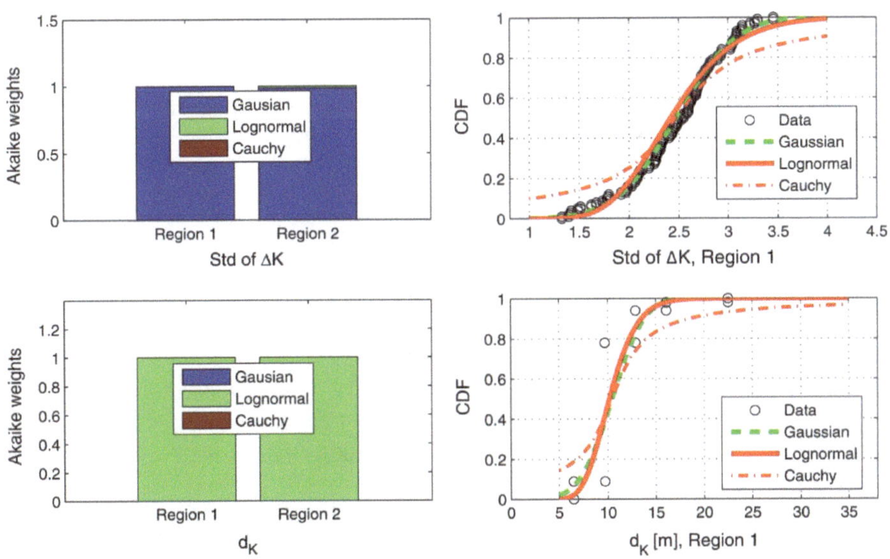

Fig. 3.23 Akaike weights of cell-to-cell distributions and the cdf plot for small-scale fading: ΔK and d_K. Reprinted from Ref. [104], Copyright 2015, with permission from IEEE

decorrelation distance d_K, which is the minimum separation distance Δd that satisfies the equation $\rho(\Delta d) \leq 0.5$. The AIC-based method and the KS test show that d_K in both regions can be modeled by the lognormal distribution. The corresponding parameters of the lognormal distributions are summarized in [104] and the values of d_K in both regions are close to each other.

3.4 Wideband Channel Characterization of High-Speed Railways

The differences between wideband and narrowband fading models is that, as the transmit signal bandwidth B increases so that $T_m \approx B^{-1}$ the approximation $u(t - \tau_n(t)) \approx u(t)$ is no longer valid, where T_m is the multipath delay spread, $u(t)$ is the equivalent low-pass signal, and $\tau_n(t)$ is the corresponding delay. With wideband signals, the received signal experiences distortion due to the delay spread of the different multipath components, so the received signal can no longer be characterized by just the amplitude and phase random processes. The most important characteristics of the wideband channel, including its time dispersion parameters, coherence bandwidth, Doppler power spectrum, coherence time, and angular domain parameters. These characteristics are described in subsequent sections.

3.4.1 Delay Characteristics

3.4.1.1 Delay Parameter Definitions

In order to compare different multipath channels for high-speed railway (HSRs), parameters which grossly quantify the multipath channel are used. The mean excess delay ($\bar{\tau}$), root mean square (RMS) delay spread (σ_τ), and excess delay (X dB) are multipath channel parameters that can be determined from a power delay profile [120]. The mean excess delay is the first moment of the power delay profile and is defined to be

$$\bar{\tau} = \frac{\sum_k a_k^2 \tau_k}{\sum_k a_k^2} = \frac{\sum_k P(\tau_k)\tau_k}{\sum_k P(\tau_k)}. \tag{3.4.1}$$

The RMS delay spread is the square root of the second central moment of the power delay profile and is defined to be

$$\sigma_\tau = \sqrt{\overline{\tau^2} - (\bar{\tau})^2}, \tag{3.4.2}$$

where

$$\overline{\tau^2} = \frac{\sum_k a_k^2 \tau_k^2}{\sum_k a_k^2} = \frac{\sum_k P(\tau_k)\tau_k^2}{\sum_k P(\tau_k)}. \tag{3.4.3}$$

It is important to note that the RMS delay spread and mean excess delay are defined from a single power delay profile which is the temporal or spatial average of consecutive impulse response measurements collected and averaged over specific HSRs.

The maximum excess delay (X dB) of the power delay profile is defined to be the time delay during which multipath energy falls to X dB below the maximum. In other words, the maximum excess delay is defined as $\tau_X - \tau_0$, where τ_0 is the first arriving signal and τ_X is the maximum delay at which a multipath component is within X dB of the strongest arriving multipath signal (which does not necessarily arrive at τ_0). The maximum excess delay (X dB) defines the temporal extent of the multipath that is above a particular threshold. The value of τ_X is sometimes called the excess delay spread of a power delay profile, but in all cases must be specified with a threshold that relates the multipath noise floor to the maximum received multipath component.

3.4.1.2 Coherence Bandwidth

It should be noted that the power delay profile and the magnitude frequency response (the spectral response) of HSRs channel are related through the Fourier transform. It is therefore possible to obtain an equivalent description of the channel in the frequency domain using its frequency response characteristics. Analogous to the delay spread parameters in the time domain, coherence bandwidth is used to characterize the channel in the frequency domain. The RMS delay spread and coherence bandwidth are inversely proportional to one another, although their exact relationship is a function of the exact multipath structure.

While the delay spread is a natural phenomenon caused by reflected and scattered propagation paths in the HSRs channel, the coherence bandwidth, B_c, is a defined relation derived from the RMS delay spread. Coherence bandwidth is a statistical measure of the range frequencies over which the channel can be a statistical "flat". In other words, coherence bandwidth is the range of frequencies over which two frequency components have a strong potential for amplitude correlation. Two sinusoids with frequency separation that greater than B_c are affected quite differently by the railway channel. If the coherence bandwidth is defined as the bandwidth over which the frequency correlation function is above 0.9, then the coherence bandwidth is approximately

$$B_c \approx \frac{1}{50\sigma_\tau}. \tag{3.4.4}$$

If the definition is relaxed so that the frequency correlation function is above 0.5, then the coherence bandwidth is approximately

$$B_c \approx \frac{1}{5\sigma_\tau}. \tag{3.4.5}$$

3.4.1.3 Current Investigations on Delay Characteristics

Channel Measurements:

Until now, there has been some wideband channel measurement for High-Speed Railways, such as

(1) The wideband measurement is carried out in 2.1 GHz with a bandwidth of 3.84 MHz in China [121].
(2) The position-based radio propagation channel for HSR is carried out by extensive measurements at 2.35 GHz with 10 MHz bandwidth in China [122].
(3) A series of broadband measurements at 950 MHz and 2,150 MHz frequency which is known as potential frequency bands for LTE-R are conducted in China [123].

Channel models:

The RMS delay spread, maximum delay, and the number of paths in different HSR scenarios are investigated in [121]. The wideband measurement is carried out in 2.1 GHz with a bandwidth of 3.84 MHz in China. It shows that most of the time (more than 80%) the RMS delay spread in plain and hilly terrains is less than 0.1 ns or almost 0 ns, which means that there is only one dominant path in channel. The cutting also has the largest RMS delay spread (0.4 ns). In the station, 95% of the RMS delay spread is less than 0.3 ns. It seems in these two scenarios the reflected and scattering components are richer. For the maximum delay, the plain and hilly terrains have the minimum value of 0.35 and 0.3 ns. The maximum delay in the station is 0.7 ns. In addition, the worst delay of 2.5 ns occurs in the cutting and the maximum number of resolved paths in the cutting is 4. The station and the hilly terrain have a similar maximum path number of 3. Only two paths can be extracted in the plain scenario. It is found that most of the time in HSRs the number of clear paths is less than six [124]. The mean and the RMS delay spread are no longer than 1.37 and 1.69 ns, respectively. In addition, the maximum delay, which is less than 6.7 ns, occurs most frequently.

The position-based radio propagation channel for HSR is carried out by extensive measurements at 2.35 GHz with 10 MHz bandwidth in China [122]. The

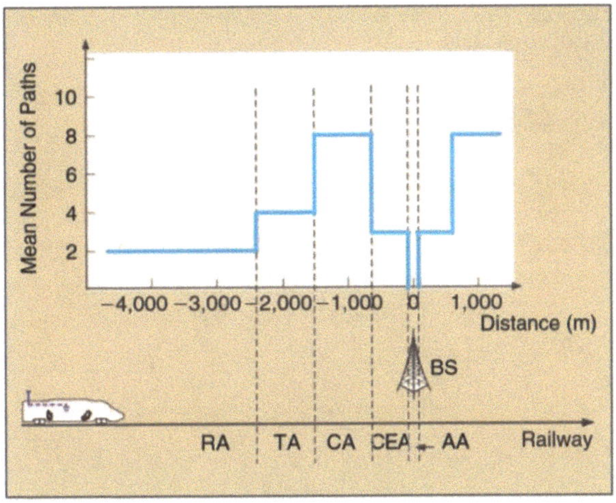

Fig. 3.24 Different subregions of HSR tracks

whole process of the train running toward the BS is divided into five subregions shown in Fig. 3.1: 1) remote area (RA) $d > 2{,}500$ m, where d is the distance between the train and the BS, 2) toward area (TA) 1,600 m $< d <$ 2,500 m, 3) close area (CA) 700 m $< d <$ 1,600 m, 4) closer area (CEA) 0 m $< d <$ 700 m, 5) arrival area (AA) $d = 0$ m (Fig. 3.24)

The RMS delay spread for RA is 0 ns with a 50% probability due to the long distance from the BS. In addition, the powers of all components are low. In TA, the RMS delay spread begins to grow. The RMS delay spread in CA lies between 0 and 1,000 ns, with the possibility of 90%, which shows the most severe multipath and the worst case for frequency selective fading. CEA has a smaller RMS delay spread than TA. There is only one specular line of sight (LOS) and no multipath can be detected in AA. Thus, the RMS delay spread and maximum delay are zero. In CA, the maximum delay takes place at 5,100 ns and has the highest probability. The value occurs in the range of 0–5,000 ns with a probability of 75%. The highest number of multipath components in CA is 12, whereas in RA it is 5. TA and CEA present similar results, with approximately one to six components. Therefore, the CA condition has the worst time dispersion. When the train is far from the BS, the signal power is low due to a large path loss. Meanwhile, the strength of the other multipath components is also low. Therefore, only a few paths are extracted. When the train continues moving, the path loss decreases. The power of the LOS path, as well as the multipath waves, begins to emerge. Therefore, more multipath components show up in this region. When the train runs near the BS, the distance between the train and BS is less than 100 m. The LOS power is so strong that it becomes the dominant propagation component. Therefore in the AA region, only one LOS path appears [125].

A series of broadband measurements at 950 MHz and 2,150 MHz frequency which is known as potential frequency bands for LTE-R are conducted in China [123]. The mean excess delay, RMS delay spread and the average number of

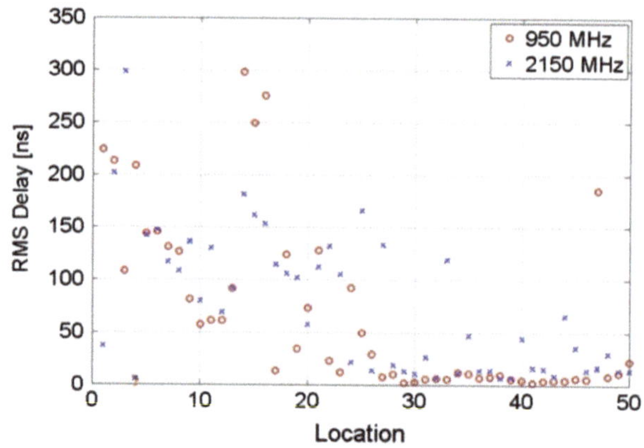

Fig. 3.25 RMS delay spread at two frequencies

multipath are summarized in Tab. I. The RMS delay at 950 MHz and 2,150 MHz are illustrated in Fig. 3.7. The value of RMS delay spread for region III and region IV lies in the range of 0–100 ns. This phenomenon is most likely due to the existence of the dominant LOS component and almost no additional multipath components can be resolved. When close to the cutting area, RMS delay spread becomes larger. The RMS delay spread is mainly distributed in the range of 100–300 ns in cutting scenarios, which reflects the severe multipath effects. In this area, there are approximately 5–6 multipath, which is due to the fact that there are richer reflected and scattered components from the cutting wall and remote hills. The RMS delay spread values of different frequencies are close (Fig. 3.25)

3.4.1.4 Future Research Directions

The delay characteristics are useful for wideband system design, especially when the LTE-R is deployed in HSRs. The OFDM is main physical technology of LTE. OFDM can also help to fight multipath fading using the cyclic prefix (CP). To realize this, the duration of CP should be longer than the maximum delay in the scenario. Otherwise, it will cause intersymbol interfere. In addition, the design of reference signal or pilot interval in the LTE broadcast channel also needs channel variation, such as coherence bandwidth, which is related to the RMS delay spread.

3.4.2 Doppler Effect

The two-path model is considered with unmodulated (sinusoidal) carrier signals firstly, though the consideration is valid for narrowband systems. The two-path model is the simplest model for explaining Doppler characteristic, i.e., Doppler shift, fading. Later, wideband channels are taken into account for its Doppler effect. Finally, the statistical parameters due to Doppler Effect are described.

3.4.2.1 The Time-Varying Two-Path Channel Model

The sample case that time-varying channel propagates over two path is considered for fundamental analysis. First, one single sinusoidal wave is considered, and the transmit signal is:

$$E_{TX}(t) \propto \cos(2\pi f_c t), \tag{3.4.6}$$

where f_c is the center frequency. As for the received signal, it is assumed as a homogeneous plane wave. If the RX moves away from the TX with speed v(the speed of movement in the direction of wave propagation is γ, see Fig. 3.24, the wave run distance between TX and RX increases with the runtime, the received signal can be described as:

$$\begin{aligned} E(t) &= E_0 \cdot \cos(2\pi f_c t - k_0[d_0 + v \cdot \cos(\gamma)t]) \\ &= E_0 \cdot \cos(2\pi t[f_c - \frac{v}{\lambda}\cos(\gamma)] - k_0 d_0), \end{aligned} \tag{3.4.7}$$

where k_0 is the wavenumber $2\pi/\lambda$ and E_0 is the signal amplitude at d_0; d_0 is the distance between TX and RX at $t = 0$. Here, we assumed that the amplitude is constant, which holds under the local area condition. Obviously, the movement of the RX leads to a shift of the received frequency, which is called the ***Doppler shift***.

In general, the Doppler Effect can be due to movements of the TX, the RX, the Interacting Objects (IOs), or any combination thereof; to simplicity, we assume only movement of the RX here. The Doppler shift is as follows, (Fig. 3.26).

$$\upsilon = -\frac{v}{\lambda}\cos(\gamma). \tag{3.4.8}$$

The ***maximum Doppler shift*** v_{max} is obtained when the direction of RX movement is aligned with the direction of wave propagation, i.e., $\gamma = 0^{\circ}$.

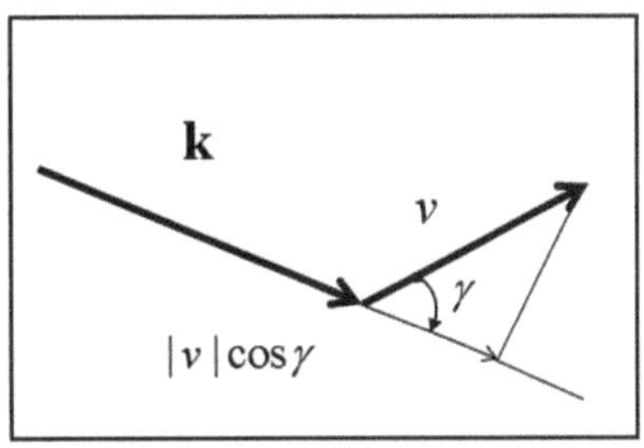

Fig. 3.26 Projection of velocity vector v onto the direction of propagation **k**

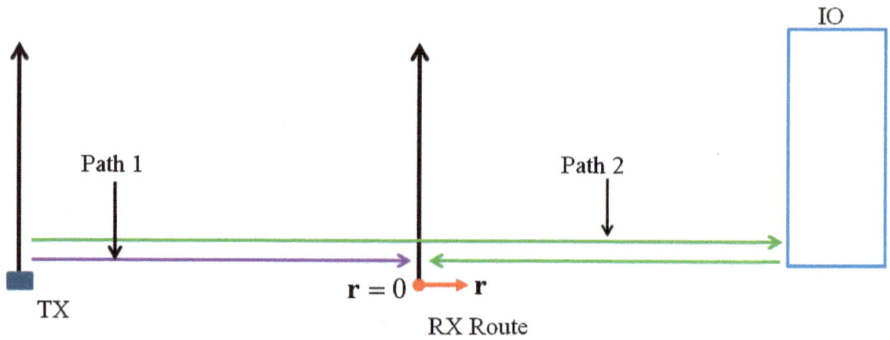

Fig. 3.27 Geometry of the time-varying two-path model

There are two points will be noted, one is that as the speed of the movement is always small compared with the speed of light, the Doppler shifts are relatively small; the other one is that Eq. (4.3) is based on several assumptions—e.g., static IOs, no double reflections on moving objects, etc.

We transform the above Eq. (4.2) into complex baseband notation,

$$E = E_0 \exp(-jk_0[d_0 + v \cdot \cos(\gamma)t]). \tag{3.4.9}$$

Note that the term $v\cos(\gamma)t$ is equal to the run distance projected onto the direction of wave propagation. For simplicity, we set $d_0 = 0$, and $k_0 v\cos(\gamma)t = \mathbf{k}_0 \cdot \mathbf{r}$. Here, $\mathbf{k}_0$ is the vector-valued wavenumber (i.e., has the absolute magnitude k_0, and is pointing into the direction of wave propagation); $\mathbf{r}$ is displacement vector pointing to the direction of movement (Fig. 3.27).

Now we consider the two paths, created by the IO in the propagation environment, see Fig. 3.25. Clearly, the two paths have different run distance, and thus different runtime:

$$\tau_1 = d_1/c_0,\ \text{and}\ \tau_2 = d_2/c_0. \tag{3.4.10}$$

Assuming that the two wave over these paths are vertically polarized, and have amplitudes E_1 and E_2 at the reference position $\mathbf{r} = 0$. We get the following expression for the superposition of two plane waves:

$$E(\mathbf{r}) = E_1 \exp(-j\mathbf{k}_1\mathbf{r}) + E_2 \exp(-j\mathbf{k}_2\mathbf{r}). \tag{3.4.11}$$

As two waves superimpose, the receive signal fluctuated for constructive and destructive interference. The fluctuation is location-dependent—i.e., $E(\mathbf{r})$ is a function of displacement vector $\mathbf{r}$. As for RX moving, it goes through a time-varying interference pattern with "mountains and valleys." We stress the fact that the interference pattern exists and not change whether or not the RX goes through it. For RX moves with high speed, more fading dips (instances of very low received power) [120] will be occurred as the spatially varying fading thus becomes time-varying fading. Hence, the fading rate, as number of fading dips per second, depends on the speed of the RX. These analyses are concentrated in time domain.

In frequency domain, since the Doppler shifts are relatively small, whether they have significant impacts on the radio link? If all multipath components (MPCs) were Doppler-shifted by the same value, the RX could maneuverable compensate for the shifts. However, the different MPCs always have different Doppler shifts according to the complex propagation scenario, even the simple two-path channel model. The MPCs with different Doppler-shifted signals leads to a random Frequency Modulation (FM) of the received signal that is awkward at RX. We also use the two-path model for analysis in Fig. 3.25. Two waves with different Doppler shifts are received by RX. The difference of the two Doppler shifts is just equal to the fading rate in time domain [125]. As we have discussed, the Doppler characteristic is a measure for the channel change rate.

3.4.2.2 Frequency Dispersion

We remain restrict our discussion to a simple scenario with a static TX and IOs, and a moving RX with speed v. Above, a sinusoidal carrier wave is considered for narrowband Doppler analysis. For a general transmit signal $s(t)$ with Fourier transform $S(f)$, we can get the receive signal as follows,

$$R(f) = hS(\alpha f),\ r(t) = \frac{h}{\alpha}s(\frac{t}{\alpha}), \quad \text{with } \alpha = 1 - \frac{v\cos(\gamma)}{c_0}, \tag{3.4.12}$$

where h is the complex attenuation factor caused by the propagation channel. In real systems, the transmit signal is band-limited around the carrier frequency, i.e., $S(f)$ is effectively zero outside a band $[f_c - B/2, f_c + B/2]$. The approximation is

$$\alpha f = f - \frac{v\cos(\gamma)}{c_0} f \approx f - \frac{v\cos(\gamma)}{c_0} f_c.$$

The accuracy increase with decreasing normalized bandwidth B/f_c. In narrowband system,

$$R(f) \approx hS(f - v),\ r(t) \approx hs(t)e^{j2\pi vt}, \tag{3.4.13}$$

where the definition for v is same as (4.3).

The Eq. (4.7) is always referred as wideband Doppler characterization; and (4.8) is often used in narrowband and approximate narrowband channel.

3.4.2.3 Stochastic Characteristic of Doppler Effect

The ***Doppler power spectrum*** is treated as a complete characterization of the Doppler Effect. The temporal statistics of fading caused by the superposition of several Doppler-shifted paths in time domain and the power for Doppler-shifted components in frequency domain both can be obtained from Doppler power spectrum.

The Doppler power spectrum in high-speed railway (HSR) is given in [126]. Different scenarios in HSR with different Doppler power spectrums are observed. For example, it is Rice Doppler Spectrum in mountain area, pure Doppler shift is in viaduct and Jakes Doppler Spectrum is found in Tunnel. Besides, the position-based Doppler shift is proposed as Doppler shift sweeps from a positive value to a negative value just in one call. Figure 3.26 depicts the measured Doppler spectrum in HSR. It is noted that a dual "Z" shape of the Doppler spectrum, which indicates a specular frequency shift.

The Doppler spectrum is given as [121],

$$S(v) = \frac{\sigma^2}{\pi\sqrt{1-(v/v_{\max})^2}} + \rho^2\delta(v - v_{\text{speclular}}), \tag{3.4.14}$$

where ρ^2 and σ^2 are power of the specular LOS component and the isotropic scattering components, respectively. The two constants satisfy $\rho^2 + \sigma^2 = 1$ and $K = 10\lg(\rho^2/\sigma^2)$ is the Ricean K-factors. And $f_{\text{speclular}}$ is the Doppler shift of the specular propagation component.

In above discussed, we showed the physical interpretation of the Doppler shift by movement. As the RX is moving, MPCs with different arrival angle (different γ) reach at the RX give rise to multiple Doppler shifts. This leads to a broadening of the received signal spectrum, i.e., ***Doppler spread***. The rms Doppler spread is

$$v_s = \sqrt{\frac{\int_{-\infty}^{\infty} S(v)v^2 dv}{P_S} - v_m^2}, \tag{3.4.15}$$

where, P_S is the integrated power,

$$P_S = \int_{-\infty}^{\infty} S_v(v)dv. \tag{3.4.16}$$

And v_m is the mean Doppler shift, defined as

$$v_m = \frac{\int_{-\infty}^{\infty} S(v)v\,dv}{P_S}. \tag{3.4.17}$$

Coherence time is the indicator of Doppler spread in time domain, and is given as

$$T_c \approx \frac{1}{f_m}, \tag{3.4.18}$$

where f_m is the maximum Doppler shift. The coherence time and Doppler spread are inversely proportional to the other. The other expressions of coherence time are given in [127]. Coherence time is actually a statistical measure of the time duration over which the channel impulse response is insignificant variant.

3.4.3 *Angular Characteristics*

3.4.3.1 Angular Parameter Definitions and Estimations

1. **Definitions**:

In multipath fading channels, the angular domain parameters refer to angle of departure (AOD) and angle of arrival (AOA) of the multipath component relative to the origin of antenna array, and the channels can be characterized in spatial domain by the angular characteristics obtained from AOD and AOA, which include:

- Power angular spectrum (PAS):
- Mean and RMS angular spread (AS);
- Angular distribution.

The PAS describes the spatial distribution of the multipath power related to the angle. Usually, the PAS can be described by uniform distribution, truncated Gaussian distribution, Laplace distribution, or Von Mise distribution.

Let $P(\theta)$ denote the signal power as a function of the angle θ. Then the mean AS can be defined as:

$$\mu_\theta = \frac{\int_{-\pi}^{\pi} \theta P(\theta) d\theta}{\int_{-\pi}^{\pi} P(\theta) d\theta}, \tag{3.4.19}$$

and the RMS AS can be expressed as:

$$\sigma_\theta = \frac{\int_{-\pi}^{\pi} (\theta - \mu_\theta)^2 P(\theta) d\theta}{\int_{-\pi}^{\pi} P(\theta) d\theta}. \tag{3.4.20}$$

By utilizing the values of mean and RMS AS, the wrapped Gaussian distribution is usually used for AOA or AOD distribution in High-Speed Railways [128].

2. **Estimations**:

In order to analyze the angular characteristics in High-Speed Railways, the estimation algorithms for angular domain parameters should be investigated. Considering the estimation of only AOA, for simplicity, the linear antenna array is chosen as an example. Assume that there are several multipath components $s_1(t), \ldots, s_D(t)$ with D different AOAs $\theta_1, \ldots, \theta_D$, and the number of antennas is M. As shown in Fig. 3.1, for a specific AOA θ, the phase difference between the adjacent antennas is $\frac{2\pi d \sin\theta}{\lambda}$, where d is the distance between the adjacent antennas and λ is the wavelength. If we take the first antenna as a reference, the phase of the m-th antenna is $\frac{2\pi m d \sin\theta}{\lambda}$. So the steering matrix can be defined as follows:

$$\mathbf{A} = [\mathbf{a}(\theta_1), \mathbf{a}(\theta_2), \ldots, \mathbf{a}(\theta_D)] = \begin{bmatrix} 1 & 1 & \cdots & 1 \\ e^{-j\frac{2\pi d \sin\theta_1}{\lambda}} & e^{-j\frac{2\pi d \sin\theta_2}{\lambda}} & \cdots & e^{-j\frac{2\pi d \sin\theta_D}{\lambda}} \\ \vdots & \vdots & \ddots & \vdots \\ e^{-j\frac{2\pi(M-1)d \sin\theta_1}{\lambda}} & e^{-j\frac{2\pi(M-1)d \sin\theta_2}{\lambda}} & \cdots & e^{-j\frac{2\pi(M-1)d \sin\theta_D}{\lambda}} \end{bmatrix}_{M\times D}, \tag{3.4.21}$$

where $\mathbf{a}(\theta_i) = [1, e^{j\frac{2\pi d \sin\theta_i}{\lambda}}, \ldots, e^{j\frac{2\pi(M-1)d \sin\theta_i}{\lambda}}]^T$, and $i = 1, 2, \ldots, D$.

So the received signal can be expressed as:

$$\mathbf{x}(t) = [\mathbf{a}(\theta_1), \mathbf{a}(\theta_2), \ldots, \mathbf{a}(\theta_D)] \cdot \begin{bmatrix} s_1(t) \\ s_2(t) \\ \vdots \\ s_D(t) \end{bmatrix} + \begin{bmatrix} n_1(t) \\ n_2(t) \\ \vdots \\ n_M(t) \end{bmatrix} = \mathbf{A} \cdot \mathbf{s}(t) + \mathbf{n}(t), \tag{3.4.22}$$

where $\mathbf{n}(t)$ is the zero-mean Gaussian noise vector. So the correction matrix for received signals can be defined as:

$$\begin{aligned} \mathbf{R_{xx}} &= E[\mathbf{x}(t) \cdot \mathbf{x}^H(t)] = E\{[\mathbf{A} \cdot \mathbf{s}(t) + \mathbf{n}(t)][\mathbf{A}^H \cdot \mathbf{s}^H(t) + \mathbf{n}^H(t)] \\ &= \mathbf{A}E[\mathbf{s}(t) \cdot \mathbf{s}^H(t)]\mathbf{A}^H + E[\mathbf{n}(t) \cdot \mathbf{n}^H(t)] \\ &= \mathbf{A}\mathbf{R_{ss}}\mathbf{A}^H + \mathbf{R_{nn}}, \end{aligned} \tag{3.4.23}$$

where $\mathbf{R_{ss}}$ is the $D \times D$ correction matrix for transmitted signals, and $\mathbf{R_{nn}}$ is the correction matrix for noise.

The purpose of different AOA estimation techniques is to define a function by using the correction matrix, and the AOA can be obtained from the maximum

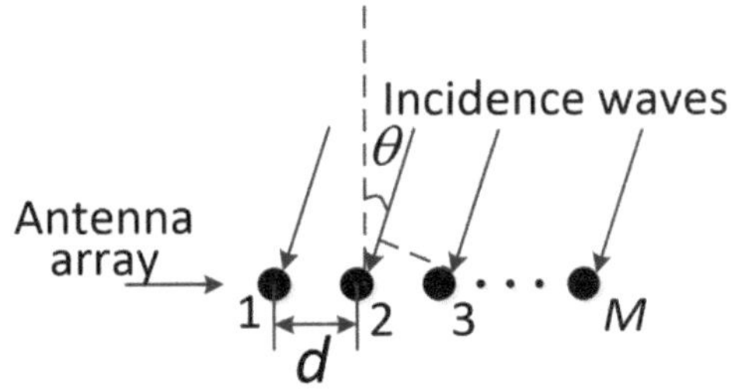

Fig. 3.28 Linear antenna array

values of this function. Different methods are utilized to define this function, such as beamforming (BF), maximum likelihood (ML), linear prediction (LP), and so on (Fig. 3.28).

Bartlett

The Bartlett function to estimate the AOAs is defined as [129]

$$P_B(\theta) = \mathbf{a}^H(\theta) \cdot \mathbf{R}_{\mathbf{xx}} \cdot \mathbf{a}(\theta). \tag{3.4.24}$$

The BF method is utilized in Bartlett estimation, but the resolution properties are limited.

MVDR (Minimum Variance Distortionless Response)

The MVDR function to estimate the AOAs is defined as [130]

$$P_C(\theta) = \frac{1}{\mathbf{a}^H(\theta) \cdot \mathbf{R}_{\mathbf{xx}}^{-1} \cdot \mathbf{a}(\theta)}. \tag{3.4.25}$$

The ML method is utilized in MVDR estimation to maximize the signal-to-interference ratio, and this method has better resolution properties than the Bartlett estimate in most cases.

LP (Linear Prediction)

The LP function to estimate the AOAs is defined as [131]

$$P_{LPm}(\theta) = \frac{\mathbf{u}_{\mathbf{m}}^{\mathbf{T}}(\theta) \cdot \mathbf{R}_{\mathbf{xx}}^{-1} \cdot \mathbf{u}_{\mathbf{m}}}{\left|\mathbf{u}_{\mathbf{m}}^{\mathbf{T}}(\theta) \cdot \mathbf{R}_{\mathbf{xx}}^{-1} \cdot \mathbf{a}(\theta)\right|^2}, \tag{3.4.26}$$

where $\mathbf{u}_{\mathbf{m}}$ is the m-th column of $M \times M$ unit matrix. The purpose of the LP method estimation is to minimize the mean squared prediction error, but the resolution properties are different when predicting signals of different elements over multiple antenna array.

MUSIC (MUltiple SIgnal Classification)

The MUSIC function to estimate the AOAs is defined as [132]

$$P_{MU}(\theta) = \frac{1}{\left|\mathbf{a}(\theta)^H \mathbf{E}_N \mathbf{E}_N^H \mathbf{a}(\theta)\right|}, \tag{3.4.27}$$

where $\mathbf{E_N} = [\mathbf{e_1}, \mathbf{e_2}, \ldots, \mathbf{e_{M-D}}]$ is the $M \times (M - D)$ matrix whose columns are the $(M - D)$ noise eigenvectors. MUSIC is one of the most popular techniques used in angle estimation. This method is applicable to antenna arrays with different geometry, but the computation and required storage are high.

ESPRIT (Estimation of Signal Parameters via Rotation Invariance Techniques)

The ESPRIT algorithm is achieved by requiring that the antenna array possess a displacement invariance [133]. ESPRIT can also be applied to antenna arrays with different geometry, and the computation and storage costs can be reduced. However, plenty of antennas are needed compared with the number of signals to detect.

SAGE (Space-Alternating Generalized Expectation–maximization)

Besides the AOA estimation techniques described above, the channel parameters can also be jointly estimated by SAGE algorithm [134].

The baseband channel impulse response (IR) of the MIMO system can be expressed as

$$\mathbf{h}(\tau) = \sum_{l=1}^{L} \alpha_l e^{-j2\pi f_c \tau_l} \mathbf{p_2}(\varphi_{2,l}) \mathbf{p_1}(\varphi_{1,l})^T \delta(\tau - \tau_l), \quad (3.4.28)$$

where α_l, τ_l, $\varphi_{1,l}$ and $\varphi_{2,l}$ donate the complex amplitude, delay, AOD and AOA of the l-th multipath component for carrier frequency f_c, respectively. The parameters which describe all the multipath components can be expressed as

$$\Theta = [\alpha_l, \tau_l, \varphi_{1,l}, \varphi_{2,l}; l = 1, \ldots, L]. \quad (3.4.29)$$

These parameters can be jointly estimated by using SAGE. This algorithm iteratively update the subset of Θ, i.e., $\theta_l = [\alpha_l, \tau_l, \varphi_{1,l}, \varphi_{2,l}]$, and finally obtain the joint ML estimation of all the parameters in Θ.

3.4.3.2 Current Investigations on Angular Characteristics

Besides the wideband channel measurement in High-Speed Railways, the multiple-input multiple-output (MIMO) technology where multiple antennas are equipped at both ends of communication link, is required if the angular domain parameters need to be extracted. The angular statistics can thus be obtained from the measurement data. Based on MIMO channel measurements or theoretical approaches, some channel models are also proposed which include the angular parameters.

Channel Measurements:

Until now, the MIMO channel measurement in High-Speed Railways is still challenging task due to the difficulties in hardware design. Only a few MIMO channel measurements in High-Speed Railways are performed:

(1) In [128], the single-input-multiple-output (SIMO) channel measurement for High-Speed Railway scenarios is conducted in the Wireless World Initiative New Radio (WINNER) project. Hence, the AOA can be obtained from the multiple receiver (Rx) side, and the angular spread (AS) is provided.
(2) The multiple-input single-output (MISO) channel measurement results are presented in [135], and the PAS is given.

Obviously, the MIMO channel measurements in High-Speed Railways are still needed for future High-Speed Railways communication system.

Channel models:

Currently, some channel models including angular parameters for High-Speed Railway scenarios are proposed.

(1) Based on the measurement results in [128], a MIMO geometry-based stochastic channel model (GSCM) for High-Speed Railway scenario is generated. As multiple antennas are only used at the Rx side, some missing parameters, such as AOD, are borrowed from other rural measurements. So the MIMO GSCM in High-Speed Railways based on measurements has not been completed. Figure 3.2 presents a diagram of GSCM in High-Speed Railways, and the cluster-based structure is suggested [128] (Fig. 3.29).
(2) Another approach is the Geometry-based deterministic model (GBDM) based on ray-tracing method [136], and the angular parameters can be directly obtained from the ray-tracing simulator. However, the angular characteristics need to be further validated by the corresponding measurement.
(3) Some theoretical MIMO channel models in High-Speed Railways are also proposed. These models assume that the effective scatterers are placed on regular shapes [137], such as one-ring, two-ring, and multiple ellipses, or at

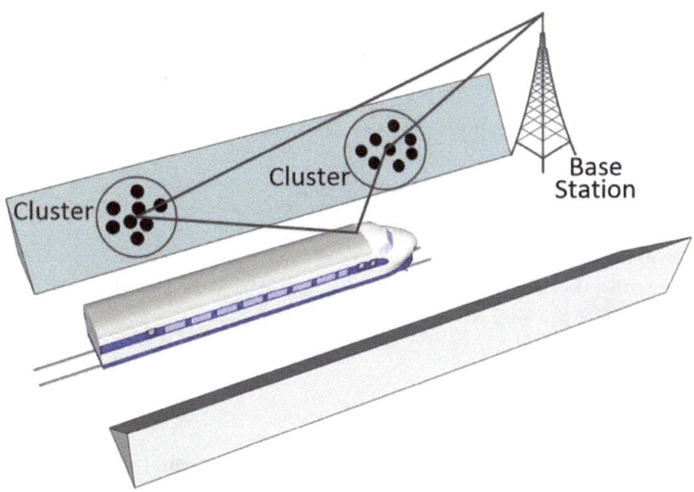

Fig. 3.29 Geometry-based stochastic channel model in High-Speed Railways

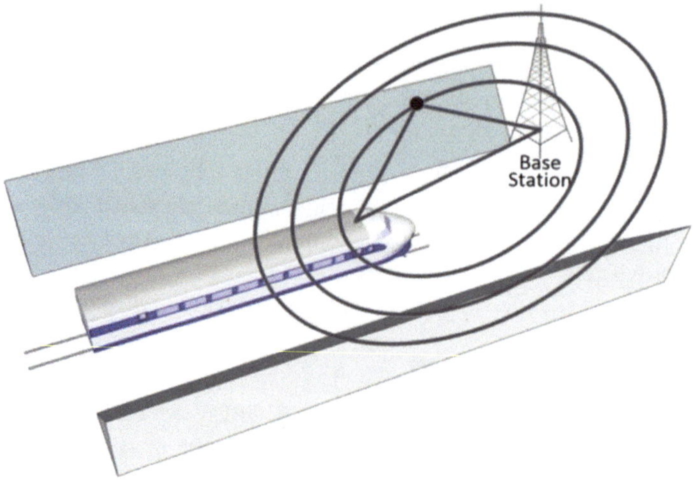

Fig. 3.30 Theoretical channel model in High-Speed Railways that effective scatterers are placed on regular shapes

random locations with certain statistical distributions [138], and the time-variant angular parameters, such as AOD and AOA are also considered. Figure 3.3 gives the diagram of theoretical MIMO channel models that the effective scatterers are placed on multiple ellipses, whereas in Fig. 3.4 effective scatterers are placed at random locations in a specific area according to the scenario. It can be seen that the angle parameters are naturally included in these models. However, the angular statistics are neither analyzed in these models, nor validated by the channel measurements or channel simulations.

In conclusion, the MIMO channel measurements in High-Speed Railways are still essential for channel modeling and validation of different channel models. The angular characteristics can thus be obtained and analyzed based on the MIMO channel measurements and models (Figs. 3.30 and 3.31)

3.4.3.3 Future Research Directions

MIMO channel measurements:

The conventional MIMO measurement can be realized in three ways:

(1) Channel measurements with parallel architecture, like a truly MIMO measurements;
(2) Channel measurement with MIMO architecture by switching antennas in time division multiplexing (TDM);
(3) Channel measurement with single-input single-output (SISO) architecture by moving antennas automatically or manually.

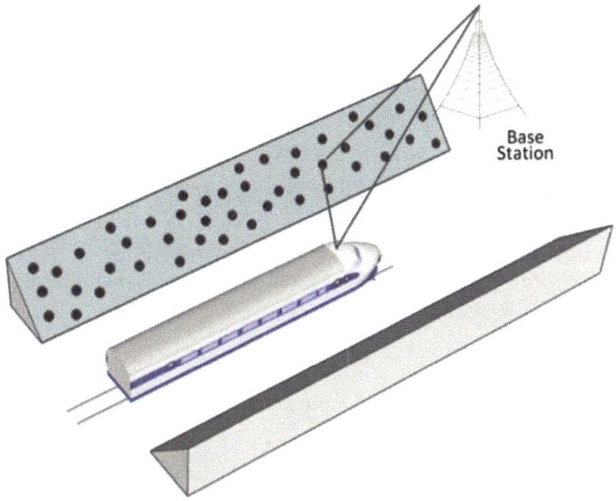

Fig. 3.31 Theoretical channel model in High-Speed Railways that effective scatterers are placed at random locations with certain statistical distributions

Due to the complexity and cost involved in designing and building a parallel architecture, the first method is hard to realize, especially for Massive MIMO. Considering the high moving speeds, the third method is impossible to utilize for High-Speed Railway scenario. So the second method is the most promising one for High-Speed Railways MIMO channel measurement, only if the time required to collecting the data of all antenna pair channels (snapshot time) is small enough to remain the channel wide-sense stationary (WSS).

Angular characteristics:

So far, in most common High-Speed Railway scenarios, the research status of wideband channel parameters is shown in Table 3.7 [139]. We can see that almost no results are reported for angular characteristics and models. Based on MIMO channel measurements in High-Speed Railways, the angular domain parameters can be obtained by various angle estimation techniques. So the angular characteristics in different High-Speed Railway scenarios, such as PAS and AS, can be introduced and summarized for future modeling. (Table 3.8)

Table 3.8 Research status of wideband channel parameters in High-Speed Railways

Scenario	Delay	Doppler Shift	AOA/AOD
Viaduct	√	√	√
Cutting	√	×	×
Rural	√	√	√
Suburban	√	√	×
In-Carriage	√	×	×
√Channel characteristic has been presented			
×Channel characteristic has not been presented			

Elevation angles:

In reality, radio waves propagate in three dimensions and scatterers are dispersed in elevation in High-Speed Railways [140], especially when the train is close to the Base Station (BS) where the impact of elevation angles should not be ignored. So the 3D channel measurements including elevation angles are necessary. Besides the GBDM by using 3D ray-tracing method, the other High-Speed Railway channel models are still generated in two dimensions. Therefore, the 3D channel models including elevation angles are also required.

3.5 Summary

To begin with, we clarify the definitions and partitions of high-speed railway propagation scenarios. Moreover, a more general concept—wide-sense V2X (WSV2X) has been formed, which has already become a regular convened session in European Conference on Antennas and Propagation (EuCAP). This concept is widely acknowledged. In this chapter, and in this manner, all the propagation scenarios of WSV2X communications are defined in detail.

The second part of the chapter presents various methods and related systems of measuring the radio channels in high-speed railway networks. We briefly review state-of-the-art high-speed railway radio channel measurement campaigns according to the scenarios, measurement equipment, measurements' setup parameters (i.e., carrier frequency, bandwidth, and antenna configuration), and estimated channel statistics. With wideband signals, the received signal experiences distortion due to the delay spread of the different multipath components, so the most important characteristics of the wideband channel, including its time dispersion parameters, coherence bandwidth, Doppler power spectrum, coherence time, and angular domain parameters, are characterized in various high-speed railway environments as well in the last part of this section.

To sum up, both narrowband and wideband radio propagation and wireless channels for railway communications are comprehensively studied in this chapter. This provides a solid basis for system design, simulation, and evaluation of various communication systems deployed in railway environments.

References

1. Ai B, He R, Zhong Z et al (2012) Radio wave propagation scene partitioning for high-speed rails. Int J Antennas Propag:1072–1075
2. Ai B (2011) Interview. Electron Lett 47(21): 1158
3. He R, Zhong Z, Ai B, Ding J (2011) An empirical path loss model and fading analysis for high-speed railway viaduct scenarios. IEEE Antennas Wirel Propag Lett 10:808–812
4. He R, Zhong Z, Ai B, Ding J (2011) Propagation measurements and analysis for high-speed railway cutting scenario. Electron Lett 47(21):1167–1168

5. Guan K, Zhong Z, Ai B, Kürner T (2014) Propagation measurements and modeling of crossing bridges on high-speed railway at 930 MHz. IEEE Trans Veh Technol 63(2):502–517
6. Guan K, Zhong Z, Ai B, Briso-Rodriguez C (2012) Propagation mechanism modelling in the near region of circular tunnels. IET Microwaves Antennas Propag 6(3):355–360
7. Guan K et al (2013) Complete Propagation Model in Tunnels. IEEE Antennas Wirel Propag Lett 12:741–744
8. He R et al (2013) Propagation channel measurements and analysis at 2.4 GHz in subway tunnels. IET Microwaves Antennas Propag 7(11):934–941
9. WINNER Group and IST-WINNER II Deliverable 1. 1. 2 v. 1.2 (2008) WINNER II channel models. Technical report IST-WINNER2
10. Vaa T, Penttinen M, Spyropoulou I (2007) Intelligent transport systems and effects on road traffic accidents: state of the art. IET Intell Transp Syst 1(2):81–88
11. IEEE Std. 802. 11–2007, Part 11: weireless lAN medium access control (MAC) and physical layer (PHY) specifications. IEEE Std. 802. 11 2007
12. ETSI ES 202 663 (2009) Intelligent transport systems, european profile standard on the physical and medium access layer of 5 GHz ITS. Draft Version 0. 0. 6
13. IEEE Std 1609. 4 (2011) IEEE standard for wireless access in vehicular environments (WAVE)-multi-channel operation
14. IEEE standard for communications-based train control (CBTC) performance and functional requirements (1999)
15. http://www.uic.asso.fr
16. ETSI ETR 300-3 (2000-02), Terrestrial trunked radio (TETRA), voice plus data (V + D), designers' guide, part 3: Direct Mode Operation (DMO), 1st edn
17. García CR, Lehner A, Strang T, Frank K (2008) Channel model for train to train communication using the 400 MHz band. In: Proceedings of the 2008 IEEE 67th vehicular technology conference, pp 3082–3086, Singapore
18. 1st year research report on railroad wireless communication system, Korea Railroad Research Institute, Seoul, Republic of Korea, 2011
19. Guan K, Zhong Z, Ai B (2011) Assessment of LTE-R using high speed railway channel model. In: Proceedings of the 3rd international conference on communications and mobile computing, pp 461–464, Qingdao, China
20. Molisch AF, Tufvesson F, Karedal J, Mecklenbrauker CF (2009) A survey on vehicle-to-vehicle propagation channels. IEEE Wirel Commun 16(6):12–22
21. Abbas T, Karedal J, Tufvesson F, Paier A, Bernadó L, Molisch AF (2011) Directional analysis of vehicle-to-vehicle propagation channels. In: Proceedings of the IEEE 73rd vehicular technology conference (VTC '11), pp 1–5, Budapest, Hungry
22. Nuckelt J, Abbas T, Tufvesson F, Mecklenbr¨auker C, Bernadó L, K¨urner T (2013) Comparison of ray tracing and channel-sounder measurements for vehicular communications. In: Proceedings of the IEEE 77th vehicular technology conference (VTC '13), pp 1–5, Dresden, Germany
23. Liu L, Tao C, Qiu J et al (2012) Position-based modeling for wireless channel on high-speed railway under a viaduct at 2.35GHz. IEEE J Sel Areas Commun 30(4):834–845
24. Wei H, Zhong Z, Guan K, Ai B (2010) Path loss models in viaduct and plain scenarios of the high-speed railway. In: Proceedings of the 5th International ICST conference on communications and networking in China (ChinaCom '10), pp 1–5, Beijing, China
25. Guan K, Zhong Z, Alonso JI, Briso-Rodnguez C (2012) Measurement of distributed antenna systems at 2.4 GHz in a realistic subway tunnel environment. IEEE Trans Veh Technol 61 (2):834–837
26. Guo Y, Zhang J, Tao C, Liu LIU, Tian L (2012) Propagation characteristics of wideband high-speed railway channel in viaduct scenario at 2. 35GHz. J Modern Transp 20(4):206–212
27. Acosta-Marum G, Ingram MA (2007) Six time- and frequency-selective empirical channel models for vehicular wireless LANs. IEEE Veh Technol Mag 2(4):4–11

28. IST WINNER II D1. 1. 2, WINNER II channel models. http:// www.ist-winner.org/WINNER2-Deliverables/D1.1.2v1.1.pdf
29. Maurer J, Fugen T, Wiesbeck W (2002) Narrow-band measurement and analysis of the inter-vehicle transmission channel at 5.2 GHz. In: Proceedings of the IEEE 55th vehicular technology conference (VTC '02), pp 1274–1278, Birmingham, UK, May 2002
30. Maurer J, Schafer TM, Wiesbeck W (2001) A realistic description of the environment for inter-vehicle wave propagation modelling. In: Proceedings of the IEEE 54th vehicular technology conference (VTC '01), pp 1437–1441, Atlantic, NJ, USA
31. Maurer J, Fugen T, Schafer T, Wiesbeck W (2004) A new inter-vehicle communications (IVC) channel model. In: Proceedings of the IEEE 60th vehicular technology conference (VTC '04), pp 9–13, Los Angeles, Calif, USA
32. Nuckelt J, Schack M, Kurner T (2011) Deterministic and stochastic channel models implementedin a physical layer simulator for Car-to-X communications. Adv Radio Sci 9:165–171
33. Guan K, Zhong Z, Ai B, Kurner T (2013) Deterministic propagation modeling for the realistic high-speed railway environment. In: Proceedings of the IEEE 77th vehicular technology conference (VTC '13), pp 1–5, Dresden, Germany
34. Jakes WC (ed) (1974) Microwave mobile communications. Wiley, New York, NY, USA
35. Akki AS, Haber F (1986) A statistical model of mobile-to-mobile land communication channel. IEEE Trans Veh Technol 35(1):2–7
36. Akki AS (1994) Statistical properties of mobile-to-mobile land communication channels. IEEE Trans Veh Technol 43(4):826–831
37. Vatalaro F (1997) Doppler spectrum in mobile-to-mobile communications in the presence of three-dimensional multipath scattering. IEEE Trans Veh Technol 46(1):213–219
38. Almers P, Bonek E, Burr A et al (2007) Survey of channel and radio propagation models for wireless MIMO systems. Eurasip J Wirel Commun Netw 2007:Article ID 19070
39. Wang C, Cheng X, Laurenson D (2009) Vehicle-to-vehicle channel modeling and measurements: recent advances and future challenges. IEEE Commun Mag 47(11):96–103
40. Chen B, Zhong Z (2012) Geometry-based stochastic modeling for MIMO channel in high-speed mobile scenario. Int J Antennas Propag 2012:Article ID184682, 6 pp
41. Karedal J, Tufvesson F, Czink N et al (2009) A geometry-based stochastic MIMO model for vehicle-to-vehicle communications. IEEE Trans Wireless Commun 8(7):3646–3657
42. MEDAV GmbH. http://www.medav.de
43. Wang CX, Ghazal A, Ai B et al (2016) Channel measurements and models for high-speed train communication systems: a survey
44. Zhou T, Tao C, Salous S et al (2015) Channel sounding for high-speed railway communication systems. IEEE Commun Mag 53(10):70–77
45. Zhou T, Tao C, Liu L et al (2012) High-speed railway channel measurements and characterizations: a review. J Modern Transp 20(4):199–205
46. Ghazal A, Wang CX, Haas H et al (2012) A non-stationary MIMO channel model for high-speed train communication systems. 2012 IEEE 75th on vehicular technology conference (VTC Spring). IEEE, 2012, pp 1–5
47. Molisch AF (2010) Wireless communications. Wiley
48. Matz G (2005) On non-wssus wireless fading channels. IEEE Trans Wirel Commun 4 (5):2465–2478
49. Bernadó R (2012) Non-stationarity in vehicular wireless channels. PhD dissertation, Technischen Universitat Wien, Vienna, Austria
50. Ai B et al (2014) Challenges toward wireless communications for high-speed railway. IEEE Trans Intell Transp 15(5):2143–2158
51. Zhou T, Tao C, Salous S, Liu L, Tian Z (2015) Channel sounding for high-speed railway communication systems. IEEE Commun Mag 53(10):70–77
52. Salous S (2013) Radio propagation measurement and channel modelling. Wiley

53. Wang C, Ghazal A, Ai B, Liu Y, Fan P (2016) Channel measurements and models for high-speed train communication systems: a survey. IEEE Commun Surv Tut (to be published)
54. Pekka K (2007) WINNER II channel models part II radio channel measurement and analysis results. IST-4-027756, WINNER II D1.1.2, v1.0
55. Parviainen R, Kyosti P, Hsieh Y, Ting P, Chiou J (2008) Results of high speed train channel measurements. In: COST 2100 TD'08, Lille, France
56. Zhao M, Wu M, Sun Y, Yu D, Di S, Zhou P, Zeng X, Ge S (2013) Analysis and modeling for train-ground wireless wideband channel of LTE on high-speed railway. In: Proceedings of the IEEE VTC'13-Spring, Dresden, Germany, June 2013, pp 1–5
57. Wei H, Zhong Z, Guan K, Ai B (2010) Path loss models in viaduct and plain scenarios of the high-speed railway. In: Proceedings of the CHINACOM'10, Beijing, China, Aug 2010, pp 1–5
58. Gao L, Zhong Z, Ai B, Xiong L (2010) Estimation of the Ricean factor in K the high speed railway scenarios. In: Proceedings of the CHINACOM'10, Beijing, China, Aug 2010, pp 1–5
59. Lu J, Zhu G, Ai B (2010) Radio propagation measurements and modeling in railway viaduct area. In: Proceedings of the IEEE WiCOM'10, Chengdu, China, Sept 2010, pp 1–5
60. He R, Zhong Z, Ai B (2010) Path loss measurements and analysis for high-speed railway viaduct scene. In: Proceedings of the IWCMC'10, Caen, France, July 2010, pp 266–270
61. Gao L, Zhong Z, Ai B, Xiong L, Zhang H (2011) Analysis and emulation of the small-scale fading characteristics in the high-speed rail scenarios. In: Proceedings of the CHINACOM'11, Harbin, China, Aug 2011, pp 1181–1185
62. Wei H, Zhong Z, Xiong L, Ai B, He R (2011) Study on the shadow fading characteristic in viaduct scenario of the high-speed railway. In: Proceedings of the CHINACOM'11, Harbin, China, Aug 2011, pp 1216–1220
63. He R, Zhong Z, Ai B, Xiong L, Wei H (2010) A novel path loss model for high-speed railway viaduct scenarios. In: Proceedings of the IEEE WiCOM'11, Wuhan, China, Sept 2011, pp 1–4
64. He R, Zhong Z, Ai B, Xiong L, Ding J (2011) The effect of reference distance on path loss prediction based on the measurements in high-speed railway viaduct scenarios. In: Proceedings of the CHINACOM'11, Harbin, China, Aug. 2011, pp 1201–1205
65. He R, Zhong Z, Ai B, Ding J (2011) An empirical path loss model and fading analysis for high-speed railway viaduct scenarios. IEEE Antennas Wireless Propag Lett 10:808–812
66. He R, Zhong Z, Ai B, Ding J (2012) Distance-dependent model of Ricean K-factors in high-speed rail viaduct channel. In: Proceedings of the IEEE VTC'12-Fall, Quebec City, Canada, Sept 2012, pp 1–5
67. He R, Zhong Z, Ai B, Ding J (2012) Measurements and analysis of short-term fading behavior for high-speed rail viaduct scenario. In: Proceedings of the IEEE international conference on communications (ICC), Canada 2012, pp 4563–4567
68. He R, Zhong Z, Ai B, Wang G, Ding J, Molisch AF (2013) Measurements and analysis of propagation channels in high-speed railway viaducts. IEEE Trans Wireless Commun 12 (2):794–805
69. Zhang B, Zhong Z, Ai B, Yao D, He R (2014) Measurements and modeling of cross-correlation property of shadow fading in high-speed railways. In: Proceedings of the IEEE VTC'14-Fall, Vancouver, Canada, Sept 2014, pp 1–5
70. Liu L, Tao C, Qiu J, Chen H, Yu L, Dong W, Yuan Y (2012) Position based modeling for wireless channel on high-speed railway under a viaduct at 2.35 GHz. IEEE J Sel Areas Commun 30(4):834–845
71. Guo Y, Zhang L, Zhang C, Tian L (2013) Correlation analysis of high speed railway channel parameters based on channel measurement. In: Proceedings of the HMWC'13, Shanghai, China, Nov 2013, pp 1–5

72. Wang Q, Xu C, Zhao M, Yu D (2014) Results and analysis for a novel 2x2 channel measurement applied in LTE-R at 2.6 GHz. In: Proceedings of the IEEE WCNC'14, Istanbul, Turkey, April 2014, pp 177–181
73. Sun R, Tao C, Liu L, Tan Z, Zhang L, Zhou T (2014) Non isotropic scattering characteristic in an alternant tree-blocked viaduct scenario on high-speed railway at 2.35 GHz. Int J Antennas Propag 2014: Article ID 642894, 9 pp. doi:10.1155/2014/642894
74. Zhou T, Tao C, Liu L, Tan Z (2014) A semi empirical MIMO channel model in obstructed viaduct scenarios on high-speed railway. Int J Antennas Propag 2014:Article ID 287159, 10 pp. doi:10.1155/2014/287159
75. Lu J, Zhu G, Briso-Rodriguez C (2011) Fading characteristics in the railway terrain cuttings. In: Proceedings of the IEEE VTC'11-Spring, Budapest, Hungary, May 2011, pp 1–5
76. He R, Zhong Z, Ai B, Ding J (2011) Propagation measurements and analysis for high-speed railway cutting scenario. Electron Lett 47(21):1167–1168
77. He R, Zhong Z, Ai B, Ding J, Yang Y (2012) Propagation measurements and analysis of fading behavior for high speed rail cutting scenarios. In: Proceedings of the IEEE GLOBALCOM'12, California, USA, Dec 2012, pp 5015–5020
78. He R, Zhong Z, Ai B, Ding J, Yang Y, Molisch AF (2013) Short-term fading behavior in high-speed railway cutting scenario: measurements, analysis, and statistical models. IEEE Trans Antennas Propag 61(4):2209–2222
79. Sun R, Tao C, Liu L, Tan Z (2013) Channel measurement and characterization for HSR U-Shape groove scenarios at 2.35 GHz. In: Proceedings of the IEEE VTC'13-Fall, Las Vegas, USA, Sept 2013, pp 1–5
80. Tian L, Zhang J, Pan C (2013) Small scale fading characteristics of wideband radio channel in the U-Shape cutting of high-speed railway. In: Proceedings of the IEEE VTC'13-Fall, Las Vegas, USA, Sept 2013, pp 1–6
81. Luan F, Zhang Y, Xiao L, Zhou C, Zhou S (2013) Fading characteristics of wireless channel on high-speed railway in hilly terrain scenario. Int J Antennas Propag 2013:Article ID 378407, 9 pp. doi:10.1155/2013/378407
82. Zhang Y, He Z, Zhang W, Xiao L, Zhou S (2014) Measurement based delay and Doppler characterizations for high-speed railway hilly scenario. Int J Antennas Propag 2014:Article ID 875345, 8 pp. doi:10.1155/2014/875345
83. Aikio P, Gruber R, Vainikainen P (1998) Wideband radio channel measurements for train tunnels. In: Proceedings of the IEEE VTC'98, Ottawa, Canada, May 1998, pp 460–464
84. Briso-Rodriguez C, Cruz JM, Alonso JI (2007) Measurements and modeling of distributed antenna systems in railway tunnels. IEEE Trans Veh Technol 56(5):2870–2879
85. Guan K, Zhong Z, Ai B, Kurner T (2014) Propagation measurements and analysis for train stations of high-speed railway at 930 MHz. IEEE Trans Veh Technol 63(8):3349–3516
86. Guan K, Zhong Z, Ai B, Kurner T (2014) Empirical models for extra propagation loss of train stations on high-speed railway. IEEE Trans Antennas Propag 62(3):1395–1408
87. Guan K, Zhong Z, Ai B, K¨urner T (2014) Propagation measurements and modeling of crossing bridges on high-speed railway at 930 MHz. IEEE Trans Veh Technol 63(8):502–517
88. Wen Y, Ma Y, Zhang X, Jin X, Wang F (2012) Channel fading statistics in high-speed mobile environment. In: Proceedings of the IEEE APWC'12, Cape Town, South Africa, Sept 2012, pp 1209–1212
89. Lu J, Zhu G, Ai B (2012) Fading analysis for the high speed railway viaduct and terrain cutting scenarios. Int J Antennas Propag 2012:Article ID 862945, 9 pp
90. Chen B, Zhong Z, Ai B (2012) Stationarity intervals of time-variant channel in high speed railway scenario. J. China Commun 9(8):64–70
91. He R, Zhong D, Ai B, Oestges C (2015) Shadow fading correlation in high-speed railway environments. IEEE Trans Veh Technol 64(7):2762–2772
92. He R, Zhong Z, Ai B, Zhang B (2014) Measurement-based autocorrelation model of shadow fading for the high-speed railways in urban, suburban, and rural environments. In: Proceedings of the APSURSI'14, Memphis, TN, USA, July 2014, pp 949–950

93. Qiu J, Tao C, Liu L, Tan Z (2012) Broadband channel measurement for the high-speed railway based on WCDMA. In: Proceedings of the IEEE VTC'12-Spring, Yokohama, Japan, May 2012, pp 1–5
94. Liu L, Tao C, Zhou T, Zhao Y, Yin X, Chen H (2012) A highly efficient channel sounding method based on cellular communications for high-speed railway scenarios. EURASIP J Wirel Commun Network 16 pp. doi:10.1186/1687-1499-2012-307
95. Yin X, Cai X, Cheng X, Chen J, Tian M (2015) Empirical geometry based random-cluster model for high-speed-train channels in UMTS networks. IEEE Trans Intell Transp Syst 16 (5):2850–2861
96. Kaltenberger F et al (2015) Broadband wireless channel measurements for high speed trains. In: ICC
97. Kita N, Ito T, Yokoyama S, Tseng M-C, Sagawa Y, Ogasawara M, Nakatsugawa M (2009) Experimental study of propagation characteristics for wireless communications in high-speed train cars. In: Proceedings of the EuCAP'09, Berlin, Germany, Mar. 2009, pp 897–901
98. Dong W, Liu G, Yu L, Ding H, Zhang J (2010) Channel properties of indoor part for high-speed train based on wideband channel measurement. In: Proceedings of the CHINACOM'10, Beijing, China, Aug 2010, pp 1–4
99. Yang J, Ai B, Zhong Z (2012) Construction and capacity analysis of high-rank LoS MIMO channels in high speed railway scenarios. Int J Antennas Propag 2012:Article ID 423759, 7 pp. doi:10.1155/2012/423759
100. He R, Zhong Z, Ai B, Ding J, Jiang W, Zhang H, Li X (2014) A standardized path loss model for the GSM-railway based high-speed railway communication systems. In: Proceedings of the IEEE VTC, pp 1–5 (to be published)
101. He R, Zhong Z, Ai B, Oestges C (2015) Shadow fading correlation in high-speed railway environments. IEEE Trans Veh Technol (to be published)
102. He R, Zhong Z, Ai B, Guan K (2016) Reducing the cost of the high-speed railway communications: from propagation channel view. IEEE Trans Intell Transp Syst (under review)
103. Molisch AF (2011) Wireless Communications, 2nd edn. Wiley, Hoboken, NJ, USA
104. He R, Ai B, Zhong Z, Molisch AF, Chen R, Yang Y (2015) A Measurement-Based Stochastic Model for High-Speed Railway Channels. IEEE Trans Intell Transp Syst 16 (3):1120–1135
105. He R, Zhong Z, Ai B, Ding J, Yang Y, Molisch AF (2013) Short-term fading behavior in high-speed railway cutting scenario: measurements, analysis, and statistical models. IEEE Trans Antennas Propag 61(4):2209–2222
106. He R, Zhong Z, Ai B, Wang G, Ding J, Molisch AF (2013) Measurements and analysis of propagation channels in high-speed railway viaducts. IEEE Trans Wirel Commun 12 (2):794–805
107. He R, Zhong Z, Ai B, Ding J, Yang Y (2012) Propagation measurements and analysis of fading behavior for high speed rail cutting scenarios. In: 2012 IEEE Global Communications Conference (GLOBECOM), Anaheim, CA, 2012, pp 5015–5020
108. Michelson D, Ghassemzadeh S (2009) Measurement and modeling of wireless channels. In: Tarokh V (ed) New directions in wireless communications research. Springer, New York
109. He R, Zhong Z, Ai B, Ding J (2012) Measurements and analysis of short-term fading behavior for high-speed rail viaduct scenario. In: 2012 IEEE international conference on communications (ICC), Ottawa, ON, 2012, pp 4563–4567
110. Guan K, Zhong Z, Ai B, Kürner T (2014) Propagation measurements and analysis for train stations of high-speed railway at 930 MHz. IEEE Trans Veh Technol 63(8):3499–3516
111. Guan K, Zhong Z, Ai B, Kürner T (2014) propagation measurements and modeling of crossing bridges on high-speed railway at 930 MHz. IEEE Trans Veh Technol 63(2):502–517

112. Guan K, Zhong Z, Alonso JI, Briso-Rodriguez C (2012) Measurement of distributed antenna systems at 2.4 GHz in a realistic subway tunnel environment. IEEE Trans Veh Technol 61 (2):834–837
113. Kozono S (1994) Received signal-level characteristics in a wide-band mobile radio channel. IEEE Trans Veh Technol 43(3):480–486
114. Gao L, Zhong Z, Ai B, Xiong L (2010) Estimation of the Ricean K factor in the high speed railway scenarios. In: Proceedings of the 2010 International ICST conference communications networking, pp 1–5
115. Parsons JD (2000) The mobile radio propagation channel. Wiley, Chichester, U.K
116. Rappaport TS (2001) Wireless communications principles and practice, 2nd edn. Prentice Hall
117. Cheng X, Wang C-X, Ai B, Aggoune H (2014) Envelope level crossing rate and average fade duration of non-isotropic vehicle-to-vehicle Ricean fading channels. IEEE Trans Intell Syst 15(1):62–72
118. Greenstein LJ, Ghassemzadeh SS, Erceg V, Michelson DG (2009) Ricean K-factors in narrowband fixed wireless channels: theory, experiments, and statistical models. IEEE Trans Veh Technol 58(8):4000–4012
119. Liu L et al (2012) Position-based modeling for wireless channel on highspeed railway under a viaduct at 2.35 GHz. IEEE J Sel Areas Commun 30(4):834–845
120. Rappaport TS (1999) Wireless communications: principles and practice. Prentice Hall, Upper Saddle River, NJ
121. Qiu J, Tao C, Liu L, Tan Z (2012) Broadband channel measurement for the high-speed railway based on WCDMA. In: Proceedings of the IEEE 75th vehicular technology conference, May 2012, pp 1–5
122. Liu L, Tao C, Qiu J, Chen H, Yu L, Dong W, Yuan Y (2012) Position-based modeling for wireless channel on high-speed railway under a viaduct at 2.35 GHz. IEEE J Sel Areas Commun 30(4):834–845
123. Ding J, Guan K, Zhang L, Yang J, Zhang B, Zhong Z, Briso C, Huang J (2016) Broadband wireless channel in composite high-speed railway scenario: measurements, analysis and modeling. Int J Antennas Propag (Submitted)
124. Wen Y, Ma Y, Zhang X, Jin X, Wang F (2012) Channel fading statistics in high-speed mobile environment. In: Proc IEEE-APS Topical conference on antennas propagation wireless communications, Sept 2012, pp 1209–1212
125. Chen B, Zhong Z, Ai B, Guan K, He R, Michelson DG (2015) Channel characteristics in high-speed railway: a survey of channel propagation properties. IEEE Veh Technol Mag 10 (2):67–78
126. Molisch AF (2007) Wireless communications. Wiley
127. Liu L, Tao C, Qiu J et al (2012) Position-based modeling for wireless channel on high-speed railway under a viaduct at 2.35 GHz. IEEE J Sel Areas Commun 30(4):834–845
128. Pekka K (2007) WINNER II channel models part II radio channel measurement and analysis results. IST-4-027756, WINNER II D1.1.2, v1.0, Sept 2007
129. Bartlett M (1961) An introduction to stochastic processes with special references to methods and applications. Cambridge University Press, New York
130. Capon J (1969) High-resolution frequence-wavenumber spectrum analysis. Proc IEEE 57 (8):1408–1418
131. Johnson D (1982) The application of spectral estimation methods to bearing estimation problems. Proc IEEE 70(9):1018–1028
132. R. Schmidt, "Multiple emitter location and signal parameter estimation," *IEEE Trans. Antenna Propag.*, vol. AP-34, no. 2, pp. 276–280, Mar. 1986
133. Roy R, Kailath T (1989) "ESPRIT-estimation of signal parameters via rotational invariance techniques. IEEE Trans Acoust Speech Signal Process 37(7):984–995
134. Fleury BH, Tschudin M, Heddergott R, Dahlhaus D, Ingeman PK (1999) Channel parameter estimation in mobile radio environments using the SAGE algorithm. IEEE J Sel Areas Commun 17(3):434–450

135. Parviainen R, Kyosti P, Hsieh Y, Ting P, Chiou J (2008) Results of high speed train channel measurements. In: COST 2100 TD'08, Lille, France, Oct 2008
136. Guan K, Zhong Z, Ai B, K¨urner T (2013) Deterministic propagation modeling for the realistic high-speed railway environment. In: Proceedings of the IEEE VTC'13-Spring, Dresden, Germany, June 2013, pp 1–5
137. Ghazal A, Wang C-X, Ai B, Yuan D, Haas H (2015) A nonstationary wideband MIMO channel model for high-mobility intelligent transportation systems. IEEE Trans Intell Transp Syst 16(2):885–897
138. Chen B, Zhong Z (2012) Geometry-based stochastic modeling for MIMO channel in high-speed mobile scenario. Int J Antennas Propag:Article ID 184682
139. Chen B, Zhong Z, Ai B, Guan K, He R, Michelson DG (2015) Channel characteristics in high-speed railway: a survey of channel propagation properties. IEEE Veh Technol. Mag 10 (2):67–78
140. Wang X, Ghazal A, Ai B, Liu Y, Fan P (2015) Channel measurements and models for high-speed train communication systems: a survey. IEEE Commun Surv Tut 99:1–1
141. WINNER Group (2007) IST-4–027756 WINNER II D1. 1. 1 V1. 1, WINNER II Interim channel models
142. Rappaport TS (1996) Wireless communications: principles and practice. Prentice Hall PTR, New Jersey

Chapter 4
Cooperation and Cognition for Railway Communications

4.1 Cooperation Scenarios

The concept of cooperation in wireless communication networks has drawn significant attention recently from both academia and industry as it can be effective in addressing the performance limitations of wireless networks due to user mobility and the scarcity of network resources.

Current advances in telecommunications, and recent trends in mobile services suggest that the future of wireless devices will be characterized by the key word multi: multi-interface, multiservice and multi-reconfigurable. On the one hand, the variety of wireless technologies available on the market is opening the door to a plethora of heterogeneous devices provided with multiple radio interfaces, and connected to the Internet through several different access technologies (e.g., Wi-Fi, Wi-Max, 3G/4G, etc.). On the other hand, novel mobile applications (e.g., location based services and mobile social networks) will coexist with the traditional Internet-based services, determining a great range of possible quality of service (QoS) requirements which must be supported by the network providers. However, the heterogeneity of devices and wireless access technologies might not always constitute a limitation, rather a potential to exploit on several scenarios like the emergency ones. Recent catastrophic events (e.g., the Japanese tsunamis or the Katrina hurricane in US) demonstrated worldwide the fragility of fixed terrestrial communication infrastructures, as well as the need for more flexible and interoperable network architectures to support reliable communication among rescue teams and survivors. Given the pervasive penetration of end users devices (e.g., smart phones), spontaneous networks constitute promising solutions to implement emergency communication systems in which heterogeneous devices share their resources (e.g., Internet connection, energy power), in order to increase the network coverage, and thus the probability to reach survivors and to coordinate with rescue teams.

Z.-D. Zhong et. al., *Dedicated Mobile Communications for High-speed Railway*,
Advances in High-speed Rail Technology, DOI 10.1007/978-3-662-54860-8_4

Cooperative communication is one of the fastest growing areas of research, and it is likely to be a key enabling technology for efficient spectrum use in future. The key idea in user cooperation is that of resource sharing among multiple nodes in a network. The reason behind the exploration of user cooperation is that willingness to share power and computation with neighboring nodes can lead to savings of overall network resources. Mesh networks provide an enormous application space for user cooperation strategies to be implemented. In traditional communication networks, the physical layer is only responsible for communicating information from one node to another. In contrast, user cooperation implies a paradigm shift, where the channel is not just one link but the network itself.

Cooperation is possible whenever the number of communicating terminals exceeds two. Therefore, a three-terminal network is a fundamental unit in user cooperation. Indeed, a vast portion of the literature, especially in the realm of information theory, has been devoted to a special three-terminal channel, labeled the relay channel. The focus of our discussion will be the relay channel, and its various extensions. In contrast, there is also a prominent portion of literature devoted to cooperation as viewed from a network-wide perspective, which we will only briefly allude to.

Communication from a single source to a single destination without the help of any other communicating terminal is called direct, single-user or point-to-point communication as shown in Fig. 4.1. User cooperation is possible whenever there is at least one additional node willing to aid in communication. The simplest and oldest form of user cooperation is perhaps multi-hopping, which is nothing but a chain of point-to-point links from the source to the destination as shown in Fig. 4.1. No matter what the channel is, there is some attenuation of the signal with distance, which makes long-range point-to-point communication impractical. This problem is overcome by replacing a single long-range link with a chain of short-range links, where at each intermediate node there is a booster or repeater to enhance signal

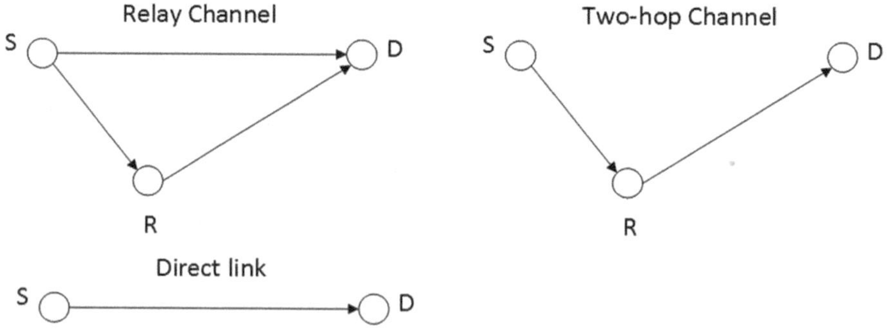

Fig. 4.1 Direct, two-hop and relay communications

quality. Multi-hopping was conceived about the same time as smoke and drum signals, therefore we do not attempt to put a time stamp on it.

More recently, the three-terminal relay channel as shown in Fig. 4.1 was introduced by van der Meulen (1968, 1971). In his original work, van der Meulen discovered upper and lower bounds on the capacity of the relay channel, and made several observations that led to improvement of his results in later years. The capacity of the general relay channel is still unknown, but the bounds discovered by van der Meulen were improved significantly by Cover and El Gamal (1979). In the interim, Sato (1976) also looked at the relay channel in the context of the Aloha system. Notably, an extensive review of results on several channels that are important to network information theory was published in van der Meulen (1977). The review summarized the state of the art at that time, but our understanding of relaying has improved considerably since then. Other important contributions of the era which contributed to the understanding of user cooperation include: Slepian and Wolf (1973), Gaarder and Wolf (1975), Cover and Leung (1981), Willems (1982), Cover (1972, 1975), Bergmans and Cover (1974), Marton (1979), Gel'fand and Pinsker (1980), Han (1981), El Gamal and van der Meulen (1981), Cover et al. (1980), Wyner (1978), Wyner and Ziv (1976).

Undoubtedly, the most prominent work on relaying to date is Cover and El Gamal (1979). Most of the results in this work have still not been superseded. In the years following Cover and El Gamal (1979), there was some interest in the relay channel, as is evident from the literature. In El Gamal and Aref (1982), the authors discovered the capacity of the semi-deterministic relay channel, where the received signal at the relay is a deterministic function of the source and relay transmissions. There was an effort to generalize the results of Cover and El Gamal (1979) to networks with multiple relays in Aref (1980), El Gamal (1981). These works also investigated deterministic relay networks with no interference, and deterministic broadcast relay networks.

The relay channel is the three-terminal communication channel shown in Fig. 4.2. The terminals are labeled the source (S), the relay (R), and the destination (D). All information originates at S, and must travel to D. The relay aids in communicating information from S to D without actually being an information source or sink. The signal being transmitted from the source is labeled X. The signal received

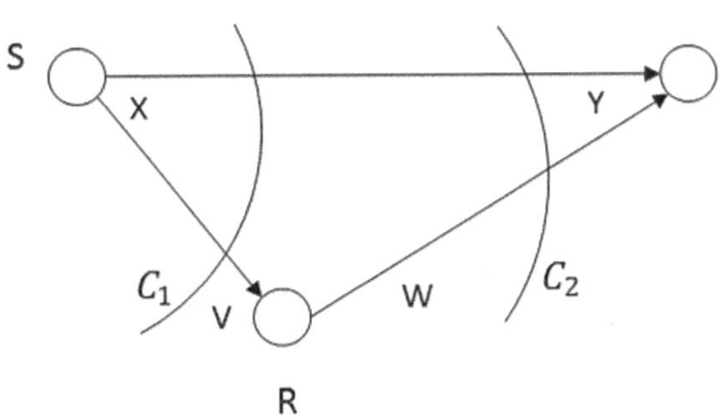

Fig. 4.2 The relay channel

by the relay is V. The transmitted signal from the relay is W, and the received signal at the destination is Y. Several notions of relaying exist in the literature. We will list the prominent ones in this section. Conceptually, information is relayed in two phases or modes: first, when S transmits and (R, D) receive, commonly called the broadcast (BC) mode; and second when (S, R) transmit and D receive, also known as the multiple access (MAC) mode. Note that this differentiation is only conceptual since it is possible for communication in both modes to take place simultaneously. We will elaborate on this a little later, but first we will enumerate four different models of relaying that can be classified based on the above two modes.

Cooperation is the process of working together, opposite of working separately in competition. Recently, such a concept has been adopted from social sciences and economics to constitute a major research area in wireless communication networks. The idea of employing cooperation in wireless communication networks has emerged in response to the user mobility support and limited energy and radio spectrum resources, which pose challenges in the development of wireless communication networks and services in terms of capacity and performance.

In the course of the development of cooperative communication, several complicating issues must be addressed, including the loss of rate to the cooperating mobile, overall interference in the network, cooperation assignment and handoff, fairness of the system, and transmit and receive requirement on the mobiles.

In a cooperative communication system, each wireless user is assumed to transmit data as well as act as a cooperative agent for another user, as shown in Fig. 4.3.

Cooperation leads to interesting trade-offs in code rates and transmit power. In the case of power, one may argue on one hand that more power is needed because each user, when in cooperative mode, is transmitting for both the users. On the other hand, the base station transmit power for both users will be reduced because of diversity.

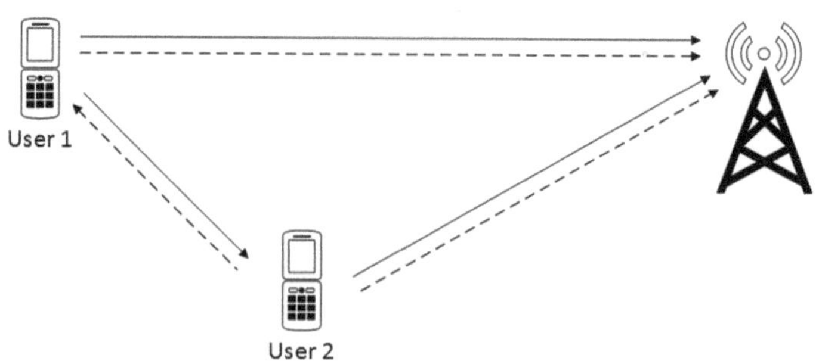

Fig. 4.3 In cooperative communication each mobile is both a user and a relay

In cooperative communication, each user transmits both his/her own bits as well as some information for his/her partner; one might think this causes loss of rate in the system. However, the spectral efficiency of each user improves because, due to cooperation diversity the channel code rates can be increased.

Generally, we can categorize three cooperation scenarios based on various studies in literature. In the first scenario, cooperation among different entities is employed to improve the wireless communication channel reliability through spatial diversity. In the second scenario, the system throughput is improved via aggregating the offered resources from cooperating entities. Finally, cooperation is used to achieve seamless service provision.

4.1.1 Improved Channel Reliability

The wireless communication channel suffers from several phenomena that decrease its reliability. These phenomena include path loss, shadowing, and fading. Cooperation in wireless networks can increase the reliability of the communications against the channel impairments. As illustrated in Fig. 4.4 for a downlink transmission from a base station to a mobile terminal, where the source node transmits its data packets toward the destination node with the help of cooperating entities. A cooperating entity is a relay node with an improved channel condition over the direct transmission channel from the source to the destination. This relay node can be a mobile terminal or a dedicated relay station as shown in Fig. 4.4.

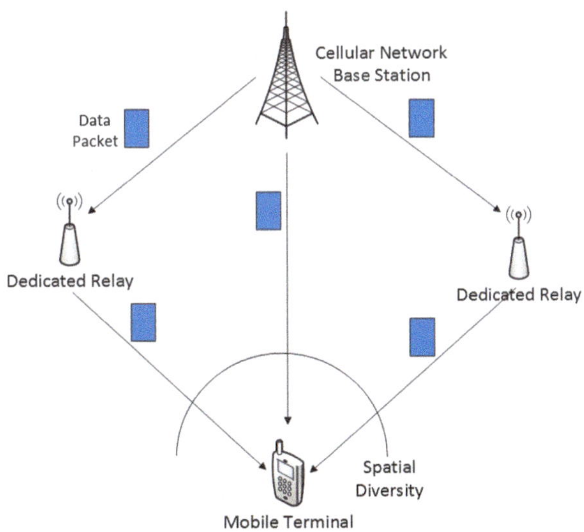

Fig. 4.4 Cooperation to improve channel reliability: spatial diversity

Taking advantage of the rich wireless propagation environment across multiple protocol layers in a network architecture offers numerous opportunities to dramatically improve network performance. Cooperative diversity systems consist of multiple nodes that share resources in order to create multiple diversity channels and thereby improve system performance, typically in terms of availability, range, and throughput.

This improved reliability can be achieved by exploiting cooperative spatial diversity. Transmitting independent copies of the signal generates diversity and can effectively combat the deleterious effects of fading. In particular, spatial diversity is generated by transmitting signals from different locations, thus allowing independently faded versions of the signal at the receiver. Cooperative communication generates this diversity in a new and interesting way. When the channel between the original source and destination is unreliable, other network entities can cooperate with the source node to create a virtual antenna array and forward the data toward the destination. Hence, different transmission paths with independent channel coefficients exist between the source and destination nodes through the cooperating entities. As a result, the destination node receives several copies of the transmitted signal over independent channels. Based on this spatial diversity, the destination can combine the data received from these entities in detection to improve the transmission accuracy.

The broadcast nature of the wireless communication medium results in interference at the different nodes in the coverage area (interference region) of each other. Such interference reduces the signal-to-interference-plus-noise ratio (SINR) at the receiving nodes and hence degrades their detection performance. Thanks to the cooperation introduced by the cooperative relays, the transmitted power from the original source can be significantly reduced due to a better channel condition of the relaying links, which greatly reduces the interference region, as illustrated in Fig. 4.5. This also helps to improve the energy efficiency of the communication system. In addition to reducing the interference region, cooperation can solve the hidden terminal problem and hence results in interference reduction.

4.1.2 Improved System Throughput

An improved system throughput can be a direct benefit from the enhanced wireless channel reliability through employing cooperative transmissions at the physical layer. In addition, cooperation can increase the achieved throughput through aggregating the offered resources from different cooperating entities. This is achieved through employing cooperative strategies at the network and transport layers. In this case, data packets are transmitted along multiple paths toward the destination. Different from the preceding cooperation scenario, the data packets transmitted through different paths are not the same copy of some transmitted signal. Instead, different transmission paths carry different data packets. This has the effect of increasing the total transmission data rate between the source and

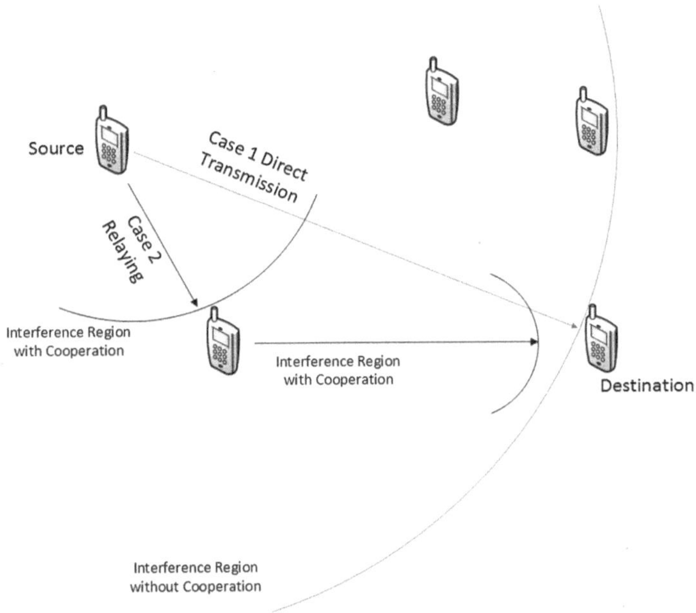

Fig. 4.5 Cooperation to improve channel reliability: interference reduction

destination nodes. In this case, the cooperating entities can be mobile terminals, base stations, or access points with sufficient resources (e.g., bandwidth), such that when these resources are aggregated, the total transmission data rate from the source to the destination can be increased. This strategy can support applications with a high required transmission rate. In Fig. 4.6, for example, resources from the cooperating cellular network and wireless local area network (WLAN) are aggregated to provide a high data rate for the mobile terminal.

A relay is said to be half-duplex when it cannot simultaneously transmit and receive in the same band. In other words, the transmission and reception channels must be orthogonal. Orthogonality between transmitted and received signals can be in time domain, in frequency domain, or using any set of signals that are orthogonal over the time–frequency plane. If a relay tries to transmit and receive simultaneously in the same band, then the transmitted signal interferes with the received signal. In theory, it is possible for the relay to cancel out interference due to the transmitted signal because it knows the transmitted signal. In practice, however, any error in interference cancelation (due to inaccurate knowledge of device characteristics or due to the effects of quantization and finite-precision processing) can be catastrophic because the transmitted signal is typically 100–150 dB stronger than the received signal. Due to the difficulty of accurate interference cancelation, full-duplex radios are not commonly used; however, advances in analog processing could potentially enable full-duplex relaying. Although early literature on information theoretic relaying was based almost entirely on full-duplex relaying, in

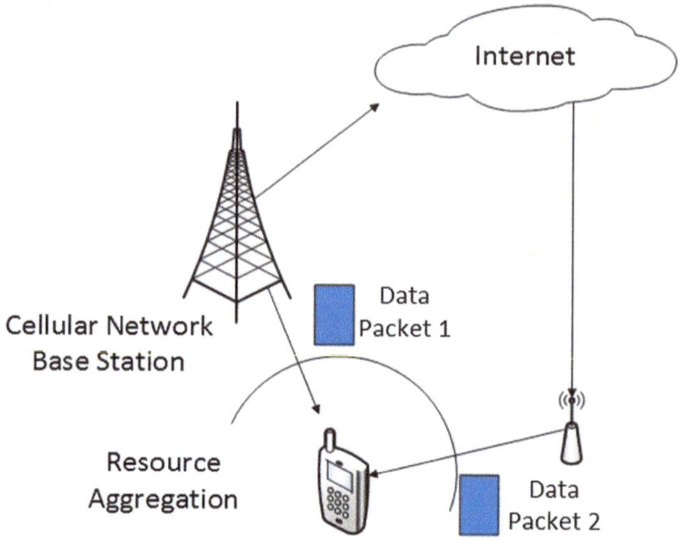

Fig. 4.6 Cooperative resource aggregation

recent years, a lot of research and especially research directed toward practical protocols has been based on the premise of half-duplex relaying.

The link performance is enhanced by conveying the information signal to the destination over more than one ideally independent fading signal paths in space, time, or frequency, through the use of a transmit antenna array. Although this is a reasonable approach for the base station, it may be impractical for the mobile unit due to the size or cost limitations which prevent the use of multiple antennas. In general, users with single antenna share their antennas which form a virtual antenna array. Each user transmits its own signals to both the destination and the partner(s). Each partner retransmits the received or some version of these signals to the destination to provide spatial diversity in a distributed fashion by means of a virtual multiple antennas transmission. The cooperation improves the robustness of the wireless system against fading and allows higher data rates.

Two main cooperation methods are amplify and forward (AF) and decode and forward (DF). In AF, cooperating user receives the noisy signal from its partner and retransmits it after amplification. Signals from the user and its partner are combined at the destination to determine transmitted data bits. In DF, the cooperating user decodes the signal received from its partner before retransmitting using the same code. Both of these methods guarantee full diversity for two-user case when the inter-user channel fading coefficients are known at the destination.

4.1.3 Seamless Service Provision

Mobile users are more sensitive to call dropping than call blocking. Call dropping interrupts service continuity for different reasons depending on the networking scenario. Cooperative strategies at the link, network, and transport layers can help to guarantee service continuity of an ongoing. In Fig. 4.7, when the service is interrupted along one path (Ch1), it still can be continued using another cooperative path (Ch2, Ch3). In this context, a cooperating entity can be a mobile terminal, base station, or access point which can create a substitute path between the source and destination nodes.

Cooperation in wireless communication networks can reduce operation costs for both mobile users and service providers. For example, it can be achieved by improving the energy efficiency, which reduces the energy costs. Another example is to extend network coverage area through relaying nodes. Such a solution is less expensive than deploying more base stations due to the high installation and maintenance costs.

4.2 Key Techniques for Cooperation

4.2.1 Relay Protocol

Spatial diversity is a well-known technique to mitigate the fading effects in a wireless channel. However, in some wireless applications, such as ad hoc networks, implementing multiple transmit and/or receive antennas to provide spatial diversity might not be possible due to the size and cost limitations. Cooperative (or relay) diversity is attractive for such networks, i.e., networks with mobile terminals having single-antenna transceivers, since it is able to achieve spatial diversity. The basic

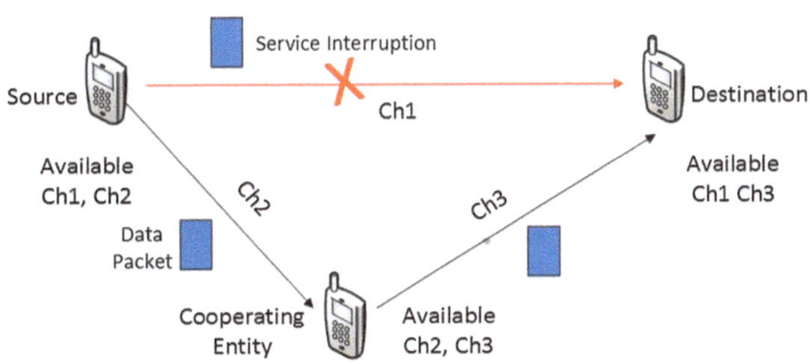

Fig. 4.7 Cooperation for seamless service provision

idea is that a source node transmits information to the destination not only through a direct link but also through the relay links.

There are at least two fundamental ideas (and a third that is practically less important) based on which the source and relay nodes can share their resources to achieve the highest throughput possible for any known coding scheme. The cooperation strategies based on these different ideas have come to be known as relay protocols.

The first idea involves decoding of the source transmission at the relay. The relay then retransmits the decoded signal after possibly compressing or adding redundancy. This strategy is known as the decode-and-forward protocol, named after the fact that the relay can and does decode the source transmission. The decode-and-forward protocol is close to optimal when the source-relay channel is excellent, which practically happens when the source and relay are physically near each other. When the source-relay channel becomes perfect, the relay channel becomes a multiple-antenna system. Some authors use the term cooperation to strictly mean the decode-and-forward type of cooperation.

The second idea, sometimes called observation, is important when the source-relay and the source–destination channels are comparable, and the relay destination link is good. In this situation, the relay may not be able to decode the source signal, but nonetheless it has an independent observation of the source signal that can aid in decoding at the destination. Therefore, the relay sends an estimate of the source transmission to the destination. This strategy is known as the estimate-and-forward (also known as compress and forward or quantize and forward) protocol.

The amplify-and-forward (also sometimes called scale and forward) protocol is a special case of the above strategy where the estimate of the source transmission is simply the signal received by the relay, scaled up or down before retransmission. A multi-antenna system is a relay channel where amplify and forward is the optimal strategy, and the amplification factor is dictated by the relative strengths of the source-relay and source–destination links.

The third idea, known as facilitation, is mostly of theoretical interest. When the relay is not able to contribute any new information to the destination, then it simply tries to stay out of the way by transmitting the signal that would be least harmful to source–destination communication.

In this chapter, we only consider a single relay helping a user (source) in the network forwarding information. A typical cooperation strategy can be modeled with two orthogonal phases, either in TDMA or FDMA, to avoid interference between the two phases:

In phase 1, a source sends information to its destination, and the information is also received by the relay at the same time.

In phase 2, the relay can help the source by forwarding or retransmitting the information to the destination.

Figure 4.8 depicts a general relay channel, where the source transmits with power p_1 and the relay transmits with power p_2. In this chapter, we will consider the

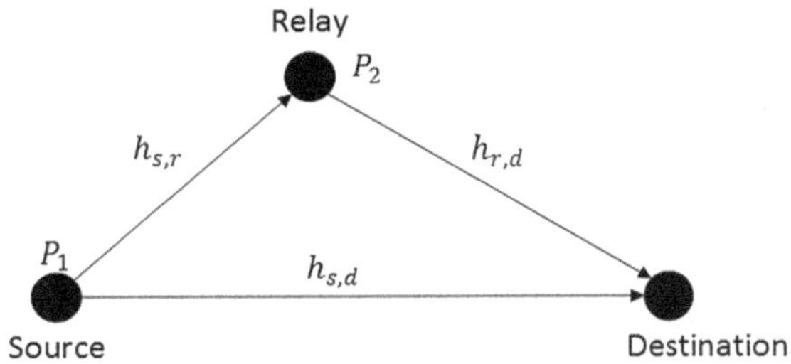

Fig. 4.8 A simplified cooperation model

special case where the source and the relay transmit with equal power P. Optimal power allocation is studied in the following chapters.

In phase 1, the source broadcasts its information to both the destination and the relay. The received signals $y_{s,d}$ and $y_{s,r}$ at the destination and the relay, respectively, can be written as

$$y_{s,d} = \sqrt{P} h_{s,d} x + n_{s,d} \tag{4.1}$$

$$y_{s,r} = \sqrt{P} h_{s,r} x + n_{s,r} \tag{4.2}$$

In which P is the transmitted power at the source and relay, x is the transmitted information symbol, and $n_{s,d}$. and $n_{s,r}$ are additive noise. $h_{s,d}$ and $h_{s,r}$ are the channel coefficients from the source to the destination and the relay, respectively. They are modeled as zero-mean, complex Gaussian random variables with variances $\delta_{s,d}^2$ and $\delta_{s,r}^2$, respectively. The noise terms $n_{s,d}$ and $n_{s,r}$ are modeled as zero-mean complex Gaussian random variables with variance N_0.

In phase 2, the relay forwards a processed version of the source's signal to the destination, and this can be modeled as

$$y_{r,d} = h_{r,d} q(y_{s,r}) + n_{r,d} \tag{4.3}$$

where the function $q(\cdot)$ depends on which processing is implemented at the relay node.

4.2.1.1 Amplify and Forward

Amplify-and-forward technique simply amplifies the signal received by the relay before forwarding it to the destination. This technique was proposed by J.N. Laneman and G.W. Wornell, and is ideal when the relay station has minimal computing power. However, one major drawback of this technique is that the noise in the signal is also amplified at the relay station, and the destination receives two independently faded versions of the signal. Figure 4.9 shows amplify-and-forward technique.

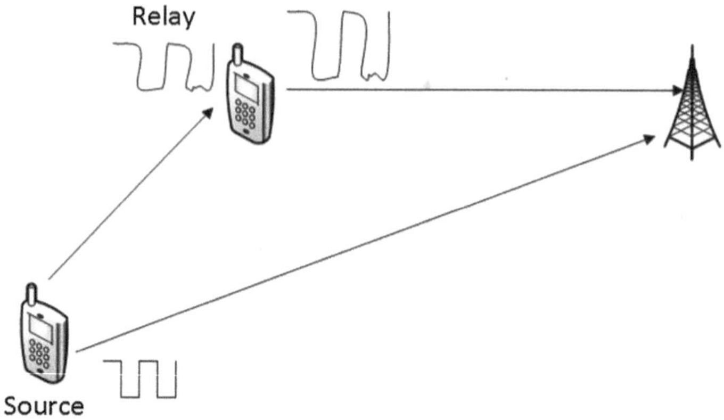

Fig. 4.9 Amplify-and-forward technique

In an amplify-and-forward relaying protocol, which is often simply called an amplify-and-forward protocol, the relay scales the received version and transmits an amplified version of it to the destination. The amplify-and-forward relay channel can be modeled as follows. The signal transmitted from the source x is received at both the relay and destination as

$$y_{s,r} = \sqrt{P}h_{s,r}x + n_{s,r} \tag{4.4}$$

$$y_{s,d} = \sqrt{P}h_{s,d}x + n_{s,d} \tag{4.5}$$

where $h_{s,r}$ and $h_{s,d}$ are the channel fades between the source and the relay and destination, respectively, and are modeled as Rayleigh flat fading channels. The terms $n_{s,r}$ and $n_{s,d}$ denote the additive white Gaussian noise with zero-mean and variance N_0. In this protocol, the relay amplifies the signal from the source and forwards it to the destination ideally to equalize the effect of the channel fade between the source and the relay. The relay does that by simply scaling the received signal by a factor that is inversely proportional to the received power, which is denoted by

$$\beta_r = \frac{\sqrt{P}}{\sqrt{P}\left|h_{s,r}\right|^2 + N_0} \tag{4.6}$$

The signal transmitted from the relay is thus given by $\beta_r y_{s,r}$ and has power P equal to the power of the signal transmitted from the source. To calculate the mutual information between the source and the destination, we need to calculate the total instantaneous signal-to-noise ratio at the destination. The SNR received at the destination is the sum of the SNRs from the source and relay links. The SNR from the source link is given by

$$SNR_{s,d} = \Gamma |h_{s,d}|^2 \tag{4.7}$$

where $\Gamma = \frac{P}{N_0}$.

In the following, we calculate the received SNR from the relay link. In phase 2, the relay amplifies the received signal and forwards it to the destination with transmitted power P. The received signal at the destination in phase 2 is given by

$$y_{r,d} = \frac{\sqrt{P}}{\sqrt{P}|h_{s,r}|^2 + N_0} h_{r,d} y_{s,r} + n_{r,d} \tag{4.8}$$

where $h_{r,d}$ is the channel coefficient from the relay to the destination and $n_{r,d}$ is an additive noise. More specifically, the received signal $y_{r,d}$ in this case is

$$y_{r,d} = \frac{\sqrt{P}}{\sqrt{P}|h_{s,r}|^2 + N_0} \sqrt{P} h_{r,d} h_{s,r} x + n^*_{r,d} \tag{4.9}$$

where

$$n^*_{r,d} = \frac{\sqrt{P}}{\sqrt{P|h_{s,r}|^2 + N_0}} h_{r,d} n_{s,r} + n_{r,d} \tag{4.10}$$

4.2.1.2 Decode and Forward

In recent years, much more research works have focused on non-coherent cooperative networks, i.e., the networks in which channel state information (CSI) is assumed to be unknown at the receivers (relays and destination). It is due to the fact that true values of the CSIs cannot actually be obtained in realistic systems. Differential phase-shift keying (DPSK), a popular candidate in non-coherent communications, has been studied for both AF and DF protocols in. However, with the DF protocol in, the authors considered an ideal case that the relay is able to know exactly whether each decoded symbol is correct or not.

The most popular method for processing the signal at the relay node is decode and forward, in this technique, the relay detects the source data, decodes, and then transmits it to the desired destination. The concept of the decode-and-forward technique is shown in Fig. 4.10.

An error correcting code can also be implemented at the relay station. This can help the received bit errors to be corrected at the relay station. However, this is only possible, if the relay station has enough computing power.

With DF, relays decode the source's messages, re-encode, and retransmit to the destination. However, it is not simple to provide cooperative diversity with the DF protocol. This is due to possible retransmission of erroneously decoded bits of the message by the relays.

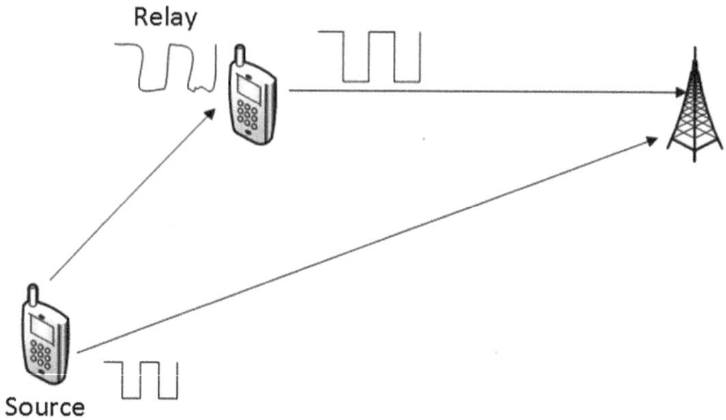

Fig. 4.10 Decode-and-forward technique

Consider a wireless network as illustrated in Fig. 4.11, where K relays help one source node to communicate with its destination. Every node has only one antenna and operates in a half-duplex mode (i.e., a node cannot transmit and receive simultaneously). The K relays communicate with the destination over orthogonal channels and the DF protocol is employed at each relay. The source, relays, and destination are denoted and indexed by node 0, node i, $i = 1, \ldots, K$, and node $K + 1$, respectively.

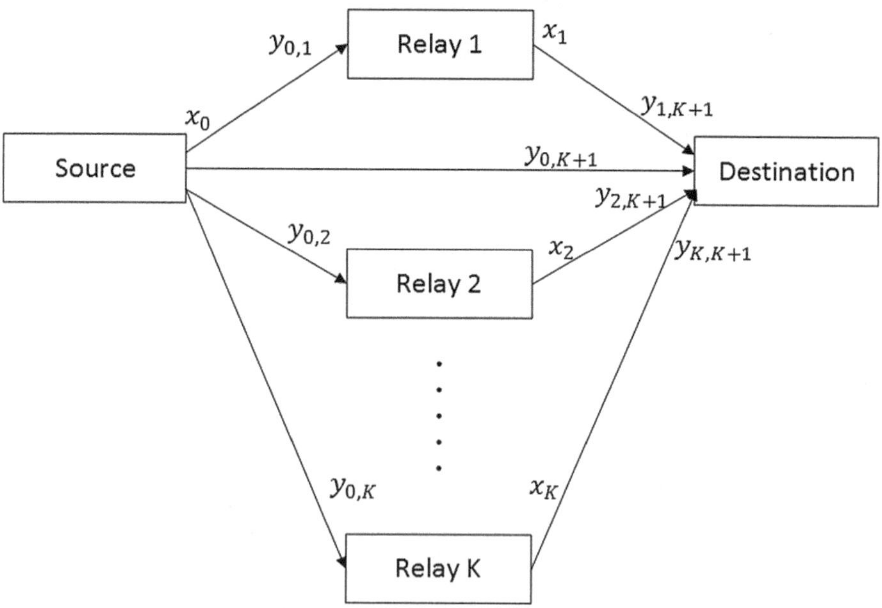

Fig. 4.11 A wireless relay network

Signal transmission from the source to destination is completed in two phases as follows: in the first phase, the source broadcasts a BFSK signal. In the baseband model, the received signals at node i are written as

$$y_{0,i,0} = (1 - x_0)\sqrt{E_0}h_{0,i} + n_{0,i,0} \tag{4.11}$$

$$y_{0,i,1} = x_0\sqrt{E_0}h_{0,i} + n_{0,i,1} \tag{4.12}$$

where $h_{0,i}$ and $n_{0,i,k}$ are the fading channel coefficients between node 0 and node i and the noise component at node i, $i = 1, \ldots, K + 1$, respectively. E_0 is the average transmitted symbol energy of the source. In (4.11) and (4.12), the third subscript $k \in \{0, 1\}$ denotes the two frequency sub-bands used in BFSK signaling. Furthermore, the source symbol $x_0 = 0$ if the first frequency sub-band is used and $x_0 = 1$ if the second frequency sub-band is used.

For the DF protocol, node i decodes the signal received from the source and retransmits a BFSK signal to the destination. If node i transmits in the second phase, the baseband received signals at the destination are given by

$$y_{i,K+1,0} = (1 - x_i)\sqrt{E_i}h_{i,K+1} + n_{i,K+1,0} \tag{4.13}$$

$$y_{i,K+1,1} = x_i\sqrt{E_i}h_{i,K+1} + n_{i,K+1,1} \tag{4.14}$$

where E_i is the average transmitted symbol energy sent by node i, and $n_{i,K+1,k}$ is the noise component at the destination in the second phase. Note that if the ith relay makes a correct detection, then $x_i = x_0$. Otherwise $x_i \neq x_0$.

The channel between any two nodes is assumed to be Rayleigh flat fading, modeled as $\mathcal{CN}(0, \sigma^2_{(i,j)})$, where i, j refer to transmit and receive nodes, respectively. The noise components at the relays and destination are modeled as *i.i.d.* $\mathcal{CN}(0, N_0)$, random variables. The instantaneous received SNR for the transmission from node i to node j is $\gamma_{i,j} = E_i\left|h_{i,j}\right|^2/N_0$ and the average SNR is $\bar{\gamma}_{i,j} = E_i\sigma^2_{i,j}/N_0$. With Rayleigh fading, the probability distribution function (pdf) of $\gamma_{i,j}$ is $f_{i,j}(\gamma_{i,j}) = \frac{1}{\bar{\gamma}_{i,j}}e^{-\frac{\gamma_{i,j}}{\bar{\gamma}_{i,j}}}$.

Another processing possibility at the relay node is for the relay to decode the received signal, re-encode it, and then retransmit it to the receiver. This kind of relaying is termed as a decode-and-forward (DF) scheme, which is often simply called a DF scheme without the confusion from the selective DF relaying scheme. If the decoded signal at the relay is denoted by $\bar{x}$, the transmitted signal from the relay can be denoted by $\sqrt{P}\bar{x}$, given that $\bar{x}$ has unit variance. Note that the decoded signal at the relay may be incorrect. If an incorrect signal is forwarded to the destination, the decoding at the destination is meaningless. It is clear that for such a scheme the diversity achieved is only one, because the performance of the system is limited by the worst link from the source-relay and source–destination. This will be illustrated through the following analysis.

Although fixed DF relaying has the advantage over AF relaying in reducing the effects of additive noise at the relay, it entails the possibility of forwarding erroneously detected signals to the destination, causing error propagation that can diminish the performance of the system. The mutual information between the source and the destination is limited by the mutual information of the weakest link between the source–relay and the combined channel from the source–destination and relay–destination. More specifically, the mutual information for decode-and-forward transmission in terms of the channel fades can be given by

$$I_{DF} = \frac{1}{2}\min\left\{\log\left(1+\Gamma\left|h_{s,r}\right|^2\right), \log\left(1+\Gamma\left|h_{s,d}\right|^2+\Gamma\left|h_{r,d}\right|^2\right)\right\} \tag{4.15}$$

where the min operator in the above equation takes into account the fact that the relay only transmits if decoded correctly, and hence the performance is limited by the weakest link between the source–destination and source–relay.

The outage probability for the fixed DF relaying scheme is given by $\Pr[I_{DF} < R]$. Since log is a monotonic function, the outage event is equivalent to

$$\min\left\{\left|h_{s,r}\right|^2, \left|h_{s,d}\right|^2+\left|h_{r,d}\right|^2\right\} < \frac{2^{2R}-1}{\Gamma} \tag{4.16}$$

The outage probability can be written as

$$\Pr[I_{DF} < R] = \Pr\left[\left|h_{s,r}\right|^2 < \frac{2^{2R}-1}{\Gamma}\right] + \Pr\left[\left|h_{s,r}\right|^2 > \frac{2^{2R}-1}{\Gamma}\right]\Pr\left[\left|h_{s,d}\right|^2+\left|h_{r,d}\right|^2 < \frac{2^{2R}-1}{\Gamma}\right] \tag{4.17}$$

Since the channel is Rayleigh fading, the above random variables are all exponential random variables with parameter one. Averaging over the channel conditions, the outage probability for decode and forward at high SNR is given by

$$\Pr[I_{DF} < R] \simeq \frac{1}{\sigma_{s,r}^2}\frac{2^{2R}-1}{\Gamma} \tag{4.18}$$

From the above, fixed relaying has the advantage of easy implementation, but the disadvantage of low bandwidth efficiency. This is because half of the channel resources are allocated to the relay for transmission, which reduces the overall rate. This is true especially when the source–destination channel is not very bad, because under this scenario a high percentage of the packets transmitted by the source to the destination can be received correctly by the destination and the relay's transmissions are wasted.

Besides the two most common techniques for relaying, there are other techniques, such as compress-and-forward cooperation and coded cooperation, which deserve some attention.

4.2.1.3 Compress-and-Forward Cooperation

The main difference between compress and forward and decode/amplify and forward is that while in the later the relay transmits a copy of the received message, in compress and forward the relay transmits a quantized and compressed version of the received message. Therefore, the destination node will perform the reception functions by combining the received message from the source node and its quantized and compressed version from the relay node.

The quantization and compression process at the relay node is a process of source encoding, i.e., the representation of each possible received message as a sequence of symbols. For clarity and simplicity, let us assume that these symbols are binary digits (bits). At the destination node, an estimate of the quantized and compressed message is obtained by decoding the received sequence of bits. This decoding operation simply involves the mapping of the received bits into a set of values that estimate the transmitted message. This mapping process normally involves the introduction of distortion (associated to the quantization and compression process), which can be considered as a form of noise.

In addition, the entropy provides a benchmark against which it is possible to evaluate the performance of source encoders. Next, and for the purpose of simplifying the presentation, let us consider that the source data is generated from a discrete memoryless source. For this setting, the entropy of the random variable being encoded at the source provides a lower bound on the average number of bits per source symbol (the source encoding rate) needed to encode the source. In this sense, the entropy provides a lower bound on the source encoding rate used at the relay node if in a peer-to-peer communication setup. The use of cooperation and the possibility of combining at the destination the messages from the source and the relay node, changes this point. Effectively, the information received at the destination from the source can be used, as side information, while decoding the message from the relay. This will allow for encoding at a lower source encoding rate.

As a final note to compress-and-forward cooperation, we note that much of the source encoding operation done at the relay falls into the realm of the set of coding techniques known as distributed source coding, Sleppian–Wolf coding, or Wyner–Ziv coding.

4.2.1.4 Coded Cooperation

Coded cooperation differs from the previous schemes in that the cooperation is implemented at the level of the channel coding subsystem. Note that both the amplify-and-forward and the decode-and-forward schemes presented earlier in this chapter were based on schemes where the relay repeats the bits sent by the source. In coded cooperation, the relay sends incremental redundancy, which when combined at the receiver with the codeword sent by the source, results in a codeword with larger redundancy.

To understand coded cooperation, consider first the operation of a typical error correcting code. The encoder for an error correcting code takes a sequence of information bearing symbols and applies mathematical operations in such a way that it generates a sequence of symbols containing not only the information present at the input sequence, but also redundant information. The redundancy in the codeword is used at the receiver to increase the chances of recovering the original information if errors have been introduced during the transmission process. While in some codewords, the information and redundancy are encoded in such a way that they can only be separated through complete decoding, in many other codes the encoding and decoding operation can be done in such a way that it is possible to add redundancy to, or remove redundancy from, the codeword in a simple manner (such as through concatenation of new redundancy symbols or deletion of selected symbols). It is this second type of code that is used in coded cooperation.

Coded cooperation is a method that integrates cooperation into channel coding. Coded cooperation works by sending different portions of each user's code word via two independent fading paths. The basic idea is that each user tries to transmit incremental redundancy to its partner. Whenever that is not possible, the users automatically revert to a noncooperative mode. The key to the efficiency of coded cooperation is that all this is managed automatically through code design, with no feedback between the users.

The users divide their source data into blocks that are augmented with cyclic redundancy check (CRC) code. In coded cooperation, each of the users' data is encoded into a codeword that is partitioned into two segments, containing N_1 bits and N_2 bits, respectively. It is easier to envision the process by a specific example: consider that the original codeword has $N_1 + N_2$ bits; puncturing this codeword down to N_1 bits, we obtain the first partition, which is a valid (weaker) codeword. The remaining N_2 bits in this example are the puncture bits. Of course, partitioning is also possible via other means, but this example serves to give an idea of the intuition behind coded cooperation.

Likewise, the data transmission period for each user is divided into two time segments of N_1 and N_2 bit intervals, respectively. We call these time intervals frames. For the first frame, each user transmits a code word consisting of the N_1-bit code partition. Each user also attempts to decode the transmission of its partner. If this attempt is successful (determined by checking the CRC code), in the second frame the user calculates and transmits the second code partition of its partner, containing N_2 code bits. Otherwise, the user transmits its own second partition, again containing N_2 bits. Thus, each user always transmits a total of $N = N_1 + N_2$ bits per source block over the two frames. We define the level of cooperation as N_2/N, the percentage of the total bits for each source block the user transmits for its partner. Figure 4.12 illustrates the coded cooperation framework.

In general, various channel coding methods can be used within this coded cooperation framework. For example, the overall code may be a block or convolutional code, or a combination of both. The code bits for the two frames may be selected through puncturing, product codes, or other forms of concatenation. To obtain the performance results given in this article, we employ a simple but very

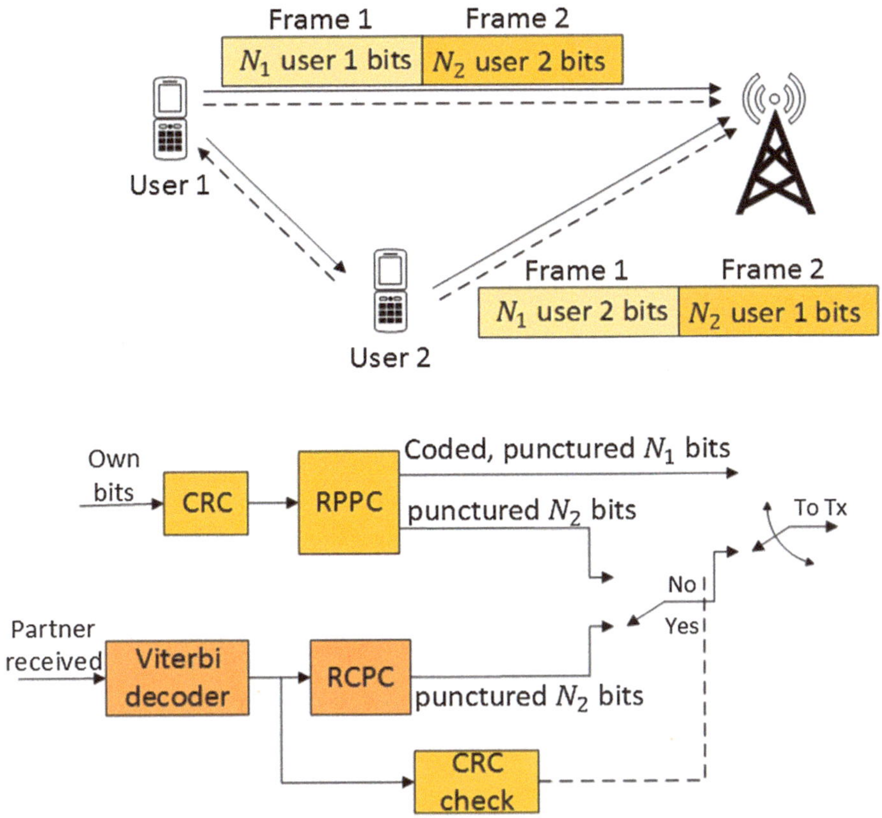

Fig. 4.12 Coded cooperation

effective implementation using rate compatible punctured convolutional (RCPC) codes. In this implementation, the code word for the first frame is obtained by puncturing a code word of length N bits to obtain N_1 code bits. The additional code bits transmitted in the second frame are those punctured to form the first frame code word.

The users act independently in the second frame, with no knowledge of whether their own first frame was correctly decoded. As a result, there are four possible cooperative cases for the transmission of the second frame: both users cooperate, neither user cooperates.

4.2.2 MIMO and Cooperative Communication

How to develop cooperative schemes to improve performance? The key lies in the advances in MIMO (multiple-input multiple-output) communication technologies. Wireless networks, very high data rates can only be expected for full-rank MIMO users. More specifically, full-rank MIMO users must be equipped multiple transceiver antennas. In practice, most users either do not have multiple antennas installed on small-size devices, or the propagation environment cannot support MIMO requirements. To overcome the limitations of achieving MIMO gains in future wireless networks, one must think of new techniques beyond traditional point-to-point communications.

Recently, there has been great interest in the use of multi-antenna physical arrays at the transmitters and/or receivers in a wireless system. Physical arrays offer space diversity to combat fading, or when sufficient knowledge of the channel conditions are available at both the transmitter and receiver, offer beamforming to combat both fading and interference from other terminals, and other wireless systems in the same band. As a result, physical arrays increase capacity and improve robustness to fading. Motivated by these possible gains, a great deal of research effort has focused on design of practical space–time codes and their associated decoding algorithms. Several studies have shown that, aside from suitable encoding and decoding algorithms, the key to leveraging spatial diversity with physical arrays is to have separation among the antennas on the order of several wavelengths of the carrier frequency so that the fading coefficients are uncorrelated. As carrier frequencies increase, this constraint becomes less restrictive; however, terminal size also decreases with time and circuit integration, thereby limiting the number of antennas that can be effectively placed in a transmitter or receiver.

For systems in which size constraints limit the number of antennas that can be placed in the transmitters or receivers, our research examines issues associated with creating a virtual array by allowing multiple users to cooperate and effectively share their antennas. Figure 4.13 compares block diagrams for physical and virtual arrays. While multi-antenna array problems are generally treated at the physical layer, virtual arrays can be dealt with at a variety of layers, including interaction across layers.

Clearly, much can be gained from comparing virtual arrays to physical arrays, as in Fig. 4.13. The performance of physical array systems provides useful performance bounds for virtual array systems. Furthermore, space–time code designs for physical arrays can be readily adapted to cooperative settings.

Figure 4.14 shows a general model for multi-antenna systems utilizing T transmit and R receive antennas. The model can be expressed in vector form as

$$y = \boldsymbol{A}x + n \tag{4.19}$$

where $\boldsymbol{A}$ is a $R \times T$ matrix, and y and n (resp. x) are column vectors of size $R \times 1$ (resp. $T \times 1$). Here, the element $[A]_{r,t} = \alpha_{r,t}$ captures the effects of multipath fading

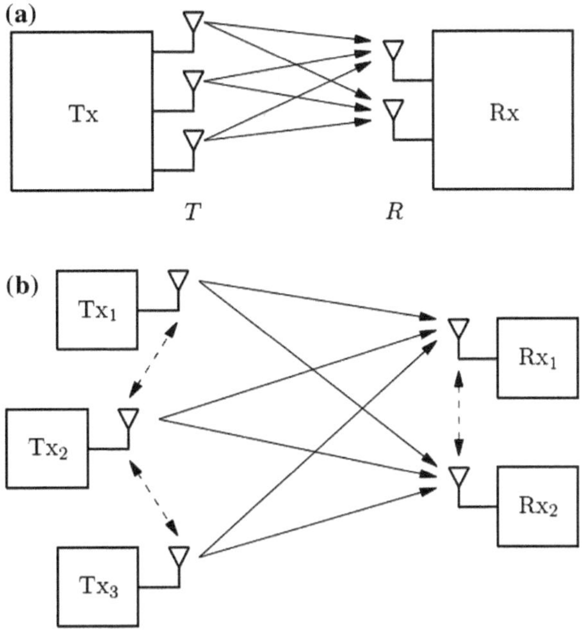

Fig. 4.13 Block diagrams relating a point-to-point physical array (**a**) and a multiuser virtual array (**b**) arising from cooperative diversity transmission

between transmit antenna t and receiver antenna r, while n_r captures the effects of receiver thermal noise and other forms of interference. Note that the multi-antenna model of Fig. 4.14 is a special case of the general wireless network consisting of a single transmitter and receiver, with vector inputs and outputs, respectively.

There has been great and growing interest in channels of the form shown in Fig. 4.14. Initially, attention focused on systems with multiple receiver antennas and their associated diversity combining algorithms, e.g., maximum ratio and selection combining, and array processing techniques, e.g., beamforming and interference mitigation, but more recently systems employing multiple transmitter antennas, possibly with multiple receiver antennas, have been emphasized. Transmit antenna arrays generally require more sophisticated algorithms than receive antenna arrays alone, both because different signals can be transmitted from the multiple antennas and because these signals super impose at the receiver antennas. Substantial energy has focused on characterizing the ultimate limits on performance for multi-antenna systems, and designing practical coding and decoding algorithms that approach these limits.

Of late, there has been substantial work characterizing the limiting performance of multi-antenna systems under a variety of fading conditions. For example, for systems without delay constraints and with sufficient fading variability (ergodicity), within the coding interval, classical Shannon theory provides the capacity of the channel. The Shannon, or ergodic, capacity for the channel model in Fig. 4.14 has been developed for several different cases of channel state information available to the transmitter and/or receiver: no channel state information; channel state

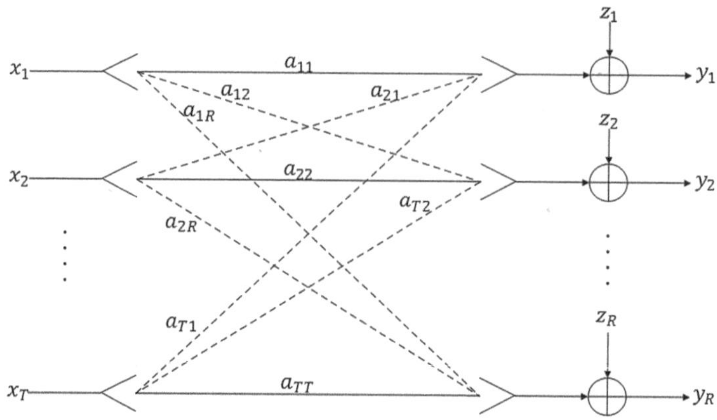

Fig. 4.14 Multi-antenna system model

information available to the receiver only; state information available to both transmitter and receiver.

The ergodic capacity results to date suggest that dramatic increases in capacity are possible using multi-antenna systems. For example, for the case of channel state information available to the receiver only, the ergodic capacity increases by $\min\{T, R\}$ b/s/Hz for each additional 3 dB of SNR, in the high SNR regime.

For the case of no channel state information at either the transmitter or receiver, the channel capacity depends upon the number of transmit and receive antennas as well as the coherence time K of the channel, defined to be the number of samples for which the channel remains constant in the assumed block fading model before it changes to another independent realization. In this case, the ergodic capacity has been shown to increase as $T'(1 - T'/K)$ bps/Hz for each additional 3 dB of SNR in the high SNR regime, where $T' = \min\{T, R, K/2\}$. The slope is maximized by employing $T = K/2$ transmit antennas, assuming $R \geq T$, and in fact degrades if more than this number of transmit antennas is utilized. In this case, the capacity increases as $T/2$ bps/Hz for each additional 3 dB of SNR.

As a point of reference, the capacity of an AWGN channel (without fading) increases by only 1 bps/Hz for each additional 3 dB of SNR in the high SNR regime. Thus, quite large spectral efficiencies can, in principle, be achieved using multi-antenna systems. Adding antenna elements, along with suitable transmitter coding and receiver processing methods, is a kin to adding cabling in a wireline setting.

For systems with tighter delay constraints, the channel may not exhibit its ergodic nature within a coding interval, so that the Shannon capacity is zero. In such cases, alternative performance metrics such as capacity-versus-outage/outage probability or delay-limited capacity can be employed to evaluate the efficacy of multi-antenna schemes.

4.2.3 Distributed Space–Time Coding

In this section, the design of distributed space–time codes for wireless relay networks is considered. The two-hop relay network model depicted in Fig. 4.15, where the system lacks a direct link from the source to destination node, is considered. Distributed space–time (space–frequency) coding can be achieved through node cooperation to emulate multiple antennas transmitter. First, the decode-and-forward protocol, in which each relay node decodes the symbols received from the source node before retransmission, is considered. A space–time code designed to achieve full diversity and maximum coding gain over MIMO channels is shown to achieve full diversity but does not necessarily maximize the coding gain if used with the decode-and-forward protocol. Next, the amplify-and-forward protocol is considered; each relay node can only perform simple operations such as linear transformation of the received signal and then amplify the signal before retransmission. A space–time code designed to achieve full diversity and maximum coding gain over MIMO channels is shown to achieve full diversity and maximum coding gain if used with the amplify-and-forward protocol.

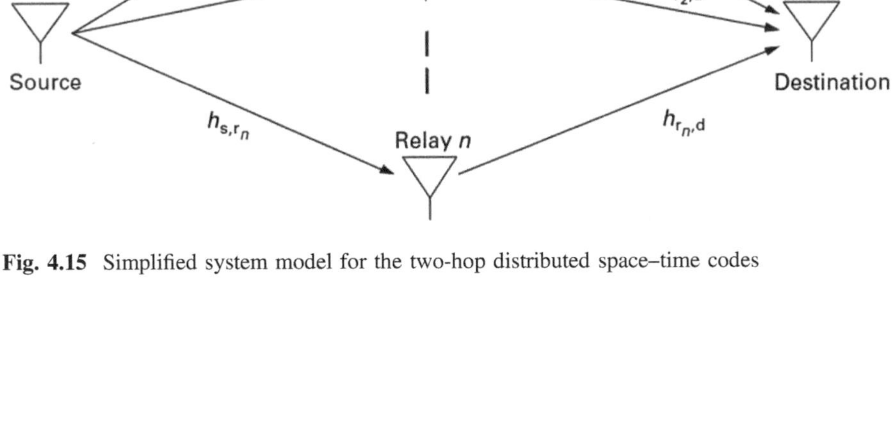

Fig. 4.15 Simplified system model for the two-hop distributed space–time codes

Next, the design of DSTC that can mitigate the relay nodes synchronization errors is considered. Most of the work on cooperative transmission assume perfect synchronization between the relay nodes, which means that the relays' timings, carrier frequencies, and propagation delays are identical. Perfect synchronization is difficult to achieve among randomly located relay nodes. To simplify the synchronization in the network, a diagonal structure is imposed on the space–time code used. The diagonal structure of the code bypasses the perfect synchronization problem by allowing only one relay to transmit at any time slot (assuming TDMA). Hence, it is not necessary to synchronize simultaneous in-phase transmissions of randomly located relay nodes, which greatly simplifies the synchronization among the relay nodes.

The aim here is to derive fractional resource allocation strategies tailored to distributed multi-hop networks utilizing estimate-and-forward (EF) protocols. Of prime interest is the derivation of fractional frame duration, power, and modulation order for each relaying stage to achieve maximum end-to-end throughput.

To this end, we will first dwell on the system model. We then derive the error rates for spatially distributed STBCs. These are eventually used to obtain resource allocation strategies, which optimize the end-to-end throughput for topologies with complete as well as partial cooperation between nodes belonging to the same relaying stage.

The general system model is a source that communicates with a destination via a given number of relays. Spatially adjacent relays are grouped into relaying virtual antenna arrays (VAAs), where we will briefly describe the functioning of the transmitting, relaying and receiving VAA stages.

The functional blocks of the transceivers forming the distributed-MIMO multistage relaying network are depicted in Fig. 4.16. The top of Fig. 4.16 relates to the source VAA containing the source; the center panel relates to an arbitrary relaying VAA tier; and the bottom relates to the destination VAA containing the destination. In this figure, each VAA tier is shown to consist of three terminals; it is, however, understood that any reasonable number of terminals can be accommodated. The coreblocks are:

- Source VAA. Specifically, the information source passes the information to a cooperative transceiver, which relays the data to spatially adjacent relays belonging to the same VAA. In contrast to other protocols dealt with in this book, this communication is assumed to happen over an air interface distinct from the interface used for interstage communication or an air interface not requiring any optimization, and is not considered further. It is also assumed that these cooperative links are error free due to the short communication distances. Each of the terminals in the VAA perform distributed space–time block encoding of the information according to some prior specified codebook. That information is then transmitted from the spatially distributed terminals after having been synchronized. Note that the problem related to synchronization is beyond the scope of this section but is increasingly dealt.

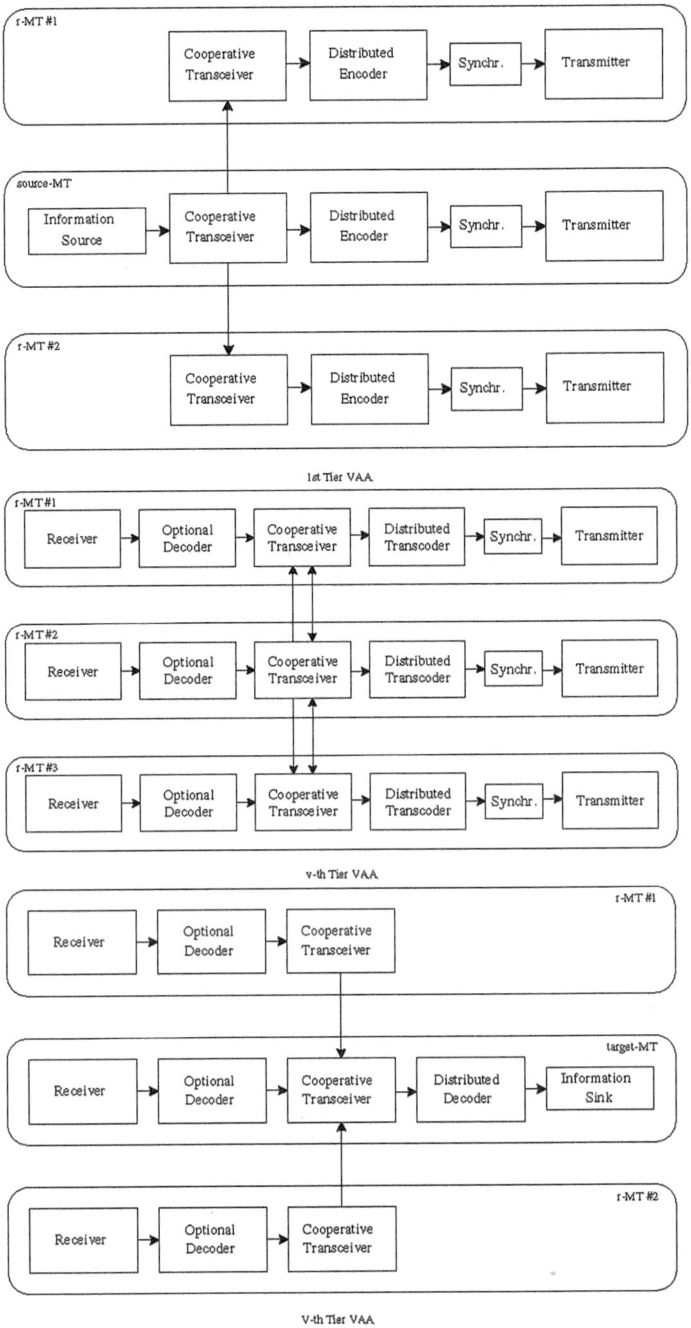

Fig. 4.16 Functional blocks of the source VAA (*top*), the vth relaying VAA (*center*) and the target VAA (*bottom*)

- Relaying VAA. Any of the relaying VAA tiers functions as follows. First, each relay within that VAA receives the data, which is optionally decoded before being passed on to the cooperative transceiver. Ideally, every terminal cooperates with every other terminal; however, any degree of cooperation is feasible. If no decoding is performed, then an unprocessed or a sampled version of the received signal is exchanged with the other relays. Note that unprocessed relaying is equivalent to transparent relaying. After cooperation, appropriate decoding is performed. The obtained information is then re-encoded in a distributed manner, synchronized and retransmitted to the subsequent relaying VAA tier.
- Destination VAA. As for the destination VAA, the functional blocks are exactly the opposite to the source VAA. All terminals receive the information, possibly decode it, then pass it onto the cooperative transceivers, which relay the data to the target terminal. The data is processed and finally delivered to the information sink.

The functional blocks of the distributed transcoder, that is encoder and decoder, are now elaborated on in more detail. To this end, the encoder and decoder are shown in Fig. 4.17 and described as follows:

- Distributed Encoder. A channel encoder within a distributed encoder does not normally differ from a non-distributed encoder; however, as has become evident throughout this book, it is generally possible and advisable to design channel codes that reflect the distributed nature of the encoding process. The role of a space–time encoder is to utilize the additional spatial dimension created by the

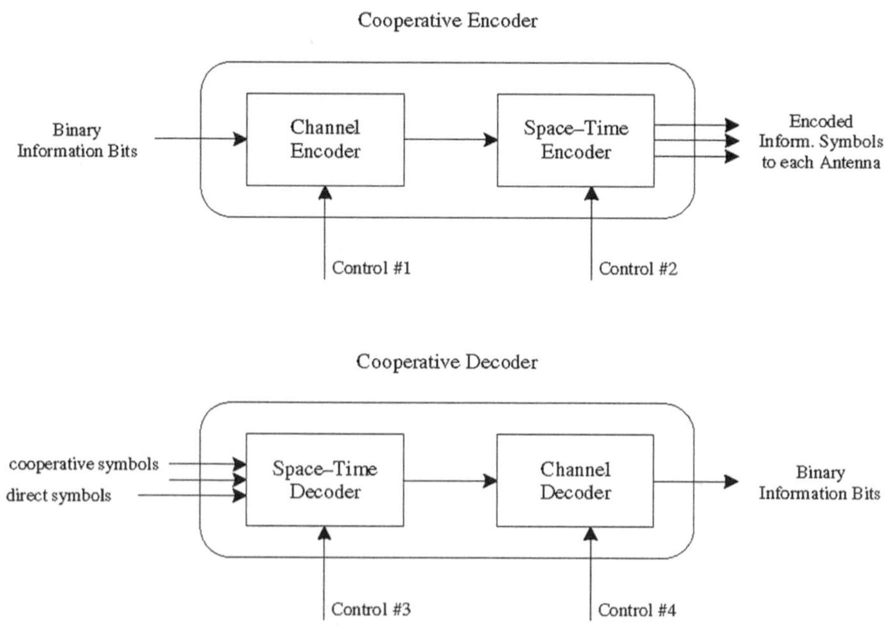

Fig. 4.17 Distributed encoder and decoder

sufficiently spaced antenna elements to increase the system performance. The functionality of distributed space–time codes (STCs) differs from a traditional deployment because only a fraction of the entire space–time codeword is transmitted from any of the spatially distributed terminals. The transmission across all terminals then yields the complete space–time codeword. Therefore, a control signal to each distributed space–time encoder is essential, as it tells each of them which fraction of the entire space–time codeword to pass onto the transmitting antenna(s). This is indicated as Control #2 in Fig. 4.17. This control information is assumed to be available to the space–time encoder, and is hence not discussed further.

- Distributed Decoder. The cooperative decoder can be realized as the inversion of all processes at the cooperative transmitter. Here, the space–time decoder is fed with the signals directly received from the available antenna(s), as well as the information received via the cooperative links from adjacent terminals. Again, a control signal is needed that specifies the type of information fed into the space–time decoder, to allow for optimum decoding. For example, the control signal could inform the decoder that the relayed signals are a one bit representation of the sampled soft information available at the respective cooperative relaying terminals. After the space–time decoding process, the information is passed on to the channel decoder, which performs the inverse process to the channel encoder. In a cooperative transcoder, the resulting binary information output may then be fed into the cooperative encoder, to get relayed to the next VAA tier.

4.2.4 Physical Layer Network Coding and Cooperative Communication

In recent years, the application of physical layer network coding (PNC) has attracted significant attention. Compared with the conventional network coding which consumes three time slots and time scheduling scheme which consumes four time slots, PNC provides a substantial throughput enhancement in two-way relay channels as it requires only two transmission time slots. For the PNC-enabled bidirectional relaying in two-way relay channel, we refer to the uplink in the first time slot as the multiple access (MAC) phase and the downlink in the second time slot as the broadcast (BC) phase.

We note that the two-way relay channel only supports two users' data exchange. As a natural and enriched extension, redesigning the PNC to accommodate the multiuser network (more than two users) is seen as an attractive topic. However, the PNC should be able to deal with interference which increases proportionally with the increase of the number of users. Moreover, the PNC should ensure that the NCS can be unambiguously decoded at each user such that the desired symbol from the other users can be extracted. One possible solution is to transmit the NCSs with larger code space. However, this sacrifices the spectral efficiency. In summary, the challenge of PNC in multiuser network is that the redesigned PNC should transmit fewer symbols but support more users.

4.2.4.1 The Two-Way Relay Channels

Wireless relaying is identified as a promising technique to offer spatial diversity and to extend the coverage of wireless networks. In a wireless relay network, the relay acts as the "intermediary" for data exchange among different users. The two-way relay channel is regarded as a classical representative of wireless relay network and has been investigated extensively in recent years. The origin of two-way relay channel can be traced to Shannon's pioneering work, where the rudiment of two-way relay channel, i.e., the two-way channel without relay was investigated. Later on, some pioneer work was led and done by Van der Meulen, Cover and El Gamal.

The two-way relay channel can be treated as a combination of two-way channel and relay network. As seen in Fig. 4.18, the two-way relay channel is a three-node linear network in which two users A and B want to exchange their data via a relay node R. The uplink of two-way relay channel, i.e., the links from the two users to the relay, can be seen as a multiple access channel (MAC) while the downlink, i.e., the links from the relay to the two users, can be seen as a broadcast channel (BC). All nodes operate in half-duplex mode and a direct link between the two users is unavailable. Similar to other types of relay network, the traditional amplify-and-forward (AF) and decode-and-forward (DF) strategies can be implemented in the two-way relay channel.

A conventional bidirectional data exchange protocol is TDMA, as shown in Fig. 4.18a. In such a protocol, each user alternately transmits their signal to the relay which avoids the co-channel interference. However, this consumes four orthogonal time slots and hence sacrifices spectral efficiency. As an alternative approach, standard network coding, as shown in Fig. 4.18b, allows the relay to

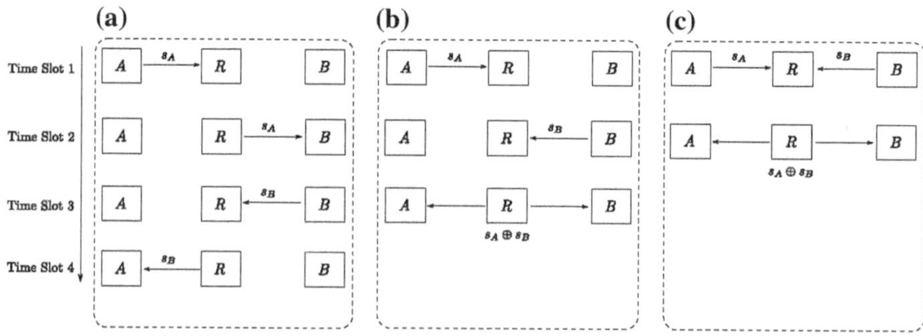

Fig. 4.18 **a** Conventional TDMA. **b** Standard NC. **c** PNC

generate the XOR combination (regarded as the network coded data) of the data from the two users and forward to them. Two users then extract the desired data by using the XOR operation on the received network coded data and their side information. However, the standard NC still requires three transmission phases as each user transmits data to the relay using different time slots.

By fully exploiting the superposition nature of electromagnetic waves, physical layer network coding (PNC) allows two users simultaneously to transmit their signals to the relay in the MAC phase, as shown in Fig. 4.18c. The relay directly maps the superimposed signal into the XOR combination of data from the two users, which is referred to as the network coded symbol. Then in the BC phase, the resulting network coded symbol is forwarded to the users. PNC provides a substantial improvement in terms of the spectral efficiency over the TDMA and standard NC protocol as it only consumes two transmission time slots. The concept of PNC is detailed in the next section.

4.2.4.2 The Multi-way Relay Channels

The natural extension of the two-way relay channel is the multi-way relay channel, as shown in Fig. 4.19. The multi-way relay channel consists of M users $(U_i, i \in \{1, 2, ..., M\})$ and a shared relay (R). All users operate in half-duplex mode and there is no direct link among users. The multi-way data exchange takes place among the users with the help of the relay. Each user expects to decode the data from all other users based on exploiting the signal received from the relay and its own side information.

The conventional data exchange protocol is the user scheduling, in which the users alternately transmit their signal to avoid co-channel interference. However, this results in a low spectral efficiency. In contrast to the scheduling approach, PNC

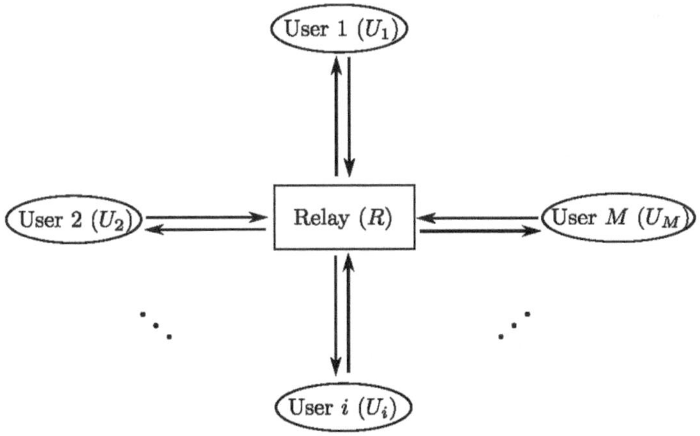

Fig. 4.19 The model of MULTI-WAY RELAY CHANNEL

allows the users to simultaneously transmit the signals in the same channel. The spectral efficiency is thus much improved. However, due to the co-channel interference in the MAC phase, the question of how the users recover their desired signal with the minimum cost is the major concern for PNC design in multi-way relay channel.

4.2.4.3 Physical Layer Network Coding

In this section, we provide the fundamental concept of PNC. Here, we show the simplest case of PNC in the two-way relay channel, where two users adopt BPSK modulation. Let $\mathcal{A}_2 = \{-1, +1\}$ denote the Gray coded BPSK alphabet. The mapping from user symbol to modulated symbol is denoted as $\mathcal{M}_B : GF(2) \rightarrow \mathcal{A}_2$. The BPSK symbols transmitted by user i, $i \in \{A, B\}$ denoted as x_i, is then given by $x_i = \mathcal{M}_B(s_i) = 1 - 2s_i$. The channel gain from user i to R is denoted as h_i. In the MAC phase, A and B simultaneously transmit their signal to R. Due to the superimposition nature of EM waves, the relay R receives

$$y_R = h_A x_A + h_B x_B + n_R \tag{4.20}$$

where n_R is the additive white Gaussian noise (AWGN) with the variance σ^2. Without loss of generality, we assume that $|h_A| \geq |h_B|$. The noiseless superimposed constellation at the relay is illustrated in Fig. 4.20, where the PNC mapping proposed is implemented. We observe that the superimposed signal is in fact mapped as the XOR combination of data from the two users, given by $s_R = s_A \oplus s_B$, where s_R denotes the network coded symbol (NCS) and $\oplus$ denotes the bit-wise XOR operation (the module-2 sum in the binary field). Let $x_{AB} \triangleq h_A x_A + h_B x_B$ denote the

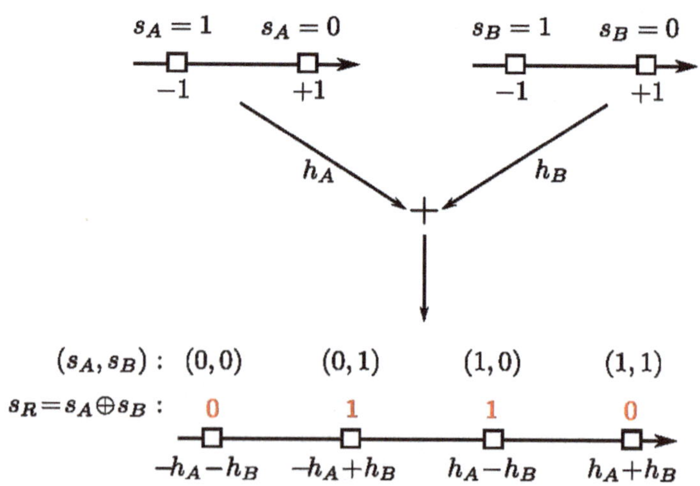

Fig. 4.20 Superimposed constellation of PNC when using BPSK

superimposed signal. Based on Fig. 4.20, the mapping from the superimposed signal into the NCS is given by

$$\begin{aligned} x_{AB} &\longrightarrow s_R \\ s.t. s_R &= s_A \oplus s_B \end{aligned} \tag{4.21}$$

Let s_R denote the alphabet of NCS. Clearly, the above mapping function results in a compression on data from the two users since the cardinality of SR is equal to that of the user alphabet. As such, BPSK can be adopted to transmit the NCS. The BPSK modulated NCS, denoted as x_R, is given by $x_R = \mathcal{M}_B(s_R)$. After receiving the NCS, each user can decode their desired symbol by using XOR operation, i.e., $\tilde{s}_B = s_A \oplus s_R$ and $\tilde{s}_A = s_B \oplus s_R$, where $\tilde{s}_i$ denotes the recovered symbol.

4.3 Signal Classification in Cognitive Radio

The development of mobile internet and internet of things (IoT) leads to a fast growing number of mobile devices, such as smart phones, tablets, and machine-to-machine (M2M) devices, which further results in an explosion in the demand for wireless communications services. The growth of the worldwide mobile traffic based on the ITU-R M.2243 report [1] is shown in Fig. 4.21. It not only stresses the importance of the growth of the traffic in the past few years, but also provides an overview of consolidated forecasts of mobile broadband traffic on a worldwide basis. In 2011, the UMTS Forum also published a report to forecast the mobile traffic for the next generation 2010–2020 [2]. According to the report, the worldwide mobile traffic will increase by a factor of 33 from 2010 to 2020, and total worldwide traffic will grow from 3.86 to 127.8 EB. This growth will come from the combination of a higher number of subscriptions and the importance of video traffic.

The above data and studies demonstrate that the data traffic on mobile networks will continue to grow explosively in the future, which brings an immerse requirement of the available frequency bands. The ITU-R M.2290 report [3] provides a global perspective on the future spectrum requirement estimate for terrestrial IMT, in which the summary of national spectrum requirement in some countries is estimated as well, as shown in Table 4.1.

All countries in Table 4.1 have a large demand for the IMT spectrum. Although the estimate of the spectrum requirements may be different from country to country, it can be concluded that countries worldwide will more or less suffer from the spectrum shortage in the future.

In 1930, the Federal Communications Commission (FCC) in the United State provides the advice of spectrum allocation and management, i.e., the use of the radio frequency spectrum is regulated by governments. Under this policy, almost all radio frequency spectrums are allocated so that only the licensed user can access to it. Hence, most of the spectrum is either unused or underutilized, which leads to an inefficient

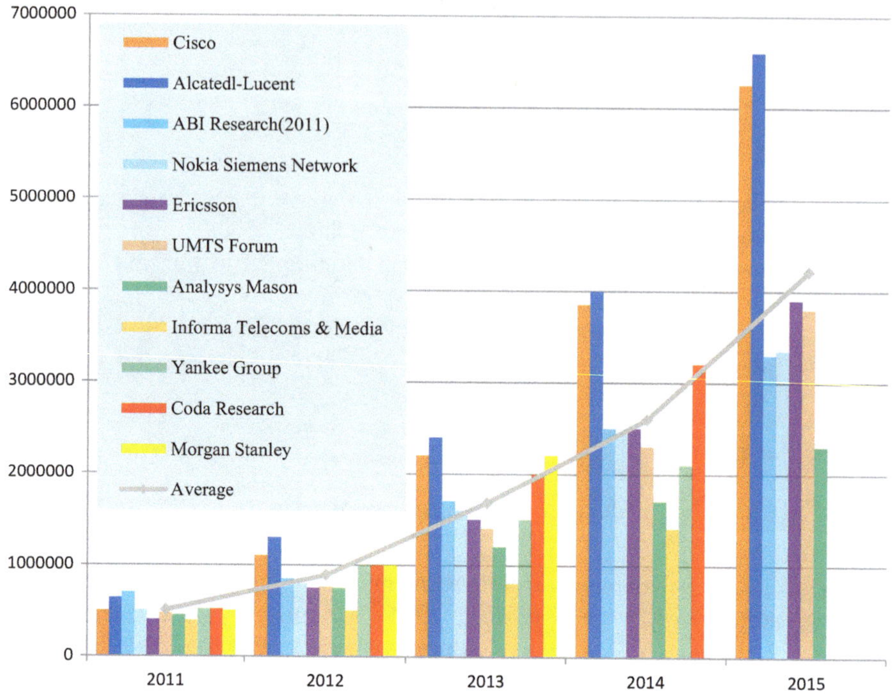

Fig. 4.21 Mobile global data traffic estimates 2011–2015 based on multiple sources. *Source* ITU-R M.2243 Report

Table 4.1 Summary of national spectrum requirement in some countries

Country	Spectrum requirement
US	Additional requirement of 275 MHz by 2014
Australia	Total requirement of 1081 MHz (Additional requirement of 300 MHz by 2020)
Russia	Total requirement of 1065 MHz (Additional requirement of 385 MHz by 2020)
China	Total requirement of 570–690 MHz (by 2015). Total requirement of 1490–1810 MHz (by 2020)
GSMA6	Total requirement of 1600–1800 MHz for some countries
India	Additional requirement of 300 MHz by 2017. Additional requirement of another 200 MHz by 2020
UK	Total requirement of 775–1080 MHz for the low demand setting. Total requirement of 2230–2770 MHz for the high demand setting

usage of the spectrum. In the fall of 2009, Shared Spectrum Company collected spectrum usage data at its spectrum observatory in the prime frequency bands between 30 MHz and 3 GHz over a three-and-a-half-day period [4]. A summary of the occupancy calculations across the bands can be found in the figure below.

It can be seen from Fig. 4.22 that there are a number of bands that have low measured spectrum occupancy, lower than 20%. Some frequency bands such as

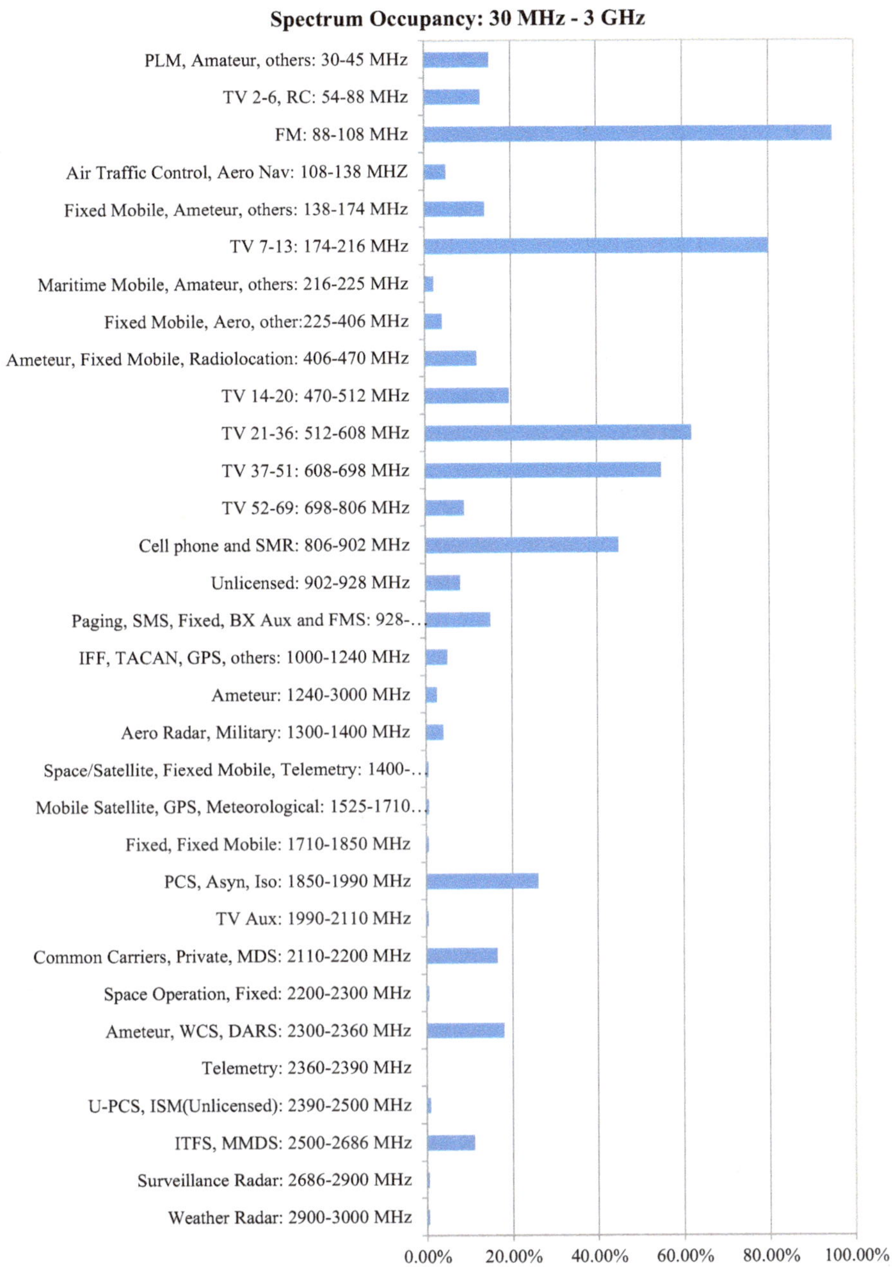

Fig. 4.22 Summary of spectrum band occupancy calculations

1400–1430 MHz (space/satellite), 1430–1520 MHz (telemetry), 1525–1710 MHz (mobile satellite/meteorological), and 1710–1850 MHz (fixed/fixed mobile) are highly underutilized or almost never used. Hence, there are many drawbacks in the current command-and-control spectrum allocation regulation. In 2002, the FCC published a three-page report to improve the conventional spectrum resource management way [5]. This report reveals the actual situation of the spectrum utilization: "In many bands, spectrum access is a more significant problem than physical scarcity of spectrum, in large part due to legacy command-and-control regulation that limits the ability of potential spectrum users to obtain such access."

The definition of cognitive radio provided by Simon Haykin is that "Cognitive radio is an intelligent wireless communication system that is aware of its surrounding environment (i.e., outside world), and uses the methodology of understanding-by-building to learn from the environment and adapt its internal states to statistical variations in the incoming RF stimuli by making corresponding changes in certain operating parameters (e.g., transmit power, carrier frequency, and modulation strategy) in real time, with two primary objectives in mind:

(1) highly reliable communications whenever and wherever needed;
(2) efficient utilization of the radio spectrum." [6]

This definition points out six characteristics of cognitive radio, i.e., awareness, intelligence, learning, adaptivity, reliability, and efficiency. All of these can be summarized as cognitive capability. Besides, cognitive radio is also endowed with reconfigurability.

The cognitive capability allows the cognitive radio to capture the information from the outside radio frequency environment, to sense the unoccupied spectrum at a given time and location, and to adaptively adjust the operation parameter such as transmit power, carrier frequency, modulation formats, etc. The cognitive radio should implement three main tasks:

(1) Radio-scene analysis, which involves the estimation of interference temperature of the radio environment, and the detection of spectrum holes;
(2) Channel identification, which encompasses the estimation of channel state information (CSI), and the prediction of channel capacity for use by the transmitter;
(3) Transmit power control and dynamic spectrum management.

Basically, task (1) and (2) are implemented in the receiver, while task (3) is carried out in the transmitter. These three tasks form a basic cognition cycle, which is shown in Fig. 4.23 as follows.

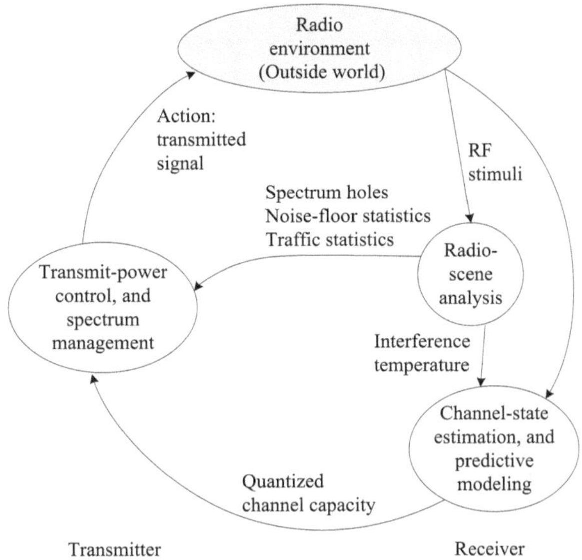

Fig. 4.23 The cognition cycle

4.3.1 Spectrum Sensing

The main idea behind cognitive radio is that the exploitation of the spectrum hole. Therefore, the spectrum sensing technique, by which the secondary user detects the presence of the primary user, provides a fundament for the secondary user to opportunistically access a spectrum without interfering the primary user. Generally, the spectrum sensing can be categorized into two classes: local spectrum sensing and cooperative spectrum sensing.

1. **Local spectrum sensing**
   Assume that the primary user is present, then, the discrete time signal received by the secondary user is given by

$$y_n = Hx_n + w_n, n = 1, \ldots, N, \tag{4.22}$$

   where x_n is the transmit signal of the primary user, w_n is the additive white Gaussian noise (AWGN) sample, $w_n \sim_{\mathrm{c}}(0, \sigma^2)$, and H is the fading coefficient. On the other hand, if the primary user is absent from the sensed spectrum, the received signal is expressed as

$$y_n = w_n, n = 1, \ldots, N. \tag{4.23}$$

   Hence, the spectrum sensing problem can be formulated as a binary hypothesis test, which the null and alternative hypothesis are stated as primary user is absent and primary user is present, respectively, i.e.,

$$\begin{cases} \mathcal{H}_0\text{: the primary user is absent, } y_n \text{ follows (4.23)} \\ \mathcal{H}_1\text{: the primary user is present, } y_n \text{ follows (4.22)} \end{cases}$$

The performance of the spectrum sensing algorithm can be evaluated by two probabilities: probability of detection P_d, and probability of false alarm P_{fa}. P_d is the probability of detecting a signal on the sensed spectrum when it truly is present, while P_{fa} is the probability of detecting a signal on the sensed spectrum when it actually is absent, i.e., $P_{fa} = P\{\mathcal{H}_1|\mathcal{H}_0\}$.
A number of algorithms have been proposed for local spectrum sensing. The most widely used algorithms include the energy detector, cyclostationary detector, matched filtering detector, waveform-based detector and radio identification-based detector, etc.

① Energy detector
Suppose that the received signal of the secondary user is given by (4.22), then, the decision metric for the energy detector can be written as

$$M = \sum_{n=0}^{N} |y_n|^2. \tag{4.24}$$

The main idea behind the energy detector is to detect the primary user's signal by comparing the output of the energy detector with a predefined threshold [7]. That is to say, the primary user is present if the energy of the received signal is larger than the threshold; otherwise, the primary user is absent.
Note that the energy detector only decides whether the primary user is present or not, the receiver does not need any prior knowledge of the primary users' signals. Furthermore, due to its low computational and implementation complexities, energy detector is the most common used spectrum sensing scheme [8–13]. However, there are some of the challenges with energy detector, such as the selection of the threshold for detecting primary users, inability to differentiate interference from primary users and noise, and poor performance under low signal-to-noise ratio (SNR). Moreover, energy detectors do not work efficiently for detecting spread spectrum signals [14].

② Cyclostationary detector
In communications system, the signal processing process such as spread spectrum, sampling, and modulation generates the periodicity in the transmit signal or in its statistics like mean and autocorrelation, which further causes the cyclostationarity feature. The cyclostationary detection exploits the cyclic correlation function to analyze whether the sensed spectrum is occupied by the primary user or not [15–17].
The cyclic spectral density (CSD) function of a received signal in (4.22) is expressed as

$$S(f,\alpha) = \sum_{\tau=-\infty}^{\infty} R_y^{\alpha}(\tau)e^{-j2\pi f\tau}, \tag{4.25}$$

where

$$R_y^{\alpha}(\tau) = E[y(n+\tau)y^*(n-\tau)e^{j2\pi\alpha n}], \tag{4.26}$$

is the cyclic autocorrelation function (CAF), and α is the cyclic frequency. Note that since noise is wide-sense stationary (WSS) with no correlation while modulated signals are cyclostationary with spectral correlation due to the signal periodicities, therefore, the cyclostationary detector can be employed to identify the type of the primary users' signals. Furthermore, it can differentiate communication systems using different modulation, coding, and multiplexing ways. However, the cyclostationary detector is computationally more complicated than the energy detector. In addition, it needs a prior knowledge of the primary user [18]. Hence, the implementation of the cyclostationary detector remains a challenge.

③ Matched filtering detector

Matched filtering algorithm detects the presence of the primary user by correlating the received signal with the known transmit signal, which can maximize the SNR of the received signal in the context of the additive noise [19]. Furthermore, with the assumption that the transmit signal is perfectly known, the matched filtering detector is known as the optimal algorithm for spectrum sensing [20]. Compared with other spectrum sensing algorithms, the matched filtering detector achieves a certain probability of false alarm in a short time. However, matched filtering requires perfect knowledge of the primary users, such as bandwidth, carrier frequency, modulation type, and so on, so that the implementation complexity is relatively large [21].

④ Waveform-based detector

Basically, the transmit signals in the actual communications system employ some patterns such as the pilot sequence and preambles, which are known to the receiver to accomplish the synchronization, estimation, or other purposes. The waveform-based detector senses the spectrum by performing the correlation between the received signal and some prior known patterns [14, 22].

Consider the received signal of the secondary user in (4.22), the decision metric of the waveform-based detector is defined as

$$M = \text{Re}\left[\sum_{n=1}^{N} y_n s_n^*\right], \tag{4.27}$$

where * denotes the conjugation. The decision is made by comparing the output decision matrix with a predefined threshold, which is similar with that of the energy detector. The main advantage of the waveform-based detector is that the detection performance outperforms the energy detector with a relatively short convergence time [14]. However, it is noted that the waveform-based detector only suits for communications systems with known signal patterns.

⑤ Radio identification based detector

The radio identification based detector detects the presence of the primary user by identifying the transmission techniques employed by the primary user, which includes two main procedures: feature extraction and classification [23–25]. Note that the radio identification based detector can obtain higher dimensional knowledge by utilizing the proper features; it can achieve high detection performance. A number of features have been proposed in the radio identification based detection. The most widely used features include channel bandwidth, center frequency, the standard deviation of the instantaneous frequency, and the maximum duration of a signal, etc. [23]. In addition, in some literature, features obtained by the energy detector and the cyclostationary detector are used for classification as well. The comparison between the above detectors is shown in the following figure.

It can be seen from Fig. 4.24 that the energy detector has the lowest computational complexity, but its probability of detection is limited as well. The waveform-based detector is more reliable and robust than other spectrum sensing methods. The matched filtering detector shares a similar

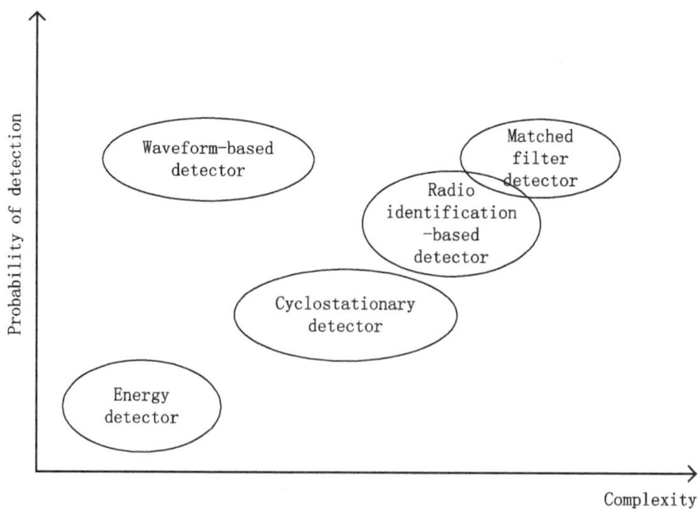

Fig. 4.24 Comparison between the main spectrum sensing algorithms

detection performance with the waveform-based detector, but suffers from a higher complexity as well. The cyclostationary and radio identification based detector have some trade-off between the detection performance and the complexity.

2. **Cooperative spectrum sensing**
In the actual cognitive radio systems, some harmful effects introduced by the path loss, shadow fading, and multipath fading channel may cause the hidden primary user problem, which is similar to the hidden terminal problem in Carrier Sense Multiple Accessing (CSMA). A simplified diagram of the hidden primary user problem is shown in Fig. 4.25, in which the two circles represent the transmission ranges of the primary user and the cognitive radio device, respectively. Due to the hidden primary user problem, the secondary user cannot detect correctly the presence of the primary user, which leads to an unwanted interference to the primary user.
In such a case, the local spectrum sensing does not work well when secondary users scanning for the primary users' transmit signal. Hence, the cooperative spectrum sensing technology is proposed to handle this problem [10, 11]. The cooperative sensing exploits the independence of the multipath fading channels and the multiuser diversity to improve the probability of detection. However, challenges of cooperative sensing include developing efficient information sharing algorithms and fusion rules, and increased complexity as well. Generally, the cooperative spectrum sensing can be divided into two classes: the centralized cooperative spectrum sensing and the distributed cooperative spectrum sensing [26].

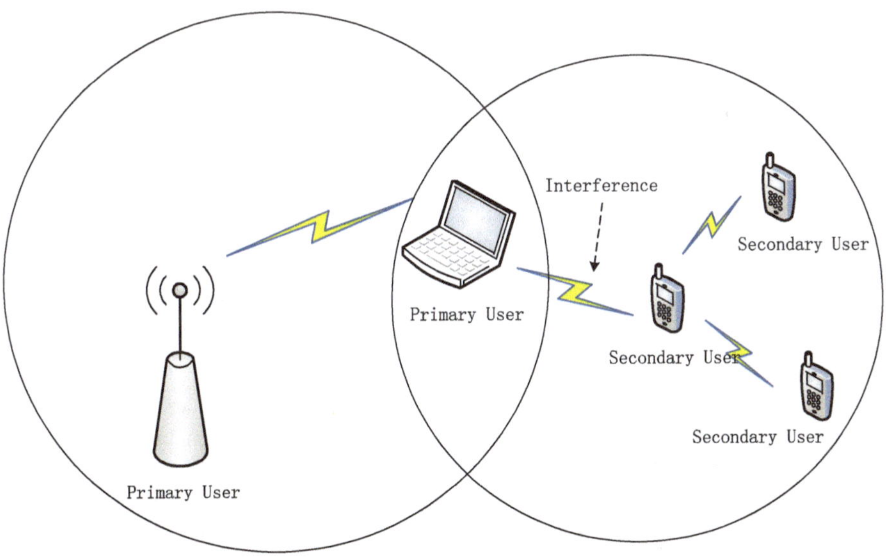

Fig. 4.25 Hidden primary user problem in the cognitive radio system

① Centralized cooperative spectrum sensing

In the centralized sensing, a fusion center (FC) collects sensing information sent by all secondary users in the network, makes global decision according to a certain fusion rule, and then broadcasts the final decision to the secondary users. The procedures of the centralized sensing are described as follows:

- Every secondary user implements the spectrum sensing locally;
- All secondary users send their (hard or soft) sensing information to the fusion center, which is generally the access point (AP) in WLAN or the base station (BS) in the cellular network;
- The fusion center employs some fusion rule to make global decision, which is sent back to all secondary users by broadcasting.

Based on the decision information type sent by the secondary user, the fusion rule can be categorized into two schemes: hard fusion rule and soft fusion rule.

Hard fusion rule: In this scheme, each secondary user detects the presence of the primary user locally and sends a one bit 0–1 hard decision to the fusion center. The main advantage of this fusion rule is that it can be easily implemented and has low processing delay. When binary decisions are reported to the fusion center, three decision rules can be exploits: AND rule, OR rule, and K-out-of-N rule. The AND rule decides that a primary user is present if all secondary users have detected the primary user. On the other hand, the OR rule decides that a primary user is present if any of the secondary users have detected the primary user. The third rule is the K-out-of-N rule that decides on the signal presence if at least K of the N secondary users have detected the primary user.

Soft fusion rule: In this scheme, the secondary users forward the entire sensing result to the fusion center without performing any local decision. The global decision is made by combining these results at the fusion center by using appropriate combining rules such as equal gain combing (EGC), selection combining (SC), maximal ratio combining (MRC), etc. The soft fusion rule provides better performance than the hard one, but it requires a larger bandwidth, and also generates more overhead than the hard fusion rule.

② Distributed cooperative spectrum sensing

In the distributed sensing, there is no centralized fusion center to make decision. The secondary users share sensing information among each other, and each of them makes its own decision based on the sensing information they have. Distributed sensing is more advantageous than centralized sensing in the sense that there is no need for a backbone infrastructure and it has reduced cost.

4.3.2 Automatic Modulation Classification

Automatic modulation classification (AMC) is a technique to identify the modulation formats from a set of predefined candidate modulations by observing the received signals. It plays a key role in cooperative as well as the noncooperative communication. For instance, in the civil communications, by utilizing the automatic modulation classification, the radio frequency spectrum authority is able to supervise and manage the frequency band to avoid the licensed spectrum being occupied by illegal users; while in the military communications, particularly in the electronic warfare, such as signal interception, reconnaissance, and electromagnetic countermeasure (ECM), the automatic modulation classification is the fundament of the subsequent signal processing procedure like demodulation and decoding. In recent year, the automatic modulation classification has drawn more attentions in cognitive radio networks and interference identification.

At receiver, the automatic modulation classification technique is an intermediate task between signal detection and demodulation, which is a prerequisite and fundament of the subsequent signal processing procedures [27]. A systematic diagram of the modulation classification is shown in Fig. 4.26. Generally, a modulation classifier includes two steps: signal preprocessing and proper design/selection of the classification algorithm. Some main tasks of the signal preprocessing part include noise reduction, estimation of signal parameters such as carrier frequency, symbol period, signal power, equalization, etc. The accuracy of the signal preprocessing affects the classification performance of the classification algorithms, and also, the levels of accuracy required by the classification performance conversely decides the precision of the preprocessing.

Various modulation classification algorithms have been proposed in the past few decades. In general, these methods can be divided into two classes: the likelihood-based (LB) [28–31] and the feature-based (FB) methods [32–36].

① Likelihood-based algorithm
The likelihood-based algorithm computes a certain form of likelihood function of the received signal, by which the modulation classification problem is modeled as a two-hypothesis classification problem, or

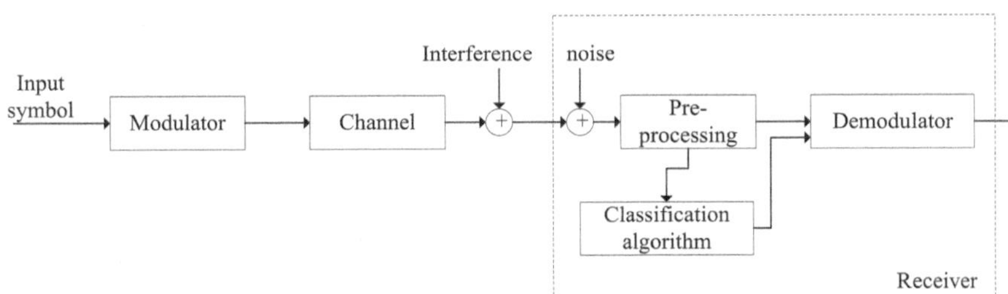

Fig. 4.26 System block diagram of automatic modulation classification

multiple binary hypothesis classification problems. The decision is made by comparing the likelihood ratio with a predefined threshold. Considering the unknown parameters chosen from the received signal, there are three potential solutions: average likelihood ratio test (ALRT), generalized likelihood ratio test (GLRT), and hybrid likelihood ratio test (HLRT).

ALRT considers the unknown parameters as random variables with certain probability density function (PDF). The likelihood function (LF) under the hypothesis H_i, representative of the ith modulation, $i = 1, \ldots, N_{\text{mod}}$, is given by

$$\Lambda_A^{(i)}[r(t)] = \int \Lambda[r(t)|\mathbf{v}_i, H_i] p(\mathbf{v}_i|H_i) \mathrm{d}\mathbf{v}_i, \tag{4.28}$$

where $\Lambda[r(t)|\mathbf{v}_i, H_i]$ is the conditional likelihood function of the noisy received signal $r(t)$ under H_i, conditioned on the unknown vector $\mathbf{v}_i$, and $p(\mathbf{v}_i|H_i)$ is the a priori PDF of $\mathbf{v}_i$ under H_i. The known PDF of $\mathbf{v}_i$ enabled us to reduce the problem to a simple hypothesis testing problem by integrating over $\mathbf{v}_i$.

GLRT treats the unknown quantity as deterministic, which is also known as the maximum likelihood (ML) method, and the LF under Hi is given by

$$\Lambda_G^{(i)}[r(t)] = \max_{\mathbf{v}_i} \Lambda[r(t)|\mathbf{v}_i, H_i]. \tag{4.29}$$

HLRT is an approach that combines ALRT and GLRT, for which the likelihood function under H_i is given by

$$\Lambda_H^{(i)}[r(t)] = \max_{\mathbf{v}_{i_1}} \int \Lambda[r(t)|\mathbf{v}_{i_1}, \mathbf{v}_{i_2}, H_i] p(\mathbf{v}_{i_2}|H_i) \mathrm{d}\mathbf{v}_{i_2}, \tag{4.30}$$

where $\mathbf{v}_i = [\mathbf{v}_{i_1}^{\dagger} \mathbf{v}_{i_2}^{\dagger}]^{\dagger}$ and $\mathbf{v}_{i_1}$ and $\mathbf{v}_{i_2}$ are vectors of unknown quantities modeled as unknown deterministics and random variables, respectively. Usually, $\mathbf{v}_{i_1}$ and $\mathbf{v}_{i_2}$ consist of parameters and data symbols, respectively. In a two-hypothesis classification problem, the decision is made according to

$$\frac{\Lambda_A^{(1)}[r(t)]}{\Lambda_i^{(2)}[r(t)] \underset{R_2}{\overset{R_1}{\gtrless}} \eta_1}, 1 = A(\text{ALRT}), G(\text{GLRT}), H(\text{HLRT}), \tag{4.31}$$

where η_1 is a threshold. The left-hand side is referred to as the likelihood ratio and the test is called average likelihood ratio test (ALRT), generalized likelihood ratio test (GLRT), and hybrid likelihood ratio test

Table 4.2 Summary of likelihood-based classifiers [27]

Author(s)	Classifier(s)	Modulations	Unknown parameter(s)	Channel
Sills [37]	ALRT	BPSK, QPSK, 16QAM, V29, 32QAM, 64QAM	Carrier phase θ	AWGN
Wei and Mendel [30]	ALRT	16QAM, V29	–	AWGN
Kim and Polydoros [38, 39]	quasi-ALRT	BPSK, QPSK	Carrier phase θ	AWGN
Huang and Polydoros [28]	quasi-ALRT	UW, BPSK, QPSK, 8PSK, 16PSK	Carrier phase θ and timing offset ε	AWGN
Sapiano and Martin [40]	ALRT	UW, BPSK, QPSK, 8PSK	–	AWGN
Long et al. [41]	quasi-ALRT	16PSK, 16QAM, V29	Carrier phase θ	AWGN
Hong and Ho [42]	ALRT	BPSK, QPSK	Signal level α	AWGN
Beidas and Weber [43]	ALRT and quasi-ALRT	32FSK, 64FSK	Phase jitter $\{\varphi_k\}_{k=1}^{K}$	AWGN
Beidas and Weber [44, 45]	ALRT and quasi-ALRT	32FSK, 64FSK	phase jitter $\{\varphi_k\}_{k=2}^{K}$ and timing offset ε	AWGN
Panagiotu et al. [29]	GLRT and HLRT	16PSK, 16QAM, V29	Carrier phase θ	AWGN
Chugg et al. [46]	HLRT	BPSK, QPSK, OQPSK	Carrier phase θ, signal power S and PSD N_0	AWGN
Hong and Ho [47]	HLRT	BPSK, QPSK	Signal level α	AWGN
Hong and Ho [48, 49]	HLRT	BPSK, QPSK	Angle of arrival $\emptyset$	AWGN
Dobre et al. [50]	HLRT	BPSK, QPSK, 8PSK, 16PSK, 16QAM, 64QAM	Channel amplitude α and phase ϑ	Flat fading
Abdi et al. [51]	ALRT and quasi-ALRT	16QAM, 32QAM, 64QAM	Channel amplitude α and phase ϑ	Flat fading

(HLRT), respectively, depending on the method employed to compute the likelihood function. The above equation can be extended straightforward to solve the multi-class problem.

Although the solutions achieved by the LB approaches are optimal, they suffer from the high computational complexity. Moreover, they are not robust to the presence of frequency or phase offsets, non-Gaussian noise, multipath channel, etc.

The summary of the likelihood-based classifiers is shown in Table 4.2.

② Feature-based algorithm

On the other hand, the feature-based approaches are based on the features extracted from the received signal. In contrast to the likelihood-based algorithms, the feature-based methods can be implemented readily and is capable to achieve nearly optimal performance with proper designed. The features used as decision metric can be instantaneous amplitude, phase and frequency, or PDF of these quantities. Among these features, the most widely adopted one is cumulant. Cumulant is a statistic of random variables, defined by the cumulant-generating function, which is the logarithm of the moment-generating function. For a complex-valued stationary random process $y(n)$, second-order moments can be defined in two different ways depending on placement of conjugation

$$\begin{aligned} C_{20} &= E[y^2(n)] \\ C_{21} &= E[|y(n)|^2]. \end{aligned} \tag{4.32}$$

Similarly, fourth-order moments and cumulants can be written in three ways. Thus, fourth-order cumulants can be defined as

$$\begin{aligned} C_{40} &= \text{cum}(y(n), y(n), y(n), y(n)) \\ C_{41} &= \text{cum}(y(n), yn), y(n), y^*(n)) \\ C_{42} &= \text{cum}(y(n), y(n), y^*(n), y^*(n)). \end{aligned} \tag{4.33}$$

The statistics in (4.32) and (4.33) are the zeroth lags of the correlations and fourth-order cumulants of $y(n)$. For zero-mean random variables w, x, y, and z, the fourth-order cumulant can be written as

$$\text{cum}(w, x, y, z) = E(wxyz) - E(wx)E(yz) - E(wy)E(xz) - E(wz)E(xy). \tag{4.34}$$

The cumulants in (4.32) and (4.33) can be estimated from the sample estimates of the corresponding moments. We assume that is zero-mean; in practice, the sample mean is removed before cumulant estimation. Sample estimates of the correlations are given by

$$\begin{aligned} \hat{C}_{21} &= \frac{1}{N}\sum_{n=1}^{N} |y(n)|^2 \\ \hat{C}_{20} &= \frac{1}{N}\sum_{n=1}^{N} y^2(n). \end{aligned} \tag{4.35}$$

This leads to the following estimates:

$$\hat{C}_{40} = \frac{1}{N}\sum_{n=1}^{N} y^4(n) - 3\hat{C}_{20}^2$$
$$\hat{C}_{41} = \frac{1}{N}\sum_{n=1}^{N} y^3(n)y^*(n) - 3\hat{C}_{20}\hat{C}_{21} \tag{4.36}$$
$$\hat{C}_{42} = \frac{1}{N}\sum_{n=1}^{N} |y(n)|^4 - |\hat{C}_{20}|^2 - 2\hat{C}_{21}^2.$$

In practice, the normalized cumulants can be estimated by

$$\tilde{C}_{4k} = \hat{C}_{4k}/\hat{C}_{21}^2. \tag{4.37}$$

A number of existing literature has exploited cumulants to implement classification for modulation formats, especially for QAM and PSK modulations. For modulation formats with lower order, the classification can be performed by using low-order cumulants, such as the fourth-order cumulant. While a higher order cumulant is used to classify higher order modulation formats with a cost of the increase of the computational complexity and delay.

The summary of the feature-based classifiers is shown in Table 4.3.

③ Hypothesis test based algorithm

Recently, a new approach to modulation classification based on the hypothesis test is proposed. F. Wang et al. proposed a fast and robust Kolmogorov–Smirnov (K–S) test based modulation classification method [75], which develops various K–S classifiers based on different decision statistics for both QAM and PSK modulations, and for various channels, such as the AWGN channel, the flat fading channel, the OFDM channel, and the channel with unknown phase and frequency offsets, as well as the non-Gaussian noise channel.

In statistics, the Kolmogorov–Smirnov (K–S) test is a nonparametric test which can be used to compare the cumulative distribution of the samples with a hypothesized cumulative distribution. The K–S test calculates the absolute distances between the empirical cumulative distribution function (CDF) of the samples and the CDF of a hypothesized distribution, choosing the largest one as the K–S test statistic. The null hypothesis can be given by

$$H_0: \qquad F_1 = F_0, \tag{4.38}$$

where F_1 and F_0 represent the empirical CDFs of the samples and of a hypothesized distribution, respectively.

Table 4.3 Summary of feature-based classifiers [27]

Author(s)	Features	Modulations	Unknown parameter(s)	Channel
Azzouz and Nandi [52, 53]	Maximum power spectral density of normalized centered amplitude, standard deviations of normalized centered amplitude, phase and frequency	2ASK, 3ASK, BPSK, QPSK, 2FSK, 4FSK	–	AWGN
Soliman and Hsue [54, 55]	Variance of the zero-crossing interval sequence, phased difference and zero-crossing interval histograms	UW, BPSK, QPSK, 8PSK, BFSK, 4FSK, 8FSK	–	AWGN
Soliman and Hsue [56–58]	PDF of phase	UW, BPSK, QPSK, 8PSK	–	AWGN
Soliman and Hsue [36, 59, 60]	Statistical moments of phase	UW, BPSK, QPSK, 8PSK	–	AWGN
Sapiano et al. [61]	DFT of phase PDF	UW, BPSK, QPSK, 8PSK	–	AWGN
Ho et al. [62, 63]	Variance of HWT magnitude, HWT magnitude and peak magnitude histograms	BPSK, QPSK, 8PSK, FSK, 4FSK, 8FSK, CP2FSK, CP4FSK, CP8FSK, MSK	–	AWGN
Hong and Ho [64]	Variance of HWT magnitude and normalized HWT magnitude	QPSK, 4FSK, 16QAM	–	AWGN
Swami and Sadler [32]	Normalized fourth-order cumulants of the received signal	BPSK, 4ASK, 16QAM, 8PSK, V32, V29, V29c	Carrier phase θ, frequency offset Δf and timing offset ε	AWGN, impulsive noise, co-channel interference
Swami et al. [65]	Normalized fourth-order cumulants of the received signal and the cost function of the alphabet-matched equalization algorithm	BPSK, 4ASK, QPSK, 16QAM, V29, V32, 64QAM	–	Frequency selective channel
Martret and Boiteau [66]	Fourth- and second-order moments of the received signal	QPSK, 16QAM	–	AWGN

(continued)

Table 4.3 (continued)

Author(s)	Features	Modulations	Unknown parameter(s)	Channel
Marchand et al. [67, 68]	Fourth- and second-order cyclic cumulants of the received signal	QPSK, 16QAM, 64QAM	–	AWGN
Spooner et al. [69–71]	Sixth-, fourth-, and second-order cyclic cumulants of the received signal	MSK, QPSK, BPSK, 8PSK, 8QAM, 16QAM, 64QAM, V29	Frequency offset Δf, excess bandwidth, symbol period T, signal amplitude α	AWGN, co-channel interference
Dobre et al. [34]	Eighth-order cyclic cumulants of the received signal	BPSK, QPSK, 8PSK, 4ASK, 8ASK, 16QAM, 64QAM, 256QAM	–	AWGN
Dobre et al. [72]	Eighth-, sixth-, and fourth-order cyclic cumulants of the received signal	4QAM, 16QAM	Carrier phase θ, phase jitter $\{\varphi_k\}_{k=1}^{K}$ and frequency offset Δf	AWGN, impulsive noise
Dobre et al. [73]	Eighth-order cyclic cumulants of the signal at the output of a selection combiner	4ASK, 8ASK, BPSK, QPSK, 16QAM, 32QAM, 64QAM,	–	Rayleigh and Ricean fading channels
Yu et al. [74]	DFT of the received signal	2FSK, 4FSK, 8FSK, 16FSK, 32FSK,	–	AWGN

Assuming that $z_1, \ldots, z_N$ is a sequence of independent and identically distributed (i.i.d.) real-valued samples. Then, the empirical CDF of $\{z_n\}$ can be calculated by

$$\hat{F}_1(z) \triangleq \frac{1}{N}\sum_{n=1}^{N} I(z_n < z), \tag{4.39}$$

where $I(.)$ is the indicator function, which equals 1 if $z_n < z$, equals 0 otherwise.

To discriminate the modulation formats by the K–S method, we need to extract a sequence of statistic $\{z_n\}$ from the received signal $\{y_n\}$ and calculate the empirical CDF $\hat{F}_1$. The decision statistic z_n can be chosen as the real and imaginary components, the magnitude or the phase of y_n. For the K possible modulation candidates $\{M_1, \ldots, M_K\}$, we can obtain the CDF F_0 of $\{z_n\}$. Therefore, the K–S statistics $\hat{D}_K$ can be written as

$$\hat{D}_k = \max_{1 \le n \le N} \left| \hat{F}_1(z_n) - F_0^k(z_n) \right|, \quad k = 1, \ldots, K. \tag{4.40}$$

The decision rule of the K–S test is to choose the modulation candidate which has the minimum K–S statistic

$$\hat{k} = \arg \min_{1 \le k \le K} \hat{D}_k. \tag{4.41}$$

It is noted that the computational complexity of the K–S test is comparable with that of the cumulant, which can be easily implemented.
Additionally, P. Urriza et al. proposed a modulation classification algorithm employing the Kuiper's hypothesis test [76]. The Kuiper's test is closely related to the K–S test. As with the K–S test, it calculates the largest discrepancy between two CDFs as the K–S statistic, while the Kuiper's test utilizes both the most positive and negative discrepancies between two CDFs as the Kuiper test statistic. Note that the Kuiper's statistic uses information on two points, it is more reliable and robust than the K–S test in classifying different modulations.
F. Wang et al. further proposed a variational-distance-based scheme for the modulation classification problem [77]. It decides on the modulation that minimizes the variational distance between the theoretical and empirical probability density of the received signal. This algorithm outperforms some existing featured-based classifiers, including the cumulant classifier, K–S classifier, and Kuiper classifier. Its computational complexity is comparable to those classifiers but it is more robust to the error in estimating the noise power.

4.3.3 Specific Emitter Identification

In typical communication systems, a receiver only focuses on the information transmitted by the emitter, but pays no attention to the emitter-specific and unintentional hardware information (referred to as the fingerprint of the emitter) of the emitter. Specific emitter identification (SEI) is the process of discriminating individual emitters by comparing the fingerprint carried by the received signal with a categorized feature set, and choosing the class that best matches the features [78].

Specific emitter identification originated in the mid-1960s, when a high-priority problem of the US government was to identify and track unique mobile emitters for targeting, and developed a method to designate a unique emitter of a given signal by using external features, namely the specific emitter identification (SEI). This technique has been studied over the past five decades, especially in government or military communications, including the signal interception, reconnaissance, and electromagnetic countermeasure (ECM). In addition, some processing systems have been designed and built to apply the SEI to signals from communication devices.

Since 1990s, the SEI technique has also been applied to commercial customers. Basically, the security of the wireless communication system today relies on bit-level cryptographic techniques and associated protocols at various levels of the data processing stack. These solutions have drawbacks, which presents a major risk to society. Standardized protections within public wireless networks are not secure enough; even if enhanced ciphering and authentication protocols exist, they have constraints and add additional costs for the users of public networks. By utilizing the specific emitter identification technique, the security approaches can be introduced into the signal processing procedure and physical layer, where the illegal and untrusted emitters are identified by extracting and analyzing the hardware features. Recently, it has become increasingly important with the advent of new technologies, such as cognitive radio. For example, the radio frequency spectrum authority is able to avoid an unlicensed secondary user to occupy the spectrum by imitating the primary user.

A typical specific emitter identification generally consists of several subsystems: signal preprocessor, feature extraction, and identification classifier. Figure 4.27 illustrates a combined specific emitter identification system. A valid SEI process relies on the feature extraction scheme to obtain features providing a wide separation between different classes.

Based on the operation mode of the emitter, SEI is applicable either to the transient signal or to the steady-state signal [79].

(1) Transient signal feature extraction
The transient signal, commonly known as the turn-on signal, provides unique and distinguishable characteristics suited for feature extraction and emitter identification [80]. Basically, the transient features are emitter-specific and consistent, which is advantageous to identification; nevertheless, they are difficult to be captured since the duration of the transient signal is extremely short. Furthermore, the transient features are easily hampered by nonideal and

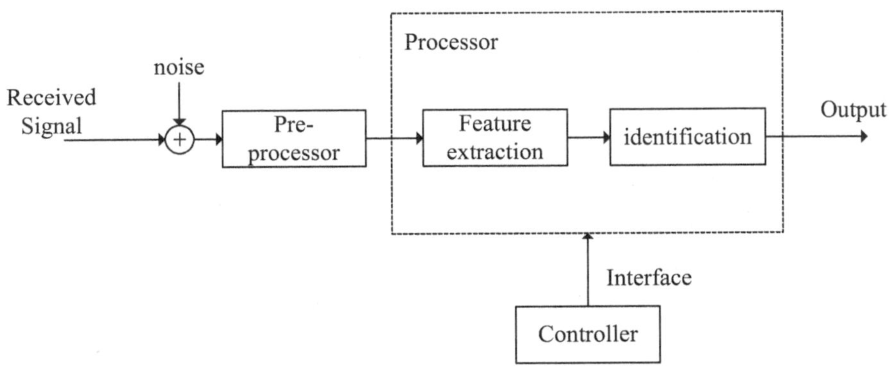

Fig. 4.27 System block diagram of specific emitter identification

complicated channel conditions, which may negatively affect the identification results.

In general, the collected transient signal contains large amount of channel noise, i.e., the starting point of the transient signal follows a sequence of noise samples. Hence, in order to measure the transient features, a foremost step is to extract the transient signal from the noise by detecting its starting and ending points, i.e., to find the exact time when the transient signal begins and ceases. However, many transients exhibit characteristics similar to noise due to their high degree of irregularity. In other words, due to the nonlinear and nonstationary nature of transmitter transients, though, the task of separating the transient from the channel noise is very difficult. D. Shaw and W. Kinsner developed an approach based on the multi-fractal analysis, which utilizes the variance fractal dimension trajectory (VFDT) to characterize the degree of irregularity along the duration of the signal [81]. The main idea behind this algorithm is that the variance fractal dimension of the noise and the transient signal differs a lot. Then, when a significant change occurs, it signifies the start of the transient. This approach is computationally efficient and can be easily implemented; however, the detection performance is sensitive to the SNR. Additionally, a predefined threshold is required to extract transient signal from the noise, therefore, an inappropriate threshold may negatively affect the detection performance as well. Besides, J Hall et al. proposed a scheme called the transient detection using phase characteristic (TDPC) [82]. This approach utilizes the change of the phase slope to identify the starting point. Since TDPC method is similar with that of the VFDT, therefore, the shortcoming of the threshold still exists in this method. In order to cope with the threshold problem, O. Ureten and N. Serinken developed a series of detection algorithm based on the Bayesian theory, including the Bayesian change point detection [83], the Bayesian step change detection (BSCD) [84], and the Bayesian ramp change Detection (BRCD) [85]. These Bayesian approaches require no prior knowledge, but are computationally more complicated than other methods.

(2) Steady-state signal feature extraction

The steady-state signal is transmitted by emitters operating under stable conditions. Although the steady-state features tend to be corrupted by the transmitted information, leading to difficulties in extracting them, the investigation of the steady-state signal has considerable practical implications, as this is easily detected and captured.

A number of feature extraction schemes have been studied in such a case, with a widely used approach relying on the time–frequency representation based feature. A time–frequency representation maps the signal onto a two-dimensional plane of time and frequency, which illustrates the temporal and spectral information simultaneously. G. Lopez-Risueno et al. propose a signal detection and identification system based on the short-time Fourier transform (STFT) [86]. However, STFT is a linear transform which cannot be adopted to analyze a nonlinear signal. A similar radar waveform identification algorithm utilizing the Wigner and Choi–Williams distributions is proposed by J. Lunden and V. Koivunen [87]. Another class-dependent scheme is proposed by B. Gillespie and L. Atlas, which employs smoothing regular quadratic time–frequency representations to extract features for radar emitter identification [88]. However, the difficulty with these quadratic time–frequency representations is the inevitable cross-terms problem. In 1998, N. E. Huang from NASA proposed a new method, namely the Hilbert–Huang transform (HHT) [89]. By using this method, any complicated data set can be decomposed into a finite number of "intrinsic mode functions" that admit well-behaved Hilbert transforms. Since the decomposition is based on the local characteristic time scale of the data, it is applicable to nonlinear and nonstationary processes. Based on this adaptive and efficient method, S. Xu et al. proposed a feature extraction algorithm employing the Hilbert spectrum.

On the other hand, the advantage of employing higher order statistics, such as cumulants, and polyspectr as identification feature is widely recognized. Among all higher order statistics, bispectra is the most commonly used. It is noticed that bispectra results is a two-dimensional matching score, which cannot be directly exploited in the signal classification and specific emitter identification. In order to solve this problem, several integrated bispectra based feature extraction algorithms have been proposed, including the radially integrated bispectra (RIB) [90], the axially integrated bispectra (AIB) [91], and the circularly integrated bispectra (CIB) [92]. The main idea behind the three algorithms is to select a certain integration path, along which the surrounding line integration is performed to turn the two-dimensional bispecta into one-dimensional vector. Note that all the three integrated bispectra-based algorithms are one-dimensional matching scores, which are feasible for signal classification and specific emitter identification. The entire identification feature extracted by the three algorithms has three key properties presented below. These properties are important to attain accurate identification performance:

- translation invariance;
- scale variance;
- phase information of the underlying signal.

① Radially integrated bispectra (RIB)

Chandran and Elgar are the first to utilize integrated bispectra in signal classification. The phase radially integrated bispectra (PRIB) is proposed as the identification feature. It is obvious that the PRIB is the phase of the integrated bispectra along radial lines passing through the origin in bi-frequency space. It has been proved that the PRIB parameters are translation invariant, amplification invariant, DC level invariant, and scale invariant. Furthermore, the PRIB approach is computationally efficient, the reasons are as follows: (i) The dimension of PRIB is equivalent to the dimension of the signal, which is much lower than the dimension of the original bispectra; (ii) the correlation between the test signals and the template signals is a one-dimensional matching score without sliding movement.

② Axially integrated bispectra (AIB)
An alternative integrated bispectrum based algorithm is proposed by Tugnait in 1994. This scheme is referred to as the axially integrated bispectra (AIB). In this algorithm, the bispectra are integrated along paths parallel to ω_1 or ω_2 axes in bi-frequency plane. It is noted that the AIB scheme retains the scale character of the received signal. However, a shortcoming of the AIB approach is that it contains less phase information than the original bispectra. This can be easily explained that in AIB algorithm, only information of a single slice $C_{3x}(0, \tau)$ is utilized, while most phase information related to $\int_{-\infty}^{\infty} C_{3x}(m, \tau)e^{-j\omega\tau}d\tau$ with $m \neq 0$ is lost.

③ Circularly integrated bispectra (CIB)
Additionally, Liao and Bao proposed an integrated bispectra based algorithm, namely the circularly integrated bispectra (CIB). Unlike the RIB and AIB, a set of concentric circles with the origin as the center are selected as the integral paths of CIB. The CIB keep the scale information of signals, which is the same as the RIB and AIB, but in contrast to the AIB, the CIB utilizes more radial slices of bispectra.

To sum up, the PRIB, RIB, AIB, and CIB methods select the integrated bispectra with the most discriminant power as the identification features of the interested signals; their difference lies in the choice of the integral paths. The performance comparison is analyzed as follows:

- The PRIB is translation invariant and retains the phase information of the signal as well. However, due to scale invariance, it loses the scale character of the signal.
- The AIB is translation invariant and have the scale variance, but it loses most of the phase information.
- The RIB and CIB are translation invariant, they keep the scale variance and the phase information of the signal as well.

4.4 Cooperation and Cognition for High-Speed Railway

With the development of high-speed railway and public growing demand on data traffic, people pay much more attention to provide high data rate and high reliable services under high mobility circumstance. Due to the higher data rate and lower system latency, long-term evolution (LTE) has been chosen as the next generation's evolution of railway mobile communication system by the International Union of Railways. However, there are still many problems to be solved in the high mobility applications of LTE, especially the higher handover failure probability, which seriously degrades the reliability of railway communication.

High-speed railway is playing an important role in mass transportation, due to its lower energy consumption, less environmental pollution, larger capacity, and higher safety features. The development of high-speed railway makes people's life more and more convenient. Meanwhile, providing high quality of service broadband communications for fast-moving users still remains unsolved, despite the fact that new solutions of incremental improvements are keeping up with this unprecedented communication requirement growth.

Recent years, the high-speed railway in China has attracted much attention and it has stepped into a new era with rapid development. In the meantime, the high-speed railway mobile communication system is required to provide broadband service for railway managers and passengers on the train. However, high mobility of trains raises some new problems and challenges to the communication between the base station and the running train, which is also referred to as the train–ground communication. First, high mobility of trains results in a very fast time-varying channel for the train–ground communication, so it is impossible for the receiver to precisely track, to capture and to estimate the instantaneous channel state information (CSI), and then to feed it back to its transmitter in real time. Second, Doppler effect caused by high mobility leads to severe inter-carrier interference, which remarkably deteriorates the information receiving performance and then leads to very poor communication quality. Third, compared with low-speed mobile scenarios, multipath fading becomes much worse in high-speed mobile scenarios, because in high mobility scenarios, wireless channels often experience more sophisticated space effect within a very short time. Therefore in order to provide higher transmission rate with good reliability for high-speed railway mobile communication system, new technologies have to be explored. Recently, some works introduced fountain codes and cooperative relay into high mobility systems.

As for cooperative relaying communications, it has been widely investigated to increase the capacity and decrease the outage probability of wireless communication system. In cooperative relaying system, relays process and forward the received data from their sources to the destinations by using some relaying protocols (e.g., decode and forward (DF) and amplify and forward (AF)), so that the destinations can receive several signal samples of each data. By using some advanced information combing methods, cooperative diversity gains can be achieved.

Due to the lower energy consumption, less environmental pollution, larger transport capacity, and more safety, railway transportation plays an important role for the development of country. China Railway has achieved remarkable successes. Nowadays, the development of high-speed railway makes people's lives more and more convenient. Meanwhile, it puts forward higher requirements on high-speed railway communication services. The existing GSM for Railway (GSM-R) network is mainly based on the second-generation Global System for Mobile Communications (GSM), and its data rate is not fast enough to meet the broadband mobile communication access and other value-added service demands of passengers. In order to provide broadband services and applications for users not only at home but also on trip, long-term evolution (LTE) has been chosen as the next generation's evolution of railway mobile communication system by International Union of Railways (UIC), which supports significant higher data rates and lower system latency.

4.4.1 Relay Selective Cooperation in Railway Network

Wireless communication for railway has drawn much attention due to the rapid development and deployment of high-speed railway in China and around the world. For a typical high-speed railway wireless communication system, there are two transmission schemes of wireless signals. One is to directly transmit signals from base station along the railway to the mobile station inside the train. The other scheme involves two phases: in the first phase signals are transmitted from the antenna of base station to the antenna of relay stations mounted to the roof of the train, and in the second phase the train antenna retransmits the received signal to the MS inside the train.

Note that this direct transmission will suffer from a carriage passing loss that is defined as the power loss when the signal traverses through the carriage, which is typically between 12 and 24 dB due to different model of the train. In contrast, the transmission based on relays will not suffer from the carriage passing loss as the relay has its antenna inside the train. However, it is well known that the relay-based scheme will decrease the channel capacity since it is relay-based half-duplex and takes two phases in constraint. On the other hand, we can simply get the diversity gain by the help of the relay. In this way, we can improve the reliability of the system.

Cooperative transmission has been proposed to improve the wireless transmission performance. Several cooperative protocol including fixed relaying, selection relaying, and increment relaying have been proposed and analyzed for Rayleigh fading channels in terms of the outage probability.

In the railway network, the fact that signal amplitude suffers loss resulted from signal traversing through the carriage which named by carriage pass loss, highly influences the reliability of the wireless communication system.

The cooperation technique also has great potential to be applied in railway networks. The existence of the carriage passing loss will significantly decrease the reliability of the railway network if only direct transmission is occupied. To overcome this shortcoming, the relay selective cooperation transmission scheme can be used to enhance the reliability. From a practical point of view, the railway network is labeled as high mobility which means any technique with unnecessary delay will not be considered. The relay can use the following three strategies to re-transmit the signal: (1) amplify and forward (AF); (2) decode and forward (DF); (3) compress and forward (CF). Here we choose DF strategy to introduce. The system is modeled as a dual-hop relay network with different distributed channels and various fading gains as shown in Fig. 4.28.

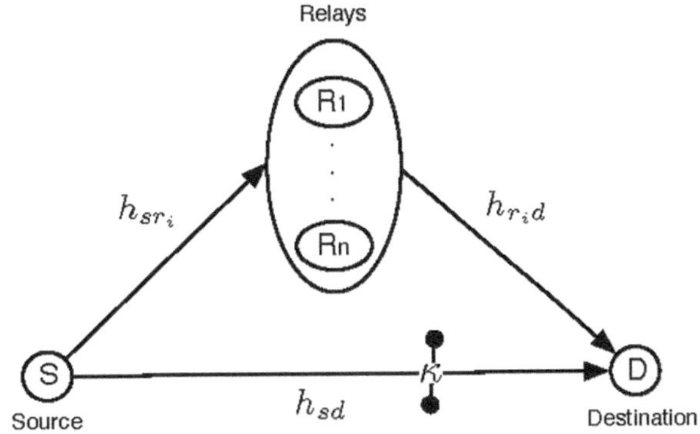

Fig. 4.28 A wireless dual-hop relay system model of high-speed railway with multiple relays

The source communicates with the destination independently, and through N half-duplex relays. Assume that source, relays, and destination have one antenna each, the channel state information (CSI) is available at destination and relays. The source transmits a signal x, which has an average power normalized to one.

For comparison purpose, the direct data transmission from source to destination is considered without relaying. Due to the carriage passing loss κ, the signal received by the destination has an attenuation. Denoting the transmit power of S by P_d, the received signal at destination can be expressed as

$$y_d = \sqrt{\kappa}\sqrt{P_d}h_{sd}x + n_{sd} \tag{4.42}$$

where h_{sd} is the fading amplitude of the channel from source to destination. Let $\gamma_d = |h_{sd}|^2/\sigma^2$ as the instantaneous signal-to-noise ratio (SNR) at the destination node. Thus, the instantaneous mutual information via direct transmission as

$$I_{sd} = \log_2(1 + |h_{sd}|^2\kappa\gamma_d) \tag{4.43}$$

N relays are assumed to assist the transmission from source to destination. In the first phase, the source broadcasts its signals to the destination and relays. We denote the set of N relays by $\mathcal{R} = r_i|i = 1, 2, \ldots, N$, where the decode-and-forward (DF) protocol is employed at relays, i.e., N relays receive the signals from source in the first phase, and then decode and forward it in the second phase. Notice that although only the DF protocol is considered, similar performance result can be obtained for other relaying protocols. Without loss of generality, those relays that succeed in decoding the source signal are represented by a set D, which called decoding set. Obviously, given N relays there are $2N$ possible subset combinations from the set $\mathcal{R}$ of N relays, the decoding set $\mathcal{R}$ is given by

$$\sum = \{\phi, D_1, D_2, \ldots, D_{2^N-1}\} \tag{4.44}$$

where ϕ denotes an empty set and D_n denotes a nonempty subset from N relays. If the decoding set is not empty, no matter whether D decodes the received signal directly from the source successfully or not, a relay will be opportunistically selected within the decoding set to forward its decoded signal if it succeeds in decoding the source's signal. The destination would combine the signal copies received during the two phases. If the decoding set is empty, the source, instead of relay, transmits new data information which is different from the one transmitted during the previous phase. We can employ a cyclic redundancy code (CRC) to determine whether CR decodes its received signal successfully or not, i.e., if the CRC checking passes, it is assumed that CR succeeds in decoding. The received signal at i relay is

$$y_{R_i} = \sqrt{P_d} h_{sr_i} x + n_{sr_i} \tag{4.45}$$

where h_{sr_i} is fading amplitude of the channel between the source and the i-th relay. Thus, the instantaneous mutual information from the source to the i-th relay I_{sr_i} is given by

$$I_{sr_i} = \frac{1}{2}\log_2\left(1 + |h_{sr_i}|^2 \gamma_d\right) \tag{4.46}$$

where the pre-log factor 1/2 is due to the half-duplex relaying constraint. During the selective process, in the information theoretic way, when the instantaneous mutual information I_{sr_i} is below the data rate R, relay r_i is deemed to decode the source signal. Thus, the event $D = \phi$ is described as

$$\frac{1}{2}\log_2\left(1 + |h_{sr_i}|^2 \gamma_d\right) < R, i \in R \tag{4.47}$$

where R denotes the set of N relays. Similarly, event $D = D_n$ can be given by

$$\frac{1}{2}\log_2(1 + |h_{sr_i}|^2 \gamma_d) < R, i \in D_n \tag{4.48}$$

$$\frac{1}{2}\log_2(1 + |h_{sr_i}|^2 \gamma_d) < R_i, i \in D_n \tag{4.49}$$

where $\tilde{D}_n = \mathcal{R} - D_n$ is the complement of D_n. Given the decoding set is D_n, an optimal relay r_i within D_n will be selected to forward its decoded signal to the destination. Assuming the transmit power of i-th relay is P_d, we can obtain the instantaneous mutual information at destination from i-th relay as

$$I_{r_i d} = \frac{1}{2}\log_2\left(1 + |h_{r_i d}|^2 \gamma_d\right), i \in D_n \tag{4.50}$$

Considering to maximum the performance at the destination, the relay with the highest instantaneous mutual information is selected to forward its decoded signal to destination. In sum, a relay within D_n that maximizes $I_{r_i d}$ is considered as the best relay, i.e.

$$\text{Best Relay} = \arg\max_{i \in D_n} I_{r_i d} = \arg\max_{i \in D_n} |h_{r_i d}|^2 \tag{4.51}$$

Without loss of generality, we denote the selected best relay as r_b. In the second phase, the best relay will forward its decoded outcome to the destination. As shown in Fig. 4.29, there are two possible cases depending on whether decoding set is empty set or not. At the destination, a signal copy with higher signal-to-noise ratio (SNR) than the other will be employed for decoding the source message. Thus, the mutual information from the source to the destination using the relay diversity transmission can be given by

$$I_{\text{relay}} = \max(\frac{1}{2} I_{sd}, \frac{1}{2} \log_2 \left(1 + |h_{bd}|^2 \gamma_d\right)) \tag{4.52}$$

where h_{bd} is fading amplitude of the channel between the best relay and the destination. If the decoding set is empty set, the source would transmit new data information. In this case, the relays would not help destination in decoding the source's signal, if it forwards an incorrect decoding results.

4.4.2 A Cooperative Handover Scheme for High-Speed Railway

Normally, the main problems caused by user's high-speed movement in cellular wireless communication system are over-frequent handover, Doppler shift and large penetration loss, among which over-frequent handover needs to be paid special attention as it seriously affects the communication quality of service (QoS) and traffic reliability. Currently only traditional hard handover scheme is supported in LTE, which encounters two challenges under high-speed movement circumstance. On the one hand, the handover delay caused by hard handover is relatively large. The high-speed train passes through the overlapping areas so fast that the handover procedure cannot be accomplished timely. On the other hand, the speed of MRS is so fast that it would miss the optimal handover position, which degrades the handover success probability. In order to overcome the challenges mentioned above, the existing handover scheme of LTE should be optimized to improve the handover success probability in high-speed movement circumstance.

Currently, more and more researches focus on the broadband communication access issues in railway communication. Coordinated multiple point transmission (CoMP) transmission and reception allows geographically separated base stations to

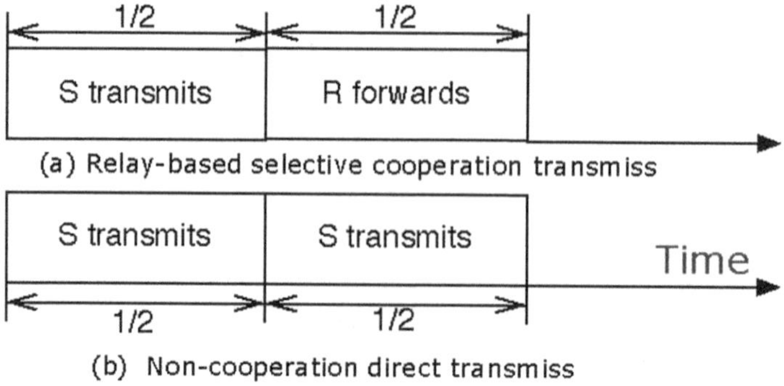

Fig. 4.29 Relay selective cooperation transmission framework: **a** relay-based selective cooperation transmission, and **b** Noncooperation direct transmission

joint sending data to one terminal and joint receiving data from one terminal, by which the intercell interference could be reduced and the system frequency spectral efficiency would be improved. CoMP can be used to solve intercell interference issues, and the CoMP schemes achieve different gains in average sector throughput and 5% edge user throughput gain as compared to that of conventional precoding scheme.

In order to achieve CoMP joint processing and transmission between the two adjacent eNodeBs along the railway track, an interference avoid co-channel deployment approach is proposed in.

As OFDM is used in LTE, the intercell interference is the main source of interference. To avoid the possible large intercell interference when train travels through the overlapping region. A frequency allocation approach for railway scenario: ignoring the reserved dedicated resources, we divided the whole frequency band into two parts which are called F1 and F2 respectively, as shown in Fig. 4.30.

By doing so the proposed co-channel network approach is given in Fig. 4.31: F1 is assigned to the up direction trains (the trains are going to the metropolis) and F2 is assigned to the down direction trains (contrary to up-direction). Compared with the existing GSM-R network whose typical frequency reuse factor is 3, the

Fig. 4.30 The frequency allocation approach

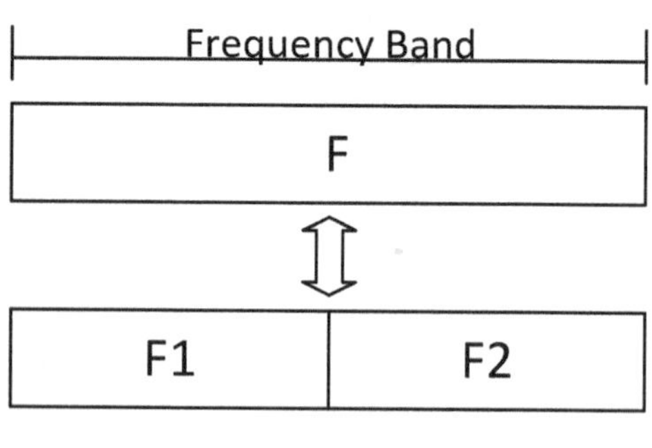

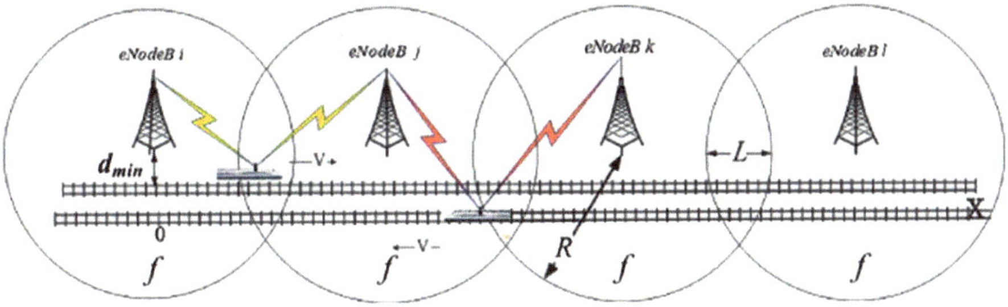

Fig. 4.31 Co-channel network approach for railway communication

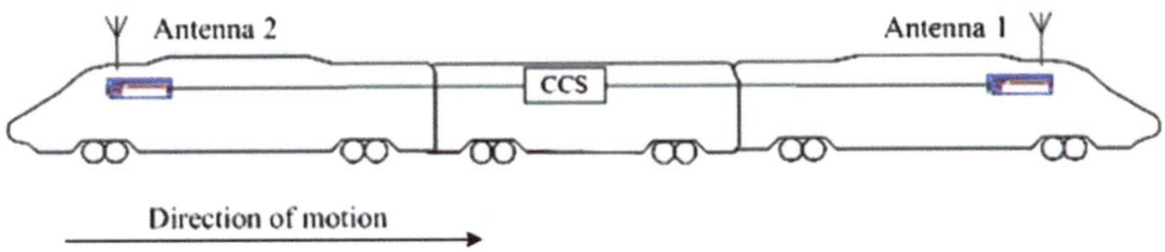

Fig. 4.32 Dual on-vehicle stations solution

proposed co-channel deployment approach can not only maximize the spectrum efficiency but also enable the interactions of adjacent eNodeBs.

For reliability and throughput, the transmission latency is detrimental. In order to eliminate the transmission delay, a dual on-vehicle stations cooperation scheme, which takes full advantage of the distributed antennas transmission and the body length of high-speed train. The mobile relay stations (MRSs) controlled by central control station (CCS) are mounted in the front and the rear of the train. Figure 4.32 shows the schematic diagram of the scheme. Antennas 1 and 2 belong to the front station and the rear station respectively. The uplink data of the users inside the train are gathered to the CCS by pico-base stations deployed on vehicle, and then the front and the rear stations transmit the gathered data to eNodeBs along the track under the control of CCS. Meanwhile, the downlink data received by the two on-vehicle stations form eNodeBs along the track are gathered to the CCS, and then the CCS forwards the collected data to the pico-base stations inside the train. With the above procedure having been done, the communication between users and eNodeBs along the track can be successfully achieved.

This scheme can solve the "processing capacity bottleneck" problem caused by the conventional single on-vehicle station scheme. Moreover, a good diversity gain would be obtained since the distance between Antennas 1 and 2 is far away enough.

The seamless soft handover scheme utilizing CoMP joint processing and transmission technology can significantly improve the handover performance when the train moves through the overlapping areas.

As shown in Fig. 4.33, as the front on-vehicle station enters into the overlapping area, the source eNodeB i activates the cooperative transmission set

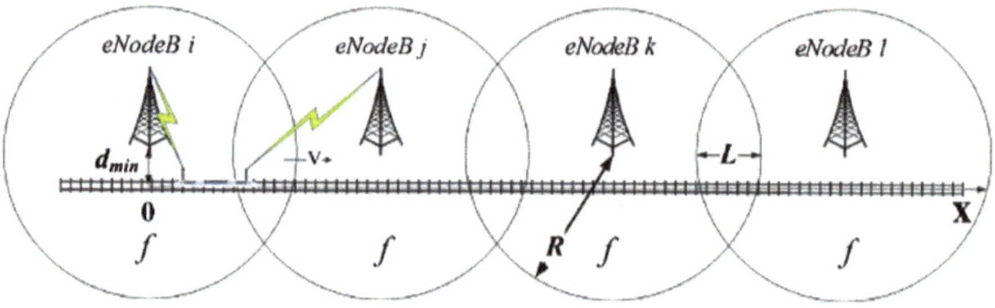

Fig. 4.33 The target eNodeB joins the cooperative set

(CTS) composed of eNodeB *i* and eNodeB *j*. The two or more eNodeBs communicating with MRSs simultaneously are called CTS. The CTS activation is based on the measurement information reported by the moving train and the position information supplied by the communication based train control system (CBTC). Once the CTS is activated, the source eNodeB *i* shares all the user plane data of users inside the train to the target eNodeB *j* by the high-speed backhaul of LTE network. The two adjacent eNodeBs both use the same frequency resource to communicate with the train. Signals from the eNodeBs in the cooperative set are in-phase superposed by precoding, which provides a diversity gain and power gain. It should be noted that CoMP CTS always contains two eNodeBs in the linear coverage topology of high-speed railway.

In high-speed environments, the Doppler effects would lead to irreducible bit error rate (BER) which is called error floor. However, according to technical specifications (TS) of LTE, the procedure of triggering handover contains three phases: the user equipments (UEs) measure the RSSI, RSRP, or RSRQ, sent the measurement reports to source eNodeB, and then the radio resource control (RRC) of source eNodeB decides whether handover is triggered or not. The 3GPP evaluation documents also point out that the handover measurement and radio link failure (RLF) only depend on the RSSI, RSRP, or RSRQ. Though the BER performance would degrade the QoS, if the RSRP remains above a certain threshold for a fixed duration, the wireless link will be reestablished and assured to complete the handover. At most of time, the high-speed train travels through the wide plain and viaduct, the line of sight (LOS) path experienced free-space loss only between MRS and BSs is available and there are few reflectors or scatterers. The major influence on wireless channel caused by relative motion between transmitter and receiver is Doppler shift instead of Doppler spread. Therefore in high-speed railway scenario, instead of considering Doppler effects which degrades BER, we only need to consider Doppler shift which would impair handover performance.

4.4.3 Cognition for High-Speed Railway

To increase quality, reliability, safety, and security of railway systems while increasing accessibility and productivity, modern railway operations rely on increasing traffic between operators' staff workstations, central databases, and also field devices widely distributed both by the trackside and on board the train. There is no single technology or standard that is universal enough to replace all the other ones while being able to support the multitude of usages and needs at the same time. As a consequence, a lot of wireless communication devices operating at different frequencies are widely deployed to address the particular needs for a railway function in a given context. The integration of all these heterogeneous wireless networks is therefore a key technical challenge to improve global efficiency of railway system.

This problem may be solved by introducing cognitive radio in high-speed railway wireless communication system, which is able to meet the needs of future wireless communication devices for control command and operational needs in the railway domain providing a continuous and always best service to the always best connected user anywhere, anytime, anyhow. Unfortunately, the high mobility leads to a difficulty in directly applying the key technologies in cognitive radio to the high-speed railway wireless communication system. The main issues are as follows [93]:

① Time-varying and nonstationary wireless channel
The time-varying and nonstationary wireless channels, resulting from the high mobility of trains, leads to a severe distortion of the transmitted signal, and further causes a degradation to the performance of signal detection and classification. In particular, for the specific emitter identification, since the fingerprint carried by the received signal is extremely subtle, the time-varying and nonstationary channel covers and contaminates the feature of the emitter, which leads to a negative effect on specific emitter identification. Meanwhile, for the spectrum sensing and modulation classification, the time-varying channel makes the channel state information more difficult to obtain, and also additionally adds computational complexity and delay of the algorithms that cope with the nonideal channel conditions.

② Severe Doppler effect
One of the problems brought by the high mobility of the railway is the severe Doppler effect, which leads to adjacent channel interference, and therefore causes negative effect on the performance of the spectrum sensing and signal classification techniques. In cognitive radio systems, spectrum sensing is the fundament of the secondary user to opportunistically access the spectrum hole. The adjacent channel interference will influence the detection accuracy of the presence of the primary user. Moreover, in order to guarantee the safety and security transmission of the control and command information, the supervisory system should classify the legal identity of the transmitted signal and interference by

using the modulation classification and specific emitter identification techniques. The adjacent interference may distort the modulation constellation and contaminate the fingerprint of the emitter, which negatively affects the classification performance.

③ Larger processing delay
The long-term evolution (LTE) mobile communication system is considered to be the natural evolution for current Global System for Mobile Communications Railways (GSM-R). As a wideband communication system, the spectrum sensing algorithms adopted in the high-speed railway wireless communication system should be wideband sensing, which directly increases the sensing delay and cannot satisfy the real-time requirement of the high mobility. Furthermore, the wideband sensing requires advanced signal processing techniques, which may further increase the computational and implementation complexity.

Therefore, the issues that the cognitive radio in high-speed railway should focus on are summarized as follows [92]:

① Spectrum sensing
The spectrum sensing techniques that taking high speed into account as well as the presence of impulsive noises in the context of multiple antennas should be considered. In particular, the decision rule of the detection needs to be designed under these nonideal channel conditions. Since no effective method has been developed so far for fast time-varying channels, a particular effort has been spent to propose an appropriate solution. Moreover, due to the efficient and real-time requirement of the high-speed railway communication system, fast and low complexity algorithms should be considered as well.

② Signal classification
For the modulation classification and specific emitter identification techniques, the most critical task is to propose robust algorithm which can effectively combat the negative effect brought by the Doppler effect and nonideal channel conditions. Particularly, more advanced signal processing theories and tools should be exploited to extract powerful and robust classification features. Furthermore, for the specific emitter identification, due to the potential framework of using the relay to avoid the penetration loss, reliable feature extraction algorithm needs to be designed to combat the negative effect introduced by the relay.
Recently, some papers or research projects present initial insights to apply cognitive radio in high-speed railway communication system. For instance, Ashwin Amanna et al. provide a future wireless communication system based on cognitive radio platform for railway. Since its wireless propagation environment is more complicated, involving characteristics like the constantly varying noise, multiple sources of interference, potential for multiple users accessing for limited spectrum,

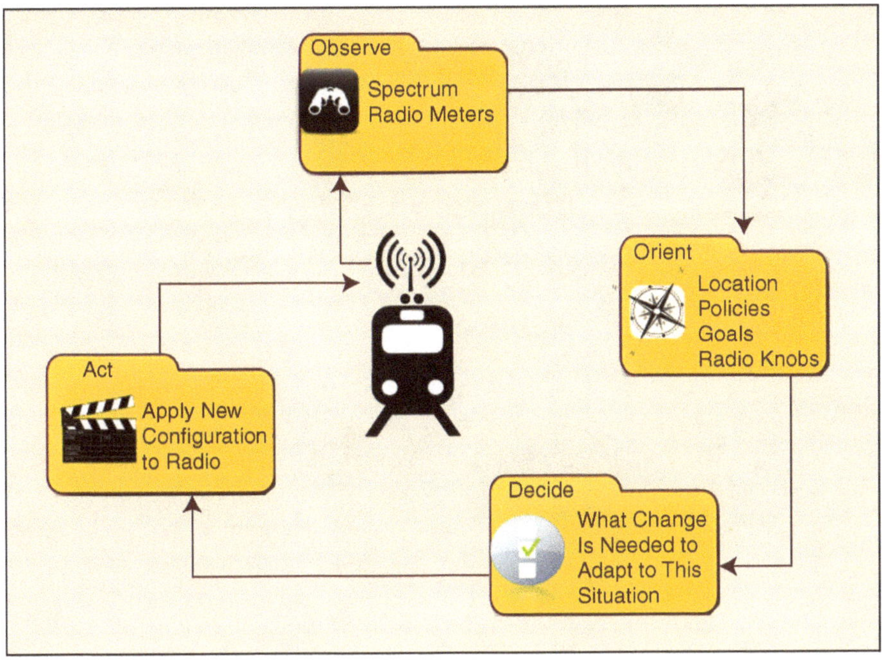

Fig. 4.34 The cognition loop of rail-CR

and unpredictable usage by user, therefore, Ashwin Amanna et al. propose a rail-CR platform, combining artificial intelligence (AI) based decision-making and learning algorithms to solve the need for adaptive communications in high-speed railway scenarios. This platform is built based on a soft-defined radio (SDR), where the cognition loop is illustrated in Fig. 4.34.

Applying cognitive radio in the railway wireless communication system provides some significant benefits, such as the improved link performance by avoiding poor channels, increased data rates on open channels, improved spectrum efficiency by allowing cognitive user to access unused spectrums, improved robustness by providing the capability of reacting to interference, and lower cost of deployment and operation.

The architecture of the rail-CR is presented as well, which contains the learning and decision-making processes. The overall procedure is shown as Fig. 4.35. This process is implemented through exploiting the artificial intelligence, including subsystems such as the case-based reasoner (CBR), optimization algorithms and policy engine. The cognition engine (CE) has a main advantage of recognizing and learning from past mistakes, and then it can improve its own decision-making ability continually.

The CBR is the first decision-making part implemented by the cognition engine, which decides the current situation according to the past history

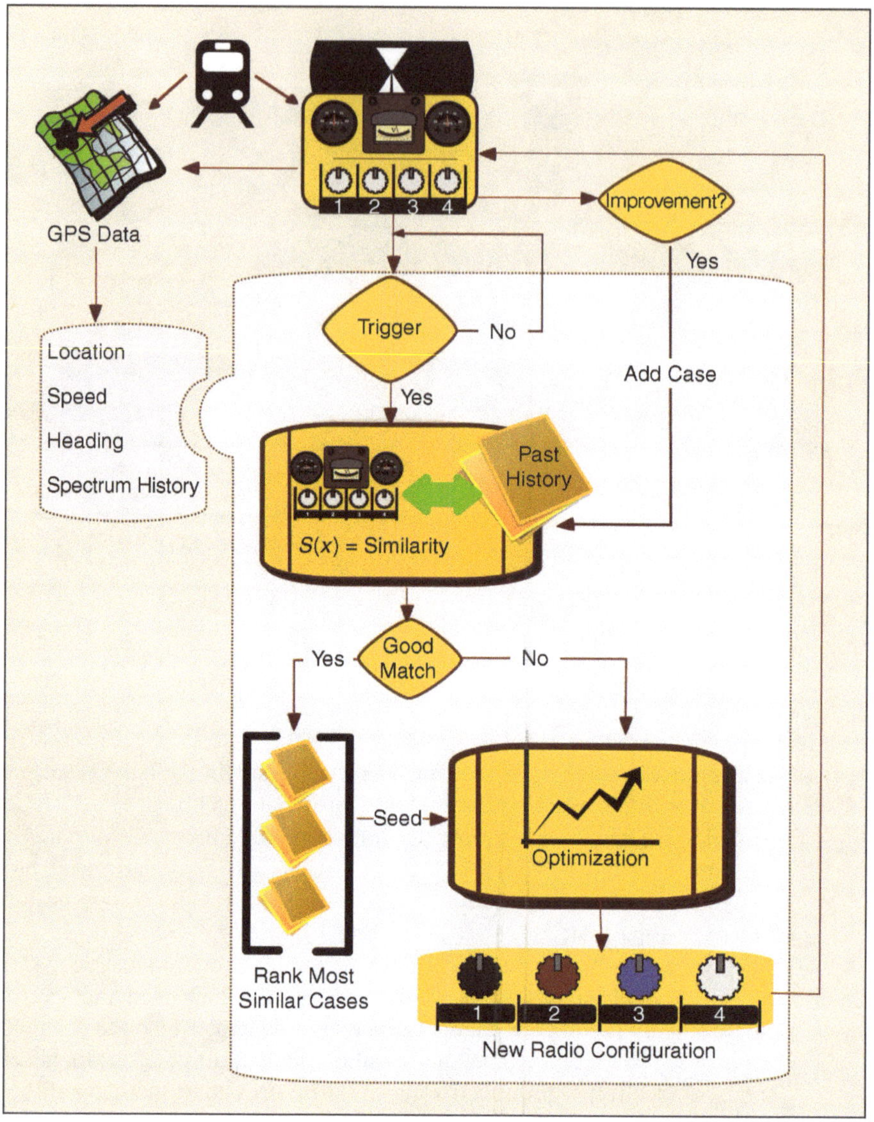

Fig. 4.35 The architecture of rail-CR

of situations, actions, and results. Each case is represented as an entry in a database, containing information about knobs, meters, and GPS position. The CBR employs a predefined function to calculate similarity and choose cases in the history data that both closely match the current situation and had significant improvements in performance as well. If an

improvement in radio performance was attained in this case, the radio knobs are selected to be adjusted in the same manner.
Then, before executing the knob changes, the entire configuration is sent through an optimization algorithm. There is a wide range of optimization algorithms, including genetic, ant colony, particle swarm, simulated annealing, and others. The cognition engine is designed in a modular fashion such that optimization algorithm can be easily switched to another method.
The policy engine is a final check to determine if the selected configuration matches the radio operation policies of the area. Policies on radio parameters (e.g., transmit power) change with region and locality. When a train is running along the border of another state or country, the policies of that region must be adhered to, even if this means a decrease in radio performance.
Furthermore, a French research project, namely CORRIDOR (COgnitive Radio for RaIlway through Dynamic and Opportunistic spectrum Reuse) is proposed recently, which evaluates the cutting-edge opportunistic air interface technologies allowing for robust low-latency links between high-speed trains and ground infrastructure. The purpose of this project is to support the main communication applications in the railway domain, including control command, close circuit television (CCTV), data for maintenance and Internet on board. Emerging techniques are proposed, implemented, and evaluated with real on-site trials, which involves carrier aggregation to exploit TV white spaces, dynamic interference management and opportunistic spectrum access, cross-layer, and handover optimization.

Several potential solutions have been presented for the cognitive radio techniques in high-speed railway as well:

① Spectrum sensing
The spectrum sensing algorithms are mainly classified into two classes: the narrowband sensing and wideband sensing. The narrowband sensing is based on sequentially or randomly detecting the narrowband channels in a wide range of spectrum. Potential solutions are described as follows: A novel predicted eigenvalue threshold-based narrowband blind spectrum sensing algorithm is proposed by Hassan et al., which achieves a good performance when compared with the conventional algorithms [94]. The entire frequency band of interest is processed when employing the wideband sensing. Besides, in the wideband case, K. Hassan proposed a combination of a nonparametric improved cooperative Welch periodogram-based spectral estimator with an optimization algorithm to accurately estimate spectral components [95]. Considering the time-varying wireless channels, caused by the high mobility of trains, affect the temporal covariance properties, a new weighted covariance

value-based spectrum sensing algorithm was proposed in such a case, which exploits the properties of time-varying channel to improve the performance.

Two narrowband algorithms have been proposed in this project: the "SPET" Algorithm ("Simplified Predicted Eigenvalues Threshold") and the "WCV-T" algorithm ("Weighted Covariance Value based method for Time-varying channel") that accounts for "Time-varying channel". Their implementations on the "Open Air Interface" are scheduled for real-time evaluation.

② Signal classification

Signal classification, especially the modulation classification, is an urgent problem in the implementation of the cognitive radio system. Hence, many researchers have proposed some algorithms which design automatic classifiers to classify different waveforms utilized in different wireless networks, i.e., blind automatic discrimination between the modulation formats exploited by the existing wireless communication standards. Furthermore, K. Hassan et al. proposed a blind modulation classification for spatially correlated MIMO systems in the context of CORRIDOR [96]. Besides, blind modulation classifier for MIMO systems in high-speed railway scenarios is presented by S. Kharbech et al. [97].

In sum, the cognitive radio for high-speed railway is an open issue recently. It is in its initial phases with several main challenges to be faced with. The key technologies are under investigation, and the architecture for cognitive radio in the high-speed railway scenario is under discussion as well.

4.5 Summary

The wireless channel is broadcast by nature. Even directional transmission is in fact a kind of broadcast with fewer recipients limited to a certain region. This implies that many nodes or users can "hear" and receive transmissions from a source and can help relay information if needed. The broadcast nature, long considered as a significant waste of energy causing interference to others, is now regarded as a potential resource for possible assistance. For instance, it is well known that the wireless channel is quite bursty, i.e., when a channel is in a severe fading state, it is likely to stay in the state for a while. Therefore, when a source cannot reach its destination due to severe fading, it will not be of much help to keep trying by leveraging repeating transmission protocols such as ARQ. If a third party that receives the information from the source could help via a channel that is independent from the source–destination link, the chances for a successful transmission would be better, thus improving the overall performance.

A wireless network system is traditionally viewed as a set of nodes trying to communicate with each other. However, from another point of view, because of the broadcast nature of wireless channels, one may think of those nodes as a set of antennas distributed in the wireless system. Adopting this point of view, nodes in the network may cooperate together for distributed transmission and processing of information. A cooperating node can act as a relay node for a source node. As such, cooperative communications can generate independent MIMO-like channel links between a source and a destination via the introduction of relay channels.

Indeed, cooperative communications can be thought of as a generalized MIMO concept with different reliabilities in antenna array elements. It is a new paradigm that draws from the ideas of using the broadcast nature of the wireless channels to make communicating nodes help each other, of implementing the communication process in a distribution fashion, and of gaining the same advantages as those found in MIMO systems. Such a new viewpoint has brought various new communication techniques that improve communication capacity, speed, and performance, reduce battery consumption and extend network lifetime, increase the throughput and stability region for multiple access schemes, expand the transmission coverage area, and provide cooperation trade-off beyond source–channel coding for multimedia communications.

The key advantages of using supportive, cooperative, or space–time relays in the system can be summarized as follows:

- Performance Gains. Large system-wide performance gains can be achieved due to path loss gains as well as diversity and multiplexing gains. These translate into decreased transmission powers, higher capacity, or better cell coverage.
- Balanced Quality of Service. While in traditional systems users at the cell edge or in shadowed areas suffered from capacity and/or coverage problems, relaying allows to balance this discrepancy and hence give (almost) equal quality of service (QoS) to all users.
- Infrastructure-Less Deployment. The use of relays allows the rollout of a system that has minimal or no infrastructure available prior to deployment. For instance, in disaster-struck areas, relaying can be used to facilitate communications even though the cellular system is nonfunctioning. For hybrid deployments, that is a cellular system coupled with relays.
- Reduced Costs. Compared to a purely cellular approach to provide a given level of QoS to all users in the cell, relaying is a more cost effective solution. In [66], however, it has also been shown that while savings are not as dramatic as hoped for, the capital and operational expenditures are generally lower when relays are used.

Some major disadvantages of using supportive, cooperative, or space–time relays in the system are given below:

- Complex Schedulers. While maintaining a single cooperative relaying link is a fairly trivial task, at system level with many users and relays this quickly becomes an arduous task. As such, relaying requires more sophisticated

schedulers since not only traffic of different users and applications needs to be scheduled but also the relayed data flows. Any gains due to cooperation at the physical layer dissipate rapidly if not handled properly at medium access and network layers.
- Increased Overhead. A full system functioning requires handovers, synchronization, extra security, etc. This clearly induces an increased overhead w.r.t to a system that does not use relaying.
- Partner Choice. To determine the optimum relaying and cooperative partner(s) is a fairly intricate task. Also, the complexity of maintaining such cooperative partnership is higher w.r.t. noncooperative relaying.
- Increased Interference. If the offered power savings are not used to decrease the transmission power of the relay nodes but rather to boost capacity or coverage, then relaying will certainly generate extra intra- and intercell interference, which potentially causes the system performance to deteriorate. An optimum trade-off needs, therefore, to be found at system level.
- Extra Relay Traffic. The relayed traffic is, from a system throughput point of view, redundant traffic and hence decreases the effective system throughput since in most cases resources in the form of extra frequency channels or time slots need to be provided.
- Increased End-to-End Latency. Relaying typically involves the reception and decoding of the entire data packet before it can be retransmitted. If delay-sensitive services are being supported, such as voice or the increasingly popular multimedia web services, then the latency induced by the decoding may become detrimental. Latency increases with the number of relays and also with the use of interleavers, such as utilized in GSM voice traffic. To circumvent this latency, either simple transparent relaying or novel decoding methods need to be used.
- Tight Synchronization. A tight synchronization needs generally be maintained to facilitate cooperation. This in turn requires expensive hardware and potentially large protocol overheads since nodes need to synchronize regularly by using some form of beaconing or other viable techniques.
- More Channel Estimates. The use of relays effectively increases the number of wireless channels. This requires the estimation of more channel coefficients and hence more pilot symbols need to be provided if coherent modulation was to be used.

From this list of advantages and disadvantages, it is obvious that many system design parameters can be traded against one another. These trade-offs are also visualized in Fig. 4.36. Some important system trade-offs are discussed below:

- Coverage versus Capacity. As already discussed in some detail, cooperative systems allow coverage to be traded against capacity or, equivalently, outage versus rate, or diversity versus multiplexing gains. Therefore, the system designer has the choice to let a relay help boost capacity or increase the coverage range. Increasing one inherently diminishes the other.

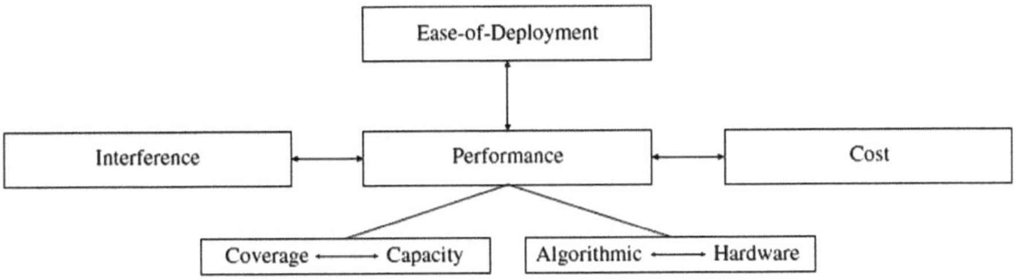

Fig. 4.36 At a given performance level, coverage can be traded capacity and algorithmic with hardware complexity. Performance can also be traded against amount of interference, ease-of-deployment and cost

- Algorithmic versus Hardware Complexity. Solving the coverage and capacity problem by means of more cellular base stations requires more complex and hence costly hardware, not to mention the expensive real estate to physically place the base stations. Relays, on the other hand, are of relatively low hardware complexity. However, the decrease in hardware complexity by using relays yields an increase in algorithmic complexity since scheduling, synchronization, handover, and other algorithms become significantly more complex. An optimum solution trading algorithmic with hardware complexity hence needs to be determined, likely leading to a hybrid deployment.
- Interference versus Performance, cooperative communications yields gains which can either be used to decrease the transmission power and hence generated interference or increase capacity/coverage. Furthermore, relaying generates extra traffic, which is an additional source of interference.
- Ease of Deployment versus Performance. Relays can be deployed in a planned and unplanned manner. In the former, the network designer optimizes the placement and parameterization of the static relay node; this is a complex task but leads to superior performance. In the latter, relays are deployed in an unplanned manner and hence can be stationary or mobile; deployment is therefore significantly simplified at the cost of decreased performance w.r.t. the planned roll-out.
- Cost versus Performance. Being a traditional trade-off, the cost of the chosen cooperative solution has a profound impact on its performance. Clearly, deploying highly complex relaying nodes that allow, say, for cooperative space–time relaying induces high costs but also improved performance.

Cooperative diversity can be used in various fields like cooperative sensing in cognitive radio, wireless ad hoc networks, wireless sensor networks, and many more. Different applications of cooperative diversity are described in the subsequent sections.

4.5.1 Cooperative Diversity in Wireless Sensor Networks

Wireless sensor networks (WSNs) are a broad class of wireless networks consisting of small, inexpensive, and energy limited devices. Due to the fact that nodes are battery powered, energy efficiency is one of the main challenges in designing wireless sensor networks. Schemes have been developed recently for energy saving of the protocol stack in specific layers. For example, multi-hop routing and clustering improve the energy efficiency of large scale WSNs. As nodes can communicate directly over small distances and have limited transmission range multi-hop routing is necessary. However, it is restricted to networks of extremely high densities. Clustering is a method of partitioning the network into local clusters, and each cluster has anode called cluster head (CH). Energy saving protocols have also been developed in the physical layer. Like all other wireless networks, wireless sensor networks suffer from the effects of fading. Cooperative diversity is a technique used to mitigate the impact of fading. This form of diversity is especially suited toward WSNs since size and power constraints restrict nodes from possessing more than one antenna. Cooperation is achieved using the simple amplify-and-forward scheme. These results can be used to predict the impact of cooperative diversity on the lifetime of sensor networks. Here different design aspects of cooperative diversity used in wireless sensor network are discussed.

The network is clustered using a distributed algorithm where CHs are selected randomly. These classes of algorithms are practical to implement in WSNs since WSNs are organized in a distributed fashion. The role of CH is evenly distributed over the network and each CH performs ideal aggregation, i.e., all cluster data is aggregated into a single packet.

4.5.2 Cooperative Diversity in Cognitive Radio

In software defined radio (SDR), the software embedded in a radio cell phone defines the parameters under which the phone should operate in real time as its user moves from one place to another. Cognitive radio (CR) is a smarter technology. Cognitive radio is a radio that is meant to be aware, sense and learn from its environment and to serve best to its user. The cognitive users are required to detect the presence of licensed (primary) users in a very short time and must vacate the band for use by primary users. Thus the main challenge in this technology is how to detect the presence of primary users.

Hence diversity gain is achieved by allowing the users to cooperate. Cooperative schemes in a TDMA system with orthogonal transmission have been recently proposed. Here different design aspects of cooperative diversity used in cognitive radio are discussed.

While many key results for cooperative communication have already been obtained, there are many more issues that remain to be addressed. An important question is how partners are assigned and managed in multiuser networks.

In other words, how is it determined which users cooperate with each other, and how often are partners reassigned? Systems such as cellular, in which the users communicate with a central base station, offer the possibility of a centralized mechanism. Assuming that the base station has some knowledge of the all the channels between users, partners could be assigned to optimize a given performance criterion, such as the average block error rate for all users in the network. In contrast, systems such as ad hoc networks and sensor networks typically do not have any centralized control. Such systems therefore require a distributed cooperative protocol, in which users are able to independently decide with whom to cooperate at any given time. A related issue is the extension of the proposed cooperative methods to allow a user to have multiple partners. The challenge here is to develop a scheme that treats all users fairly, does not require significant additional system resources, and can be implemented feasibly in conjunction with the system's multiple access protocol. Laneman and Wornell have done some initial work related to distributed partner assignment and multiple partners, and additional work by others is ongoing.

Another important issue is the development of power control mechanisms for cooperative transmission. Work thus far generally assumes that the users transmit with equal power. It maybe possible to improve performance even further by varying transmit power for each user based on the instantaneous uplink and inter-user channel conditions. Furthermore, power control is critical in CDMA-based systems to manage the near–far effect and minimize interference. Therefore, power control schemes that work effectively in the context of cooperative communications have great practical importance.

For the coded cooperation method, a natural issue is the possibility of designing a better coding scheme. Examples are given using RCPC codes. Turbo codes are applied to the coded cooperation framework. Both of these coding schemes were originally developed for noncooperative systems. An interesting open problem is the development of design criteria specifically for codes that optimize the performance of coded cooperation.

Among other interesting contributions to cooperative communication are space–time cooperative signaling, as well as new work on the relay channel, including interesting adaptive scenarios. There are also many other interesting developments.

Cognitive radio is an exciting emerging technology that has the potential of dealing with the stringent requirement and scarcity of the radio spectrum. Such revolutionary and transforming technology represents a paradigm shift in the design of wireless systems, as it will allow the agile and efficient utilization of the radio spectrum by offering distributed terminals or radio cells the ability of radio sensing, self-adaptation, and dynamic spectrum sharing. Cooperative communications and networking is another new communication technology paradigm that allows distributed terminals in a wireless network to collaborate through some distributed transmission or signal processing so as to realize a new form of space diversity to combat the detrimental effects of fading channels.

Spectrum utilization can be improved significantly by allowing a secondary user to utilize a licensed band when the primary user (PU) is absent. Cognitive radio (CR), as an agile radio technology, has been proposed to promote the efficient use of the spectrum. By sensing and adapting to the environment, a CR is able to fill in spectrum holes and serve its users without causing harmful interference to the licensed user. To do so, the CR must continuously sense the spectrum it is using in order to detect the reappearance of the PU. Once the PU is detected, the CR should withdraw from the spectrum so as to minimize the interference it may possibly cause. This is a very difficult task, as the various PUs will be employing different modulation schemes, data rates, and transmission powers in the presence of variable propagation environments and interference generated by other secondary users. Another great challenge of implementing spectrum sensing is the hidden terminal problem, which occurs when the CR is shadowed, in severe multipath fading or inside buildings with a high penetration loss while a PU is operating in the vicinity.

Cooperative communications is an emerging and powerful solution that can overcome the limitation of wireless systems. The basic idea behind cooperative transmission rests on the observation that, in a wireless environment, the signal transmitted or broadcast by a source to a destination node, each employing a single antenna, is also received by other terminals, which are often referred to as relays or partners. The relays process and retransmit the signals they receive. The destination then combines the signals coming from the source and the partners, thereby creating spatial diversity by taking advantage of the multiple receptions of the same data at the various terminals and transmission paths. In addition, the interference among terminals can be dramatically suppressed by distributed spatial processing technology. By allowing multiple CRs to cooperate in spectrum sensing, the hidden terminal problem can be addressed. Indeed, cooperative spectrum sensing in CR networks has an analogy to a distributed decision in wireless sensor networks, where each sensor makes a local decision and those decision results are reported to a fusion center to give a final decision according to some fusion rule. The main difference between these two applications lies in the wireless environment. Compared to wireless sensor networks, CRs and the fusion center (or common receiver) are distributed over a larger geographic area. This difference brings out a much more challenging problem to cooperative spectrum sensing because sensing channels (from the PU to CRs) and reporting channels (from the CRs to the fusion center or common receiver) are normally subject to fading or heavy shadowing.

With fast and agile sensing ability, CR can opportunistically fill in spectrum holes to improve the spectrum occupancy utilization. However, once the PU returns to access the licensed band, the CR should immediately stop operating in the PU licensed band. This fast switching off of the CR can guarantee minimum interference to the primary system. However, from the point of view of the cognitive system, the interruptive transmissions will lead to a discontinuous data service and intolerable delay. To cope with this problem, a cognitive relay network in which distributed cognitive users collaborate with each other so that they can share their distinct spectrum bands. By utilizing a cognitive space–time–frequency (STF) coding in the cognitive relay network, seamless data transmission within the cognitive system can also be realized.

As the demand for additional bandwidth continues to increase, spectrum policy makers and communication technologists are seeking solutions for the apparent spectrum scarcity. Meanwhile, measurement studies have shown that the licensed spectrum is relatively unused across many time and frequency slots. To solve the problem of spectrum scarcity and spectrum underutilization, the use of CR technology is being considered because of its ability to rapidly and autonomously adapt operating parameters to changing requirements and conditions. Recently, the FCC has issued a Notice of Proposed Rulemaking regarding CR that requires rethinking of the wireless communication architectures so that emerging radios can share spectrum with PUs without causing harmful interference to them.

Cognitive radio is an intelligent wireless communication system that is aware of its surrounding environment (i.e., its outside world), and uses the methodology of understanding-by-building to learn from the environment and adapt its internal states to statistical variations in the incoming radio frequency (RF) stimuli by making corresponding changes in certain operating parameters (e.g., transmit power, carrier frequency, and modulation strategy) in real time, with two primary objectives in mind: (1) Highly reliable communications whenever and wherever needed; and (2) Efficient utilization of the radio spectrum. More specifically, the CR technology will enable the users to:

- Determine which portions of the spectrum are available and detect the presence of licensed users when a user operates in a licensed band (spectrum sensing);
- Select the best available channel (spectrum management);
- Coordinate access to this channel with other users (spectrum sharing);
- Vacate the channel when a licensed user is detected (spectrum mobility).

IEEE has also endeavored to formulate a novel wireless air interface standard based on CR. The IEEE 802.22 working group aims to develop wireless regional area network physical (PHY) and medium access control (MAC) layers for use by unlicensed devices in the spectrum allocated to TV bands.

Traditional wireless networks have predominantly used direct point-to-point or point-to-multipoint (e.g., cellular) topologies. In contrast to conventional point-to-point communications, cooperative communications and networking allow different users or nodes in a wireless network to share resources and to create collaboration through distributed transmission/processing, in which each user's information is sent out not only by the user but also by the collaborating users. Cooperative communications and networking is a new communication paradigm that promises significant capacity and multiplexing gain increase in wireless networks. It also realizes a new form of space diversity to combat the detrimental effects of severe fading.

There are mainly three relaying protocols: amplify and forward (AF), decode and forward (DF), and compress and forward (CF). In AF, the received signal is amplified and retransmitted to the destination. The advantage of this protocol is its simplicity and low cost implementation. But the noise is also amplified at the relay. In DF, the relay attempts to decode the received signals. If successful, it re-encodes the information and retransmits it. Lastly, CF attempts to generate an estimate of the received signal. This is then compressed, encoded, and transmitted in the hope that the estimated value may assist in decoding the original codeword at the destination.

Cooperative techniques have already been considered for wireless and mobile broadband radio and also have been under investigation in various IEEE 802 standards. The IEEE 802.11 standard is concerned with wireless local area networks (WLANs) in unlicensed bands in indoor environments. A recent evolution of IEEE 802.11 using mesh networking, i.e., 802.11s is considering the update of 802.11 MAC layer operation to self-configuration and multi-hop topologies. The mesh point that has the ability to function as the 802.11 access point collects the information about the neighboring mesh points, communicating with them and forwarding the traffic. The IEEE 802.16 standard is an orthogonal frequency-division multiplexing (OFDM), orthogonal frequency-division multiple access (OFDMA), and single-carrier-based fixed wireless metropolitan area network in licensed bands of 10–66 Hz. As an amendment of 802.16 networks, IEEE 802.16j is concerned with multi-hop relay to enhance coverage, throughput, and system capacity.

Wireless communications technologies have seen a remarkably fast evolution in the past two decades. Each new generation of wireless devices has brought notable improvements in terms of communication reliability, data rates, device sizes, battery life, and network connectivity. In addition, the increase homogenization of traffic transports using Internet Protocols is translating into network topologies that are less and less centralized. In recent years, ad hoc and sensor networks have emerged with many new applications, where a source has to rely on the assistance from other nodes to forward or relay information to a desired destination.

Such a need of cooperation among nodes or users has inspired new thinking and ideas for the design of communications and networking systems by asking whether cooperation can be used to improve system performance. As a result, a new communication paradigm arose, which had an impact far beyond its original applications to ad hoc and sensor networks.

Cognitive radio is a novel technology that can potentially improve the utilization efficiency of the radio spectrum. Cooperative communications can play a key role in the development of CR networks.

4.5.3 Summary of Cognitive Radio

In contrast to the traditional communication systems, cognitive radio is a noncooperative system, where the (secondary users') receivers in cognitive radio are not capable of obtaining full and perfect prior knowledge of the (primary users') transmitters, such as the transmit power, carrier frequency, and modulation format.

Furthermore, the cognitive radio allowing unlicensed secondary users to opportunistically access the licensed spectrum, which may present a major risk to system security. In such cases, signal classification techniques play a key role in signal processing at the receiver and in avoiding malicious attack.

The bottleneck of the high-speed railway wireless communication systems is the scarcity of the available frequency band. Cognitive radio can reuse the licensed spectrum by allowing secondary users to communicate over an unused spectrum in specific time and location, which provides a realistic and feasible option to solve the spectrum scarcity problem in high-speed railway wireless communication. However, the high mobility leads to a difficulty in directly applying the key technologies in cognitive radio to the high-speed railway wireless communication system. The main issues including the time-varying channels and Doppler effect should be considered in the cognition techniques for high-speed railway.

References

1. Radiocommunication Sector for International Telecommunication Union (2011) The feasibility of sharing sub-bands between oceanographic radars operating in the radiolocation service and fixed and mobile services within the frequency band 3–50 MHz, ITU-R, Geneva, Switzerland, Rep. M.2234, Nov 2011
2. Universal Mobile Telecommunications System Forum (2011) Mobile traffic forecasts 2010-2020 report, Rep. 44, Jan 2011
3. Radiocommunication Sector for International Telecommunication Union (2013) Future spectrum requirements estimate for terrestrial IMT, ITU-R, Geneva, Switzerland, Rep. ITU-R M.2290-0, Dec 2013
4. Shared Spectrum Company (2010) General survey of radio frequency bands—30 MHz to 3 GHz, Rep. no. 20100323, Sep 2010
5. Federal Communications Commission (2002) Spectrum policy task force, Rep. ET Docket no. 02-135, Nov 2002
6. Haykin S (2005) Cognitive radio: Brain-empowered wireless communications. IEEE J Sel Areas Commun 23(2):201–220
7. Urkowitz H (1967) Energy detection of unknown deterministic signals. In: Proceedings of the IEEE 55:523–531
8. Shankar S, Cordeiro C, Challapali K (2005) Spectrum agile radios: utilization and sensing architectures. In: Proceedings of the IEEE international symposium on new frontiers in dynamic spectrum access networks, Baltimore, Maryland, USA, Nov 2005, pp 160–169
9. Yuan Y et al (2007) KNOWS: Cognitive radio networks over white spaces. In: Proceedings of the IEEE international symposium on new frontiers in dynamic spectrum access networks, Dublin, Ireland, Apr 2007, pp 416–427
10. Ganesan G, Li Y (2005) Agility improvement through cooperative diversity in cognitive radio. In: Proceedings of the Global Conference (GLOBECOM), St. Louis, Missouri, USA, Nov/Dec 2005, 5:2505–2509
11. Cabric D, Tkachenko A, Brodersen R (2006) Spectrum sensing measurements of pilot, energy, and collaborative detection. In: Proceedings of the IEEE international conference on communications (ICC), Washington, D.C., USA, Oct 2006, pp 1–7
12. Digham F, Alouini M, Simon M (2003) On the energy detection of unknown signals over fading channels. In: Proceedings of the IEEE international conference on communications (ICC), Seattle, Washington, USA, May 2003, 5:3575–3579

13. Papadimitratos P, Sankaranarayanan S, Mishra A (2005) A bandwidth sharing approach to improve licensed spectrum utilization. IEEE Commun Mag 43(12):10–14
14. Tang H (2005) Some physical layer issues of wide-band cognitive radio systems. In: Proceedings of the IEEE international symposium on new frontiers in dynamic spectrum access networks, Baltimore, Maryland, USA, Nov 2005, pp 151–159
15. Khambekar N, Dong L, Chaudhary V (2007) Utilizing OFDM guard interval for spectrum sensing. In: Proceedings of the IEEE wireless communication and networking conference on Hong Kong, Mar 2007, pp 38–42
16. Ghozzi M, Marx F, Dohler M, Palicot J (2006) Cyclostationarity based test for detection of vacant frequency bands. In: Proceedings of the IEEE international conference on Cognitive Radio Oriented Wireless Networks and Communication (CROWNCOM), Mykonos Island, Greece, June 2006
17. Lunden J, Koivunen V, Huttunen A, Poor HV (2007) Spectrum sensing in cognitive radios based on multiple cyclic frequencies. In: Proceedings of the IEEE international conference on Cognitive Radio Oriented Wireless Networks and Communication (CROWNCOM), Orlando, Florida, USA, July/Aug 2007
18. Oner M, Jondral F (2004) Cyclostationarity based air interface recognition for software radio systems. In: Proceedings of the IEEE radio and wireless conference on Atlanta, Georgia, USA, Sept 2004, pp 263–266
19. Proakis JG (2001) Digital Communications, 4th ed. McGraw-Hill, 2001
20. Tandra R, Sahai A (2005) Fundamental limits on detection in low SNR under noise uncertainty. In: Proceedings of the IEEE international conference on wireless networks, Communication and Mobile Computing, Maui, HI, June 2005, 1:464–469
21. Cabric D, Mishra S, Brodersen R (2004) Implementation issues in spectrum sensing for cognitive radios. asilomar conference on signals, systems and computers, Pacific Grove, California, USA, Nov 2004, 1:772–776
22. Sahai A, Tandra R, Mishra SM, Hoven N (2006) Fundamental design tradeoffs in cognitive radio systems. In: Proceedings of international workshop on technology and policy for accessing spectrum, Aug 2006
23. Yucek T, Arslan H (2006) Spectrum characterization for opportunistic cognitive radio systems. In: Proceedings of the IEEE military communication conference, Washington, D.C., USA, Oct 2006, pp 1–6
24. Vardoulias G, Faroughi-Esfahani J, Clemo G, Haines R (2001) Blind radio access technologydiscovery and monitoring for software defined radio communication systems: problems andtechniques. In: Proceedings of international conference on 3G mobile communication technologies, London, UK, Mar2001, pp 306–310
25. Mehta M, Drew N, Vardoulias G, Greco N, Niedermeier C (2001) Reconfigurable terminals: anoverview of architectural solutions. IEEE Commun Mag 39(8):82–89
26. Akyildiz I et al (2006) Next generation/dynamic spectrum access/cognitive radio wireless networks:A survey, computer networks journal (Elsevier), Sep 2006
27. Dobre O et al (2007) Survey of automatic modulation classification techniques: classical approachesand new trends. IEET Commun 1(2):137–156
28. Huang C, Polydoros A (1995) Likelihood methods for MPSK modulation classification. IEEETrans Commun 43(2/3/4):1493–1504
29. Panagiotou P, Anastasopoulos A, Polydoros A (2000) Likelihood ratio tests for modulation-classification. In: Proceedings of the IEEE MILCOM 2:670–674
30. Wei W, Mendel J (2000) Maximum-likelihood classification for digital amplitude-phasemodulations. IEEE Trans Commun 48(2):189–193
31. Yucek T, Arslan H (2004) A novel sub-optimum maximum-likelihood modulation classificationalgorithm for adaptive OFDM systems. In: Proceedings of the IEEE WCNC 2:739–744
32. Swami A, Sadler B (2000) Hierarchical digital modulation classification using cumulants. IEEETrans Commun 48(3):416–429

33. Wu H, Saquib M, Yun Z (2008) Novel automatic modulation classification using cumulantfeatures for communications via multipath channels. IEEE Trans Wireless Commun 7(8):3098–3105
34. Dobre O, Bar-Ness Y, Su W (2003) Higher-order cyclic cumulants for high order modulationclassification. In: Proceedings of the IEEE MILCOM 1:112–117
35. Dobre O, Bar-Ness Y, Su W (2004) Robust QAM modulation classification algorithm using cyclic cumulants. In: Proceedings of the IEEE wireless communications and networking conference, 2:745–748
36. Soliman S, Hsue S (1992) Signal classification using statistical moments. IEEE Trans Commun 40(5):908–916
37. Sills J (1999) Maximum-likelihood modulation classification for PSK/QAM. In: Proceedings of the IEEE MILCOM, pp 57–61
38. Kim K, Polydoros A (1988) Digital modulation classification: the BPSK versus QPSK case. In: Proceedings of the IEEE MILCOM, pp 431–436
39. Polydoros A, Kim K (1990) On the detection and classification of quadrature digital modulations in broad-band noise. IEEE Trans Commun 38(8):1199–1211
40. Sapiano P, Martin J (1996) Maximum likelihood PSK classifier. In: Proceedings of the ICASSP, pp 1010–1014
41. Long C, Chugg K, Polydoros A (1994) Further results in likelihood classification of QAM signals. In: Proceedings of the IEEE MILCOM, pp 57–61
42. Hong L, Ho K (2003) Classification of BPSK and QPSK signals with unknown signal level using the Bayes technique. In: Proceedings of the IEEE ISCAS, IV.1–IV.4
43. Beidas B, Weber C (1995) Higher-order correlation-based approach to modulation classification of digitally frequency-modulated signals. IEEE J Sel Areas Commun 13 (1):89–101
44. Beidas B, Weber C (1998) Asynchronous classification of MFSK signals using the higher order correlation domain. IEEE Trans Commun 46(4):480–493
45. Beidas B, Weber C (1996) Higher-order correlation-based classification of asynchronous MFSK signal. In: Proceedings of the IEEE MILCOM, pp 1003–1009
46. Chugg K, Long C, Polydoros A (1995) Combined likelihood power estimation and multiple hypothesis modulation classification. In: Proceedings of the ASILOMAR, pp 1137–1141
47. Hong L, Ho K (2000) BPSK and QPSK modulation classification with unknown signal level. In: Proceedings of the IEEE MILCOM, pp 976–980
48. Hong L, Ho K (2001) Modulation classification of BPSK and QPSK signals using a two element antenna array receiver. In: Proceedings of the IEEE MILCOM, pp 118–122
49. Hong L, Ho K (2002) An antenna array likelihood modulation classifier for BPSK and QPSK signals. In: Proceedings of the IEEE MILCOM, pp 647–651
50. Dobre O et al (2004) On the classification of linearly modulated signals in fading channel. In: Proceedings of the CISS Conference, Princeton, Mar 2004
51. Abdi A et al (2004) Modulation classification in fading channels using antenna arrays. In: Proceedings of the IEEE MILCOM, 2004
52. Azzouz E, Nandi A (1996) Automatic modulation recognition of communication signals, Kluwer Academic, 1996
53. Azzouz E, Nandi A (1996) Procedure for automation recognition of analogue and digital modulations. IEE Proc Commun 143(6):259–266
54. Hsue S, Soliman S (1989) Automatic modulation recognition of digitally modulated signals. In: Proceedings of the IEEE MILCOM, pp 645–649
55. Hsue S, Soliman S (1990) Automatic modulation classification using zero crossing. IEE Proc F, Radar Signal Process, 137(6):459–464
56. Yang Y, Liu C (1998) An asymptotic optimal algorithm for modulation classification. IEEE Commun Lett 2(5):117–119
57. Yang Y, Soliman S (1991) Optimum classifier for M-ary PSK signals. In: Proceedings of International Conference on Communications (ICC), pp 1693–1697

58. Yang Y, Soliman S (1997) A suboptimal algorithm for modulation classification. IEEE Trans Aerosp Electron Syst 33(1):38–45
59. Yang Y, Soliman S (1991) Statistical moments based classifier for MPSK signals. In: Proceedings of the Global Conference (GLOBECOM), pp 72–76
60. Yang Y, Soliman S (1995) An improved moment-based algorithm for signal classification. Signal Process, 43(3):231–244
61. Sapiano P, Martin J, Holbeche R (1995) Classification of PSK signals using the DFT of phase histogram. In: Proceedings of ICASSP, pp 1868–1871
62. Ho K, Prokopiw W, Chan Y (1995) Modulation identification by the wavelet transform. In: Proceedings of the IEEE MILCOM, pp 886–890
63. Ho K, Prokopiw W, Chan Y (2000) Modulation identification of digital signals by the wavelet transform. IEE Proc Radar Sonar Navig 47(4):169–176
64. Hong L, Ho K (1999) Identification of digital modulation types using the wavelet transform. In: Proceedings of the IEEE MILCOM, pp 427–431
65. Swami A, Barbarossa S, Sadler B (2000) Blind source separation and signal classification. In: Proceedings of the ASILOMAR, pp 1187–1191
66. Martret C, Boiteau D (1997) Modulation classification by means of different order statistical moments. In: Proceedings of the IEEE MILCOM, pp 1387–1391
67. Marchand P, Lacoume J, Le Martret C (1997) Classification of linear modulations by a combination of different orders cyclic cumulants. In: Proceedings of Workshop on HOS, pp 47–51
68. Marchand P, Lacoume J, Le Martret C (1998) Multiple hypothesis classification based on cyclic cumulants of different orders. In: Proceedings of ICASSP, pp 2157–2160
69. Spooner C, Brown W, Yeung G (2000) Automatic radio-frequency environment analysis. In: Proceedings of ASILOMAR, pp. 1181–1186
70. Spooner C, Brown W, Yeung G (2000) Automatic radio-frequency environment analysis. In: Proceedings of ASILOMAR, pp 1181–1186
71. Spooner C (2001) On the utility of sixth-order cyclic cumulants for RF signal classification. In: Proceedings of ASILOMAR, pp 890–897
72. Dobre O, Bar-Ness Y, Su W (2004) Robust QAM modulation classification algorithm based on cyclic cumulants. In: Proceedings of WCNC, pp 745–748
73. Dobre O et al (2005) Selection combining for modulation recognition in fading channels. In: Proceedings of IEEE MILCOM, pp 2499–2505
74. Yu Z, Shi Y, Su W (2003) M-ary frequency shift keying signal classification based on discrete Fourier transform. In: Proceedings of IEEE MILCOM pp 1167–1172
75. Wang F, Wang X (2010) Fast and robust modulation classification via Kolmogorov-Smirnov test. IEEE Trans Commun 58(8):2324–2332
76. Urriza P et al (2011) Computationally e_cient modulation level classification based on probability distribution distance functions. IEEE Commun Lett 15(5):476–478
77. Wang F, Chan C (2012) Variational-distance-based modulation classifier. In: Proceedings of the IEEE international conference, pp 5635–5639
78. Talbot KI, Duley PR, Hyatt MH (2003) Specific emitter identification and verification. Technol Rev J pp 113–133
79. Xu S (2007) On the identification technique of individual transmitter based on signalprints, Ph.D. dissertation, Huazhong University of Science and Technology, Wuhan, China
80. Choe H, Poole C, Yu A (1995) Novel identification of intercepted signals from unknown radio transmitters. In: Proceedings of the SPIE, pp 504–516
81. Shaw D, Kinsner W (1997) Multifractal modeling of radio transmitter transients for classification. in WESCANEX 97: communications, power and computing. conference proceedings, IEEE, pp 306–312
82. Hall J, Barbeau M, Kranakis E (2006) Detecting rogue devices in bluetooth networks using radio frequency fingerprinting. In:Proceedings of the IASTEDinternational conference on communications and computer networks, pp 1–6

83. Ureten O, Serinken N (1999) Bayesian detection of radio transmitter turn-on transients. In: Proceedings of the IEEE EURASIP workshop on NSIP, pp 830–834
84. Ureten O, Serinken N (1999) Detection of radio transmitter turn-on transients. Electronic Lett 35(23):1996–1997
85. Ureten O, Serinken N (2005) Bayesian detection of Wi-Fi transmitter RF fingerprinting. Electronics Lett 41(6):73–374
86. Lopez-Risueno G, Grajal J, Sanz-Osorio A (2005) Digital channelized receiver based on time-frequency analusis for signal interception. IEEE Trans Aerosp Electron Syst 41(33):879–898
87. Lunden J, Koivunen V (2007) Automatic radar waveform recognition. IEEE J Sel Topics Signal Process 1(1):124–136
88. Gillespie B, Atlas L (2001) Optimizing time-frequency kernels for classification. IEEE Trans Signal Process 49(3):485–496
89. Huang NE et al (1998) The empirical mode decomposition and the Hilbert spectrum for nonlinear and non-stationary time series analysis. Proc R Soc London Ser A 454(1971):903–995
90. Chandran V, Elgar S (1993) Pattern recognition using invariants defined from higher order spectra to one-dimensional inputs. IEEE Trans Signal Process 41(1):205–212
91. Tugnait J (1994) Detection of non-gaussian signals using integrated polyspectrum. IEEE Trans Signal Process 42(12):3137–3149
92. Liao X, Bao Z (1998) Circularly integrated bispectra: novel shift invariant features for high resolution radar target recognition. Electronics Lett 34(19):1879–1880
93. Amanna A et al (2010) Future wireless communication systems for railways. IEEE Veh Technol Mag 5(3):82–89
94. Hassan K et al (2012) Predicted eigenvalue threshold based spectrum sensing with correlated multiple antennas. In: Proceedings of Vehicular Technology Conference (VTC Spring), 2012 IEEE 75th (pp. 1–5)
95. Hassan K et al (2012) Non-parametric multiple-antenna blind spectrum sensing by predicted eigenvalue threshold. In: Proceedings of Communications (ICC), 2012 IEEE International Conference on, pp. 1634–1629
96. Hasssan K et al (2010) Blind modulation identification for MIMO systems. In: Proceedings of the global communications conference, 2010, pp. 1–5
97. Kharbech S et al (2014) Blind digital modulation identification for time-selective MIMO channels. IEEE Wireless Commun Lett 3(4):373–376

Chapter 5
Resource Management for High-Speed Railway Mobile Communications

5.1 Introduction

For the past two decades, intelligent transportation systems (ITS) have emerged as an efficient way of improving the performance of transportation systems. As an essential element of ITS, high-speed railway (HSR) has been developed rapidly as a fast, convenient, and green public transportation system and would become the future trend of railway transportation worldwide. For instance, a high-speed rail plan has been outlined in America and the length of HSR lines in China will reach 18,000 km by 2020. With the continuous construction of HSR in recent years, the issue of train operation safety has attracted more and more attention. The train operation control system plays a key role in train operation safety and is regarded as the nerve center of the HSR system. A standard has been set up for the train operation control system, which is known as European Train Control System (ETCS) [1]. In order to make ETCS work better and create a digital standard for railway communications, a dedicated mobile communication system called the global system for mobile communications for railway (GSM-R) has been proposed by International Union of Railway (UIC) [2].

GSM-R has been widely used in HSR communications and can maintain a reliable communication link between the train and the ground. However, GSM-R has some major shortcomings, such as insufficient capacity, low network utilization, and limited support for data services. A broadband wireless communication system for HSR called long-term evolution for railway (LTE-R) has been presented and determined in the 7th World Congress on High-Speed Rail [3]. Broadband wireless communications can enhance the train operation by allowing an operation center to monitor real-time train-related data information, such as safety information and track diagnostic information. In addition to the train control data transmission, LTE-R is also expected to provide passenger services such as Internet access and high-quality mobile video broadcasting [4]. With the benefit of it, passengers can treat their journey as a seamless extension of their working or leisure environment.

Z.-D. Zhong et. al., *Dedicated Mobile Communications for High-speed Railway*,
Advances in High-speed Rail Technology, DOI 10.1007/978-3-662-54860-8_5

To improve the capacity for wireless communications on the train, the future HSR communication networks are expected to be heterogeneous with a mixture of different networks and radio access technologies that can be simultaneously accessed by hundreds of users on the train [5]. For instance, the heterogeneous network architecture can be considered as a combination of satellite network, cellular network, and wireless data network, where the advantage of each access network can be taken into consideration. This architectural enhancement along with the advanced communication technologies such as multiple-input multiple-output (MIMO), orthogonal frequency-division multiplexing (OFDM), and radio over fiber (RoF), will provide high aggregate capacity and high spectral efficiency. Nevertheless, the demand for HSR wireless communications is increasingly growing, for example, the estimated wireless communication requirement could be as high as 65 Mbps per train [6]. To further relieve the contradiction between the increasing demand and limited bandwidth of HSR wireless communications, it is necessary to implement radio resource management (RRM) to improve resource utilization efficiency and ensure quality of service (QoS) requirements. However, the traditional RRM methods (e.g., handover, power control, and resource allocation) for common cellular communications may not be efficient in HSR wireless communications due to the following reasons, which are closely related to the characteristics of HSR scenario.

First, high mobility. The dramatic increase of train speed will cause frequent handover. Given a cell size of 1–2 km, a high-speed train of 350 km/h experiments one handover every 10–20 s. To solve the frequent handover problem is one of the main functions of RRM in HSR wireless communications. Moreover, the fast relative motion between the ground and the train will lead to large Doppler shift and small coherence time. The maximum speed of HSR in China is currently 486 km/h, which induces a Doppler shift of 945 Hz at 2.1 GHz. Thus, when implementing resource allocation for HSR communications, it is necessary to consider the fast-varying channel and inter-carrier interference (ICI) especially for OFDM technology.

Second, unique channel characteristics. The moving train encounters diverse scenarios (e.g., cuttings, viaducts, and tunnels) with different channel propagation characteristics [7], which causes that a single-channel model could not depict features of HSR channels accurately. It brings a big challenge to RRM schemes, which should be adaptive to diverse scenarios along the rail with different channel models. Furthermore, the line-of-sight (LOS) component is much stronger than the multipath components especially in viaduct scenario, which implies that the propagation loss mainly depends on the distance between the base station (BS) and the train. Since the distance varies with the train's position, the power control along the time has a large influence on system transmission performance.

Finally, heterogeneous QoS requirements. Many types of services with heterogeneous QoS requirements and priorities will be supported on the train. The QoS performance in HSR wireless communications will be degraded because of high mobility and unique channel characteristics, especially for real-time services and critical core services that are critical for the train operation. In order to improve system performance and satisfy heterogeneous QoS requirements, it is critical to design effective RRM schemes and resource optimization methods for multiple services transmission in HSR communications.

All these unique characteristics make it challenging to facilitate RRM design for HSR wireless communications. Thus, a new look into the RRM problem in HSR communications is urgently required, where the network architecture and unique characteristics of HSR scenario should be fully taken into consideration. This chapter provides a comprehensible survey on HSR wireless communications from the perspective of RRM and optimization design. Our goal is to present a detailed investigation and thorough discussion of current state-of-the-art RRM schemes for HSR wireless communications, as well as provide a better understanding of the RRM research challenges and open issues in HSR wireless communications.

5.2 Overview and Survey

With the growing demand for more and more QoS features and multiservice support in future HSR communication systems, RRM has become crucial and attracted great attention. RRM is the process of developing decisions and taking actions to optimize the system resource utilization. In particular, RRM consists of four key elements: admission control, mobility management, power control, and resource allocation [8]. Each element has a corresponding function with a common objective of achieving better system performance. Compared with traditional cellular communications, supporting multiservice transmission under HSR scenarios introduces some new challenging issues to these RRM elements. In this section, we provide a comprehensive state-of-the-art survey on RRM schemes for HSR wireless communications, with a focus on specific solutions for each element in detail.

5.2.1 Admission Control

Admission control is an essential tool for congestion control and QoS provisioning by restricting the access to network resources. Generally, the admission control function has two considerations: the remaining network resources and the QoS guarantees. In an admission control mechanism, a new access request can be accepted if there are adequate free resources to meet its QoS requirement without violating the committed QoS of the accepted accesses. Thus, there is a fundamental trade off between the QoS level and the resource utilization. To solve this trade-off, admission control has been extensively studied in common communication networks, and different aspects of admission control design and performance analysis have been surveyed in [9]. However, the admission control problem in HSR wireless communications is more sophisticated due to the following reasons.

On one hand, high mobility brings some challenges to the implementation of admission control schemes, such as the fast-varying fading channel and frequent handover. The fast-varying characteristics of the wireless channel will cause channel estimation error, and further lead to admission control schemes inaccurate. For frequent handover, since available handover time is short and handover connections almost arrive in batches, it is critical to consider the handover connections when implementing admission control.

On the other hand, multiple services with heterogeneous QoS requirements are supported on the train. The prioritization and fairness among different services should be taken into account in assigning admission control schemes. For example, higher priority should be given to safety-related services while the other services are delivered using the remaining network sources.

The above reasons make it difficult to directly apply the common admission control schemes into HSR communications. The investigation on novel and proper admission control schemes is of great importance. In the following, we provide an overview of the existing research on admission control for HSR wireless communications, including the classification, the detailed description, and the comparisons of different admission control schemes.

5.2.2 *Level-Based Admission Control*

Based on the admission level, the admission control schemes can be classified into the call-level admission control and the packet-level admission control. Traditional admission control schemes only consider call-level performance and are mainly designed for circuit-switched GSM-R system [10–14]. Since LET-R will become a packet-switched system, the packet-level features could be explored to improve the system performance [14]. The difference between the call-level admission control and packet-level admission control is as follows. At call level, each call is characterized by its arrival rate and holding time. If the system has enough resource, the new-call request will be accepted, otherwise, the call will be rejected. The packet-level features are characterized by the QoS profile that describes the packet arrival rate, packet queueing delay and packet loss ratio requirement. For the elastic traffic transmission, the system simply drops the excess packets based on the system status and the minimum QoS guarantee. To the best of our knowledge, the call-level admission control and packet-level admission control are investigated separately. In the packet-switched LTE-R system, the packet-level dynamics are central to the call-level performance, and thus both the packet level and call level should be considered in admission control schemes.

5.2.3 Handover-Based Admission Control

Compared with the conventional mobile communications, handover is more frequent in HSR mobile communications. Thus, to reduce the handover failure probability and prevent the handover connections from being rejected, it is critical to consider the handover connections when implementing admission control. Some existing works have studied the admission control schemes associated with handover in high-speed mobile scenarios. A spring model-based admission control scheme is proposed in [10], where every existing service is considered as a spring. For admitting handover services, bandwidth resource borrowing strategy is provided by compressing the spring as long as the lowest QoS can still be guaranteed. Performance metrics such as dropping probability and blocking probability are analyzed based on the birth–death process. In [11], a mobility-aware call admission control algorithm is proposed in mobile hotspots, where a handover queue is involved to reduce the handover-call dropping probability. The guard channels are dynamically assigned for handover calls depending on the vehicular mobility. By means of Markov chains, the proposed algorithm is evaluated in terms of new-call blocking probability, handover-call dropping probability, handover-call waiting time in the queue, and channel utilization. The literature [12] proposes joint admission control and handover policies in an integrated satellite-terrestrial architecture, and the proposed policies aim to increase both the user satisfaction and the resource utilization. Another handover-based admission control scheme is proposed in [13], where a position-based factor is introduced to reserve more resources to accept handover calls. With the help of the handover location information, the resource reservation scheme divides the system resources for handover calls and new calls, respectively.

In the above schemes [10–13], the handover services and new services are considered when making admission control decisions. Actually, the services can dynamically change their modulation and coding schemes (MCS) in HSR wireless communications. When the adopted MCS changes from high spectrum efficiency to low spectrum efficiency, the occupied physical bandwidth will increase. Then, besides handover services and new services, MCS changed service may be also dropped. In [14], the main potential origins of service dropping are classified into two types: MCS changed service dropping and handover service dropping. A cross-layer admission control scheme with adaptive resource reservation is proposed to reduce the service dropping probability, where the influences of MCS change and time-varying channels are taken into account.

5.2.4 Priority-Based Admission Control

Different types of services will be transmitted between the train and ground. Generally, different services have different priorities and bandwidth requirements.

Thus, the service priority should be considered in admission control schemes for HSR wireless communications. In [15], an admission control scheme with the complete-sharing resource allocation model is proposed for LTE-R system to maximize the number of admitted services while guaranteeing their related QoS. The proposed admission control scheme gives high priority to on-going services and guarantees the optimal bandwidth resource allocation. Zhao et al. [16] proposes an effective admission control scheme for HSR communications with MIMO antennas, where both new services and handover services are considered. Handover services will be admitted first. The reason is that dropping an ongoing service during the handover process will bring about more serious results than blocking a new service. For different handover services to be admitted, voice has the highest priority while data has the lowest. After all the handover services are processed, the new services will be considered in the same way as the handover services. A joint optimal design of admission control and resource allocation is considered in [17] for multimedia service delivery in HSR wireless networks. Different types of services are assigned to different utility functions that represent different service priorities. The considered problem aims at maximizing the total utility while stabilizing all transmission queues under the average power constraint. A threshold-based admission control scheme is proposed and the service with high priority gets large average throughput.

Based on the survey on admission control research, we can make some conclusions as follows. First, since LTE-R is designed as a packet-switched communication system, it is necessary to conduct admission control at packet level, which needs further investigations. Second, it is serious to drop ongoing services during the handover process and critical core services during the trip. Thus, the handover services and critical core services should have high priority to access the networks, where the resource reservation approach is also required. Third, the handover-based admission control and priority-based admission control should be combined together so as to deal with handover services and new critical core services reasonably. At the same time, their advantages can be exploited jointly. Finally, adaptive admission control schemes with low complexity are critical to suit the requirements such as frequent handover, quick decision-making duration and fast-varying wireless channel in HSR environments.

5.2.5 Resource Allocation

Resource allocation, as a critical part of RRM, plays an important role in enhancing the data transmission efficiency and improving the QoS performance in HSR communications. A great variety of resource allocation schemes have been proposed in the literature, aiming at sharing the limited network resources while satisfying the heterogeneous QoS requirements. Due to the unique characteristics of HSR environments, the existing resource allocation schemes for general wireless networks cannot be directly applied to HSR wireless networks. Moreover, the dynamic characteristics such as time-varying wireless channels and random packet/service arrivals, should be incorporated into the resource allocation schemes. In this subsection, the resource allocation schemes for HSR wireless communications are systematically surveyed from three aspects: interference-aware resource allocation, QoS-aware resource allocation, and dynamic cross-layer resource allocation.

5.2.6 Interference-Aware Resource Allocation

Resource allocation is an effective way to reduce the effect of interference and further improve the communication performance. Generally, the interference in HSR wireless communications mainly comes from two sources: ICI for OFDM technology and interference from two-hop links. The ICI caused by Doppler shift, is obviously not negligible and may degrade the system performance. Thus, the resource allocation problem in HSR communications with OFDM technology has attracted great research interest. For example, [18] focuses on the joint sub-carrier and power allocation problem with adaptive modulation and coding in high-speed mobile environments. The established optimization problem, which aims to minimize the overall transmit power while satisfying all the user requirements, is a nonlinear programming. Considering that the exact expression of ICI term is complicated, the statistical mean value of the ICI power is utilized to simplify the complexity. In addition, [16, 19] study the multidimensional resource allocation problem for HSR downlink communications with OFDM technique and MIMO antennas. The objective is to maximize the throughput under the total transmit power constraint and ICI. In order to reduce computational complexity, the sub-optimal solution is obtained by using quantum-behaved particle swarm optimization.

The above studies mainly focus on solving the resource allocation problems with ICI under one-hop architecture in HSR communications. Since the scarcity of spectrum nowadays, it is difficult to allocate dedicated spectrum for the two-hop links. Consequently, the spectrum reuse is inevitable and the resource sharing will lead to the interlink interference. There are also some study works on resource allocation for the two-hop communications. Qiu et al. [20] formulates a joint optimization problem of sub-carrier allocation, sub-carrier pairing, and power allocation in two-hop links. Due to the complicated expression of ICI, the joint optimization problem is decomposed into several subproblems and an iterative algorithm is proposed to solve the non-convex problem. In [21], the resource allocation problem in two-hop links is formulated as a mixed-integer nonlinear programming. Since the optimization problem is NP-hard, an alternative heuristic resource allocation scheme is proposed and then the optimal power is determined to maximize the system sum throughput. In addition, the performance analysis of HSR wireless communications is studied in [22] to better understand the effect of the interlink interference.

5.2.7 QoS-Aware Resource Allocation

It is a natural requirement for HSR wireless networks to provide end-to-end QoS to the services, with the purpose of guaranteeing the safe operation of trains and improving the quality of passenger experience. Data transmission reliability is critical in HSR communications, and is often evaluated by bit error rate (BER). With the speed increases, the communications will suffer from high BER, which may cause signaling error, retransmission, and energy waste. Thus, the BER requirement should be considered for resource allocation design in HSR communications. The work [23] tries to solve a resource allocation optimization problem, which minimizes the total transmit power while considering BER that partially depends on the modulation and coding scheme. For improving the good put of HSR communications, a link adaptation scheme is proposed in [24] under the condition of guaranteeing prescribed BER target. Multimedia entertainment is an important application in which a data stream often contains the packets with different BER requirements. For example, the video stream encoded with scalability contains the base layer packets with high BER requirement and enhancement layer packets with low BER requirement. To achieve efficient resource allocation, [25] develops a resource allocation approach by considering multiple BER requirements for different types of packets in one data stream. In order to simplify the complexity of resource allocation, a proper number of contiguous sub-carriers are grouped into chunks and the spectrum is allocated chunk by chunk.

Service transmission delay is another key QoS parameter, which directly affects perceived QoS of real-time service. The communication delay of train control services also affects the track utilization and speed profile of high-speed trains [26]. HSR services such as critical core services have a high demand for transmission delay. Thus, delay requirements should be fully taken into consideration when implementing resource allocation among the services. To better describe delay requirements, an interval-based service request model is formulated in [27], where each service has to be delivered within its given lifetime. The corresponding resource allocation problem aims to maximize the total weights of the fully completed requests and incomplete requests do not yield any revenue. As an extension, [28, 29] build more reasonable service delivery models for on-demand data packet transmission to high-speed trains, where the deadline constraints are built on each data packet rather than the whole service. Compared with the deterministic deadline constraint case, [30, 31] focus on the average delay constraint for the service transmission. The resource allocation problem is formulated as a constrained Markov decision process (MDP) and the corresponding online resource allocation algorithm is proposed.

5.2.8 Cross-Layer Dynamic Resource Allocation

Cross-layer design is a well-known approach to achieve QoS support. In cross-layer design, the challenges from the physical wireless channel and the QoS requirements are taken into account so that the resource allocation can be adapted to meet the service requirements for the given channels and network conditions. It should also consider the dynamic characteristics in HSR communication systems, such as time-varying wireless channels and random service arrivals. Thus, to enhance the efficiency of resource utilization and improve the QoS performance, it is necessary to implement dynamic resource allocation in a cross-layer way.

A cross-layer dynamic resource allocation framework is with respect to the application (APP) layer, medium access control (MAC) layer, and physical (PHY) layer. At the PHY layer, the channel state information (CSI) allows an observation of good transmission opportunity. At the MAC layer, the queue state information (QSI) provides the urgency of data packets. At the APP layer, service characteristics information (SCI) offers the service characteristics, e.g., packet arrival rate and rate-utility relationship. The control actions, which include power control action and resource allocation action, should be decided dynamically based on the PHY layer CSI, the MAC layer QSI, and the APP layer SCI. Specifically, the power control action decides the wireless link capacity, i.e., the total resource allocated to the services. The resource allocation action decides how many resources should be allocated to each service.

Based on the above framework, [31] investigates the downlink resource allocation problem in relay-assisted HSR communication systems, taking account of bursty packet arrivals, and delay performances. The considered problem is formulated as an infinite-horizon average cost constrained MDP, where the control actions depend on both the CSI and the QSI. The objective is to find a policy that minimizes the average end-to-end delay through control actions under the service delivery ratio constraints. A joint admission control, power control, and resource allocation problem is investigated in [17]. A dynamic resource allocation algorithm is proposed to maximize the system utility while stabilizing all transmission queues. In addition, [32] presents the resource allocation schemes for voice over Internet protocol (VoIP) traffic under HSR scenarios, in which the 2-state VoIP traffic model based on Markov chain is considered. An adaptive resource allocation scheme is developed based on the idea of packet bundling to increase the spectral efficiency and reduce the outage probability.

A summary of the discussed resource allocation schemes for HSR communications is provided. The existing schemes mainly focus on interference awareness, QoS requirements, and cross-layer dynamic design, which are consistent with the characteristics of HSR communications. Moreover, these three categories of resource allocation schemes correspond to different resource allocation problems. For the interference-aware resource allocation, the optimization problems are mainly built on the ICI and interlink interference. The complicated expression of the interference term makes the formulated problems non-convex, and some heuristic algorithms are proposed to obtain the suboptimal solutions. In the QoS-aware resource allocation, the formulated problems consider the reliability or delay requirements. The reliability is often evaluated by BER, which acts as a constraint and is affected by the applied modulation and coding scheme. The delay requirements are typically represented by the deterministic deadline constraint or average delay constraint. For the cross-layer dynamic resource allocation, the considered problems are dynamic optimization problems with cross-layer design. Stochastic optimization theory and MDP theory are used to effectively solve them. Finally, we point out that the combination of different resource allocation schemes would be an interesting direction for future research. For example, a practical and challenging problem is resource allocation for the delay-aware data transmission using OFDM technology in the two-hop wireless links.

5.2.9 Power Control

Compared with the conventional communication systems, there are three unique features in HSR communication systems [33], i.e., the deterministic moving direction, relatively steady moving speed, and the accurate train location information. The data transmission rate is highly determined by the transmit power and the distance between BS and the train, thus these features make it necessary and feasible to implement power control along the time. To achieve different optimization objectives under average power constraint, four power allocation schemes are proposed in [34], i.e., constant power allocation (CPA), channel inversion power allocation (CIPA), water-filling power allocation (WFPA), and proportional fair power allocation (PFPA). Figure 5.1 provide the comparisons of the power allocation results and the corresponding transmission rate results for these four schemes. The advantage and disadvantage of these schemes can be observed from

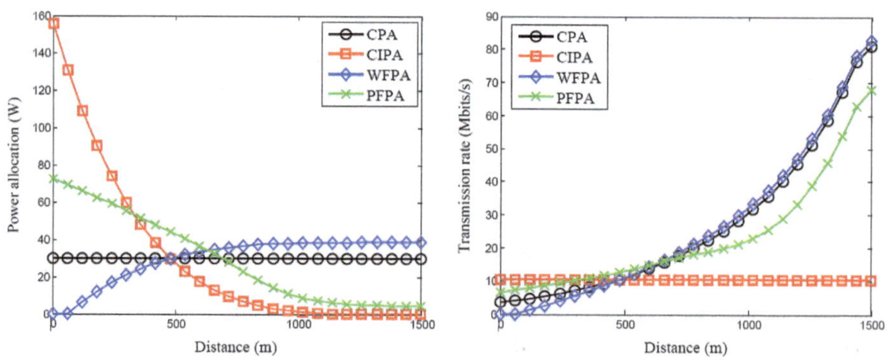

Fig. 5.1 Comparisons of four power allocation scheme. BS is located at 0, the cell radius R is 1500 m, and average power is 30 W

these two figures. For the sake of convenience in the engineering implementation, a constant power is allocated along the rail in CPA scheme while it ignores the variation of channel gain and results in the great unfairness in term of transmission rate. In order to provide a stable transmission rate and achieve the best fairness along the rail, the CIPA scheme spends much power to compensate those bad channel states when the train is far from the BS. Similar to the traditional water-filling method, the WFPA scheme can maximum the total transmission rate within one BS, whereas the services will generally suffer from starvation when the train is near the cell edge. In addition, the PFPA scheme can achieve a trade off between the total transmission rate and the fairness along the time. As an extension of [35], the work [36] investigates the utility-based resource allocation problem, which can jointly take into account power allocation along the time and packet allocation among the services.

The works [35, 36] pursue some system optimization objectives under the power constraint. On the other hand, some works consider the power allocation problem from the perspective of energy efficiency, mainly focusing on how to match the data arrival process and the time-varying channel for HSR communications. Specifically, [37] provides a novel method to minimize the total transmit power for data uplink transmission under a certain deadline constraint by exploiting the future channels. In [38], the authors study the optimal power allocation policy under given delay constraints in uplink transmission. It shows that there are two trade-offs in the transmission model, one is between the average transmit power and the delay constraint, and the other is between the average transmit power and the train velocity. Inspired by the unique spatial–temporal characteristics of wireless channels along the rail, [39] presents a novel energy-efficient and rate-distortion optimized approach for uploading video streaming. Although the above studies are useful for the optimal design of HSR wireless communications, they only take account of the time-varying channel state while do not consider the dynamic characteristics of the service or packet arrivals, which causes that the above power allocation schemes are not practical.

Dynamic power control is necessary to improve the performance of HSR communication systems, where the transmit power should be adaptive to the time-varying channel and dynamic service arrival. Considering the power constraint in HSR communications, the work [17] investigates a joint admission control and resource allocation problem. A dynamic power control and resource allocation algorithm is proposed to maximize the system utility while stabilizing all transmission queues. Different from [17], the work [40] studies the delay-aware multi-service transmission problem in HSR communication systems, with a focus on how to implement power control and resource allocation to guarantee the delay requirements under power constraint.

5.3 Resource Allocation and Power Control

In this section, we investigate the utility-based resource allocation problem at a base station in high-speed railway (HSR) wireless networks, jointly taking into account the power allocation along the time and the packet allocation among services. The problem to maximize the total utility under the average power constraint is formulated as a mixed-integer nonlinear programming (MINLP) problem. Through the integer constraint relaxation, the MINLP problem can be simplified into a convex optimization problem. The detailed analysis reveals that the relaxed problem can be equivalently decomposed into power allocation problem along the time and packet allocation problem among services, which can reduce the problem size. When the optimality of the relaxed problem is achieved, the power allocation along the time and the packet allocation along the time for each service are both proportionally fair. Since the integer relaxation causes a non-integer solution not implementable in practice, a greedy algorithm is proposed to obtain a near optimal integer solution of the MINLP problem. Finally, the performance of the proposed algorithm is analyzed by simulations under realistic conditions for HSR wireless networks.

5.3.1 System Model

A two-hop HSR wireless network architecture is considered, as shown in Fig. 5.2. The base station (BS) deployed along the rail line is connected to the backbone

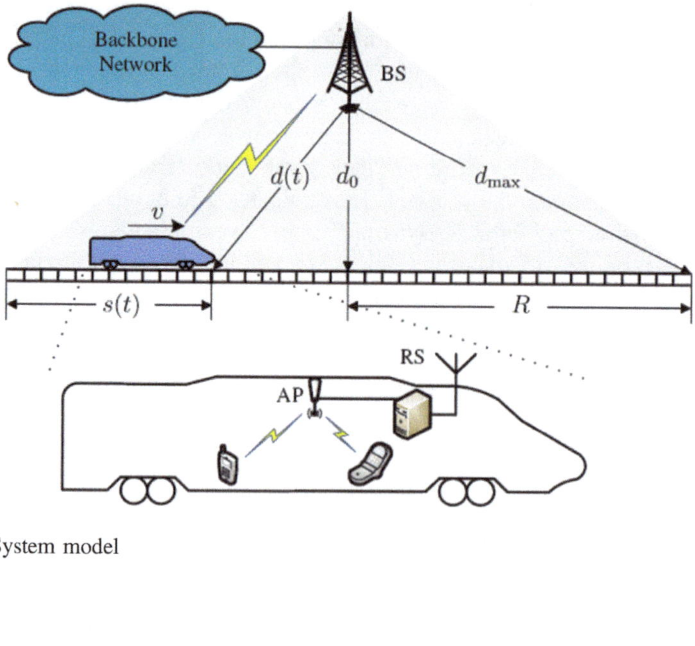

Fig. 5.2 System model

network via a wire line link. The relay station (RS) with powerful antennas installed on the top of the train is used for communicating with the BS. The RS is further connected to the access point (AP) which can be accessed by the users inside the train. Thus, there is a two-hop wireless link, consisting of the BS-RS link and the AP-Users link, which has several advantages. First, it is RS not each user to implement the handover procedure, which can achieve better handover performance and reduce the drop-off rate significantly. Second, with this two-hop wireless link, signals do not penetrate into the carriage, thus the large penetration loss can be dramatically reduced. Finally, since the users are nearly stationary with respect to the AP, the AP-Users link can provide a stable and high-speed wireless data transmission.

We consider the downlink data transmission in this two-hop architecture. The AP-Users link inside the carriage can provide a large data transmission rate by using wireless local area network (WLAN) technologies, while the BS-RS link suffers from the fast-varying wireless channel and may become the bottleneck in this architecture. Therefore, the transmission in the BS-RS link will be mainly considered with the assumption for the downlink data always being successfully received when delivered to RS.

5.3.2 Time-Distance Mapping

We consider a train travels at a constant speed υ through a single cell with radius R. The total time the train spends is $T_{\text{all}} = 2R/\upsilon$, which is divided into slots of equal duration T_s. Then the total number of slots is given by $T = T_{\text{all}}/T_s$, where we assume that T_{all} can be exactly divided by T_s and T is even. Without loss of generality, we assume that the train goes into and out of the cell coverage at slot 0 and slot T, respectively. The traveled distance until slot t is given by $s(t) = \upsilon t T_s$. A time-distance mapping function $d(t)$ is defined as the distance between BS and RS at slot t, i.e., $d(t) : [0, T] \to [d_0, d_{\max}]$, where $d_{\max} = \sqrt{R^2 + d_0^2}$ and d_0 is the distance between the BS and the rail line as shown in Fig. 5.2. The mapping function $d(t)$ can be expressed by

$$d(t) = \sqrt{|s(t) - R|^2 + d_0^2}, \quad t \in [0, T], \tag{5.1}$$

where we assume that the distance $d(t)$ does not change within slot t since T_s is very small. There are two inherent properties about the mapping function $d(t)$: the train is closest to the BS at slot $T/2$, i.e., $d(T/2) = d_0$; For any slot $t \in [0, T]$, we have $d(t) = d(T - t)$ due to the distance symmetry.

5.3.3 BS-RS Link Capacity

In HSR wireless networks, the channel condition in BS-RS Link is fast varying due to the large-scale fading and small-scale fading. Confirmed by engineering measurements [6], the line-of-sight path in BS-RS Link is available at most time, and the effect of large-scale fading is more obvious than that of small-scale fading. Therefore, we ignore the small-scale fading and assume that the channel condition variation results only from the fast-varying distance between BS and RS. As shown in [35], power allocation along the travel time plays a key role in improving the performance of HSR wireless networks. We denote $P(t)$ as the transmit power of the BS at slot t, which is limited by the average value P_{av}.

Given $P(t)$ and $d(t)$, the received signal-to-noise ratio (SNR) by RS at slot t can be expressed by

$$\mathrm{SNR} = \frac{P(t)}{WN_0 d^{\alpha}(t)} = \frac{P(t)}{N(t)}, \tag{5.2}$$

where $N(t) = WN_0 d^{\alpha}(t)$, W is the system bandwidth, N_0 is the noise power spectral density and α is the pathloss exponent. The instantaneous transmission rate in the BS-RS link at slot t is

$$R(t) = W \log_2\left(1 + \frac{P(t)}{N(t)}\right) \quad \text{bits/s}. \tag{5.3}$$

Suppose that a packet is the fundamental unit of transmission, which has equal size of L bits, hence the link capacity $C(t)$ at slot t can be denoted as the maximum number of packets, which can be expressed by

$$C(t) = \lfloor \tilde{C}(t) \rfloor = \lfloor R(t)T_s/L \rfloor, \tag{5.4}$$

where $\tilde{C}(t) = R(t)T_s/L$ and $\lfloor x \rfloor = \max\{n \in \mathbb{Z} | n \leq x\}$.

5.3.4 Utility-Based Resource Allocation

Assume that there are K types of services with infinite packets to be transmitted from BS to RS and the service set is denoted by $\mathcal{K} \triangleq \{1, \ldots, K\}$. To allocate the network resources based on the services' types, utility-based resource allocation can be employed. For any service, the utility grows as the allocated data rate increases. On one hand, equal data rate allocation does not provide equal utility, which is interpreted as equal service satisfaction. On the other hand, to achieve equal utility, the different data rates should be allocated to the services according to their types, which results in utilizing the network resources more efficiently. Thus, we consider utility-based resource allocation instead of rate-based resource allocation in this section.

Suppose in general that service k maintains an increasing and concave function $U_k(v_k)$ as its utility function, which indicates a service's degree of satisfaction on the allocated v_k packets. Instead of maximizing network throughput performance, our objective is to maximize the overall network utility, which is the summation of all services' utility functions. The utility function for service k can be defined as

$$U_k(v_k) = \begin{cases} \omega_k \frac{v_k^{1-\alpha}}{1-\alpha}, & \alpha \geq 0, \alpha \neq 1, \\ \omega_k \ln(v_k), & \alpha = 1, \end{cases} \tag{5.5}$$

where α is a parameter dictating the shape of the utility function and ω_k represents the weight of service k. When the weights of all services are same, e.g., $\omega_k = 1$, the optimization objective can be specialized into different cases according to different values of α. We choose $U_k(v_k) = \omega_k \ln(v_k)$ to obtain weighted proportionally fair resource allocation, where ω_k is assumed to be integer for $k \in \mathcal{K}$ in this section.

5.3.5 *Problem Formulation*

We develop a mathematical formulation of the optimal resource allocation problem in HSR wireless networks. Let $\mathbf{v}(t) = [v_1(t), \ldots, v_K(t)]^T$ represent the packet allocation vector at slot t, where $v_k(t)$ denotes the number of packets allocated to service k at slot t. The optimization problem consists in maximizing the BS utility to find the optimal power allocation along the time and the optimal packet allocation among services, and two necessary constraints are added: (i) The BS has an average power constraint along the time. (ii) The total number of allocated packets is no more than the link capacity at any slot. Thus, the utility-based resource allocation optimization problem is formulated as

$$\begin{aligned} (\mathbf{P1})\ \text{maximize} \quad & \sum_{t=0}^{T} \sum_{k \in \mathcal{K}} \omega_k \ln(v_k(t)) \\ \text{subject to} \quad & \frac{1}{T+1} \sum_{t=0}^{T} P(t) \leq P_{\text{av}}, \\ & 0 \leq \sum_{k \in \mathcal{K}} v_k(t) \leq C(t), \quad \forall t \in [0,T], \\ \text{variables} \quad & v_k(t) \in \mathbb{N}, P(t) \geq 0, \forall k \in \mathcal{K}, t \in [0,T]. \end{aligned} \tag{5.6}$$

The problem P1 is a mixed-integer nonlinear programming (MINLP) problem, including $T+1$ continuous variables $P(t)$ and $K(T+1)$ integer variables $v_k(t)$, which is in general NP-hard. The main difficulty of analyzing problem P1 comes from the integer nature of $v_k(t)$. To significantly improve the computational efficiency and obtain some engineering insights for solving the MINLP problem, we adopt integer constraint relaxation for problem P1, where $\mu_k(t) \in \mathbb{Q}^+$ substitutes the constraint $v_k(t) \in \mathbb{N}$ and $\tilde{C}(t)$ substitutes $C(t)$. As a result, the problem P1 is simplified into a relaxed problem P2 as follows:

$$\begin{aligned}
(\textbf{P2})\ \text{maximize} \quad & \sum_{t=0}^{T} \sum_{k \in \mathcal{K}} \omega_k \ln(\mu_k(t)) \\
\text{subject to} \quad & \frac{1}{T+1} \sum_{t=0}^{T} P(t) \leq P_{\text{av}}, \\
& 0 \leq \sum_{k \in \mathcal{K}} \mu_k(t) \leq \tilde{C}(t), \quad \forall t \in [0, T], \\
\text{variables} \quad & \mu_k(t) \geq 0, P(t) \geq 0, \forall k \in \mathcal{K}, t \in [0, T].
\end{aligned} \tag{5.7}$$

Notice that the optimal solution of problem P2 provides an upper bound to that of problem P1 since the constraints in problem P2 are looser than those in problem P1. There are totally $(K+1)(T+1)$ continuous variables in problem P2, where T is typically of the order of $10^4 - 10^5$. Standard convex optimization tools such as CVX can be employed to solve P2, however, the computational complexity is very high due to the large size of the problem. In order to obtain a low-complexity and effective algorithm for problem P2, we carry out the problem transformation. Before we present the solution for problem P2, we consider the problem decomposition to determine some characteristics which will be useful in understanding the structure of problem P2 better.

By decoupling of the optimization variables in the second constraint of (5.7), the problem P2 can be decomposed into two subproblems: (i) power allocation along the time (PAT): how to implement power allocation along the time under the average power constraint at the BS? (ii) packet allocation among services (PAS): how to allocate resources to multiple services at each slot by the given power allocation? Next, we will discuss these two subproblems separately.

5.3.6 PAT Problem

We investigate the power allocation problem along the time under the average power constraint at the BS. Since the channel state in BS-RS link is time-varying, to achieve different optimization objectives, four power allocation schemes have been proposed in [35].

Constant power allocation (CPA) The most straightforward scheme is the constant power allocation, where BS maintains a constant transmit power at all times, i.e., $P(t) = P_{\text{av}}$. Thus,

$$\tilde{C}(t) = \frac{T_s W}{L} \log_2\left(1 + \frac{P_{\text{av}}}{N(t)}\right), \quad \forall t \in [0, T] \tag{5.8}$$

Channel inversion power allocation (CIPA) It tries to maintain a constant link capacity $\tilde{C}(t)$ at the BS all the time. Therefore, based on (5.3) and (5.4), the ratio of $P(t)$ to $N(t)$ is a constant for all slots. Without loss of generality, we suppose that $P(t) = k_0 N(t)$. And then by solving $\sum_{t=0}^{T} P(t) = (T+1)P_{\text{av}}$, we have $k_0 = \frac{(T+1)P_{\text{av}}}{\sum_{t=0}^{T} N(t)}$ and

$$\tilde{C}(t) = \frac{T_s W}{L} \log_2(1 + k_0), \quad \forall t \in [0, T]. \tag{5.9}$$

Water-filling power allocation (WFPA) To maximize the total link capacity at the BS, we formulate the following optimization problem

$$\begin{aligned} &\text{maximize} && \sum_{t=0}^{T} \tilde{C}(t) \\ &\text{subject to} && \tfrac{1}{T+1} \sum_{t=0}^{T} P(t) \le P_{\text{av}}, \\ &\text{variables} && P(t) \ge 0, t \in [0, T], \end{aligned} \tag{5.10}$$

whose solution can be obtained by water-filling scheme.

Proportional fair power allocation (PFPA) To achieve a trade-off between the total link capacity and the fairness along the time, a proportional fair power allocation optimization problem is formulated as

$$\begin{aligned} &\text{maximize} && \sum_{t=0}^{T} \ln \tilde{C}(t) \\ &\text{subject to} && \tfrac{1}{T+1} \sum_{t=0}^{T} P(t) \le P_{\text{av}}, \\ &\text{variables} && P(t) \ge 0, t \in [0, T], \end{aligned} \tag{5.11}$$

whose ϵ-optimal solution can be obtained by the proposed algorithm in [35].

5.3.7 *PAS Problem*

The packet allocation problem among services is studied under the link capacity constraint by fixing the power allocation at all slots, which can be obtained according to the power allocation schemes. This setup is less complicated compared to our more general model, and its solution can provide us with some insights for solving the problem P2.

Given the fixed power allocation $P(t)$ at any slot t, the link capacity $\tilde{C}(t)$ can be computed by (5.4). The problem P2 can be divided into $T+1$ packet allocation problems and the problem at any slot $t \in [0, T]$ can be given by

$$\begin{aligned} &\text{maximize} && \sum_{k\in\mathcal{K}} \omega_k \ln(\mu_k(t)) \\ &\text{subject to} && 0 \le \sum_{k\in\mathcal{K}} \mu_k(t) \le \tilde{C}(t), \\ &\text{variables} && \mu_k(t) \ge 0, \forall\, k \in \mathcal{K}, \end{aligned} \tag{5.12}$$

which is a convex optimization problem. By applying the Karush–Kuhn–Tucker (KKT) conditions, we obtain the closed-form optimal solution of (5.12) and an important structural characteristic in the following lemma.

Lemma 1 *For the optimal solution vector* $\boldsymbol{\mu}^*(t) = [\mu_1^*(t), \mu_2^*(t), \ldots, \mu_K^*(t)]^T$ at any slot $t \in [0, T]$, *there must be the case that*

$$\frac{\mu_i^*(t)}{\omega_i} = \frac{\mu_j^*(t)}{\omega_j}, \quad \forall\, i,j \in \mathcal{K}, \tag{5.13}$$

and the closed-form optimal solution of (5.12) *is*

$$\mu_k^*(t) = \frac{\omega_k \tilde{C}(t)}{\sum_k \omega_k}, \quad \forall\, k \in \mathcal{K}. \tag{5.14}$$

Proof The proof of Lemma 5.1 is provided in Appendix A.

Based on (5.13) in Lemma 5.1, there is an inherent relationship among the optimal solutions at any slot. More resources are allocated to the service with a larger weight. In particular, for any service k, the allocated resource $\mu_k^*(t)$ at slot t is proportional to its weight ω_k. For simplicity of expression in the sequel, we introduce a virtual variable at any slot $t \in [0, T]$, which is given by

$$x(t) = \frac{\tilde{C}(t)}{\sum_k \omega_k}. \tag{5.15}$$

From (5.15), we can see that $x(t)$ is only determined by the power allocation $P(t)$ at any slot t. By plugging (5.15) into (5.14), the optimal solution of (5.12) can be rewritten by

$$\mu_k^*(t) = \omega_k x(t), \forall\, k \in \mathcal{K}, t \in [0, T]. \tag{5.16}$$

Thus, if the optimal power allocation solution in problem P2 can be obtained, then the optimal $x^*(t)$ and the optimal packet allocation solution in problem P2 can be calculated by (5.15) and (5.16), respectively.

5.3.8 Problem Transformation

To reduce the computational complexity, we consider the problem transformation for problem P2, where the number of the optimization variables dramatically decreases from $(K+1)(T+1)$ to $\frac{T}{2}+1$. Based on bisection search method, a greedy algorithm with low complexity is proposed for solving problem P2.

First, since Lemma 5.1 provides the necessary condition for the optimal solutions of (5.12) and problem P2, based on (5.16), the resource allocation variables $\mu_k(t)$ at slot t can be substituted by one single variable $x(t)$. Then, the objective function of problem P2 can be simplified into

$$\sum_{t=0}^{T}\sum_{k\in\mathcal{K}}\omega_k \ln(\mu_k(t)) = \sum_{t=0}^{T}\sum_{k\in\mathcal{K}}(\omega_k(\ln(\omega_k)+\ln(x(t)))) = \alpha+\gamma\sum_{t=0}^{T}\ln(x(t)), \tag{5.17}$$

where $\alpha(T+1)\sum_k(\omega_k(\ln(\omega_k)))$ and $\gamma=\sum_k\omega_k$ are both constant. Similarly, for the second constraint of (5.7), $\sum_k\mu_k(t)=\gamma x(t)$. Thus, the problem P2 can be transformed into

$$\begin{aligned}
(\mathbf{P3})\ \text{maximize}\quad & \sum_{t=0}^{T}\ln(x(t)) \\
\text{subject to}\quad & \frac{1}{T+1}\sum_{t=0}^{T}P(t)\le P_{\text{av}}, \\
& \gamma x(t)\le\tilde{C}(t)=\frac{T_sW}{L}\log_2\left(1+\frac{P(t)}{N(t)}\right),\quad \forall t\in[0,T], \\
\text{variables}\quad & x(t)\ge0, P(t)\ge0, \forall k\in\mathcal{K}, t\in[0,T].
\end{aligned} \tag{5.18}$$

Lemma 2 *Suppose that the optimal solution of problem P3 exists, the optimal solution provides proportionally fair resource allocation along the time for each service.*

Proof The proof of Lemma 5.2 is provided in Appendix B.

After the problem transformation, the total number of variables decreases from $(K+1)(T+1)$ to $2(T+1)$, and hence the computational complexity is dramatically reduced when K is large. Based on the investigation on problem P3, the total number of variables can be further reduced to $T+1$ as shown below.

It is easy to show that at the optimality of problem P3, the two constraints in (5.18) are both tight, otherwise, one can increase the value of $x(t)$ and $P(t)$ such that the objective function can be further maximized. Thus, we have

$$\frac{1}{T+1}\sum_{t=0}^{T} P(t) = P_{\mathrm{av}}, \tag{5.19}$$

and

$$\gamma x(t) = \tilde{C}(t) = \frac{T_s W}{L \ln 2} \ln\left(1 + \frac{P(t)}{N(t)}\right), \quad \forall t \in [0, T] \tag{5.20}$$

Based on (5.20), there exits a one-to-one correspondence established between $x(t)$ and $P(t)$, which is expressed by

$$x(t) = \eta \ln\left(1 + \frac{P(t)}{N(t)}\right), \tag{5.21}$$

where $\eta = \frac{T_s W}{\gamma L \ln 2}$. Thus, plugging (5.19) and (5.21) into problem P3 yields

$$\begin{aligned} (\mathbf{P4})\ \text{maximize} \quad & \sum_{t=0}^{T} \ln\left(\eta \ln\left(1 + \frac{P(t)}{N(t)}\right)\right) \\ \text{subject to} \quad & \sum_{t=0}^{T} P(t) = (T+1)P_{\mathrm{av}}, \\ \text{variables} \quad & P(t) \geq 0, \forall t \in [0, T] \end{aligned} \tag{5.22}$$

Lemma 3 *The optimal solution of problem P4 is the same as that of the PFPA problem.*

Proof The proof of Lemma 5.3 is provided in Appendix C.

Based on the Lemma 5.3, the optimal solution of problem P4 provides proportionally fair power allocation along the time. Furthermore, we can observe that the problem P2 can be equivalently decomposed into two subproblems: problem P4 and PAS problem, which are corresponding to power allocation problem along the time and packet allocation problem among services, respectively. Thus, we can solve the problem P4 at first, and then by using the power allocation results, the packet allocation solution can be obtained by using (5.16) and (5.21).

To solve the problem P4 effectively, the following lemma allows us to further reduce the computational complexity based on the distance symmetry at the base station.

Lemma 4 *In the optimal solution vector* $\mathbf{P}^* = [P^*(0), \ldots, P^*(T)]$, *there exists a symmetry on the optimal solution, i.e.,* $P^*(t) = P^*(T-t), \forall t \in [0, T]$.

Proof The proof of Lemma 5.4 is provided in Appendix D.

As a consequence of Lemma 5.4, the problem P4 can be simplified into the power allocation problem from slot 0 to slot $T/2$, which is labeled as P5.

$$
\begin{aligned}
(\mathbf{P5})\ \text{maximize}\quad & \sum_{t=0}^{T/2} g(P(t)) \\
\text{subject to}\quad & \sum_{t=0}^{T/2} P(t) = \left(\frac{T}{2}+1\right)P_{\text{av}}, \\
\text{variables}\quad & P(t) \geq 0, \forall t \in [0, T/2],
\end{aligned}
\tag{5.23}
$$

where $g(P(t)) = \ln\left(\eta \ln\left(1+\frac{P(t)}{N(t)}\right)\right)$ and the total number of variables decreases nearly half from $T+1$ to $\frac{T}{2}+1$.

The problem P5 is convex optimization problem, which can be solved by CVX. In addition, since problem P5 has a similar structure to the PFPA problem, the proposed algorithm in [35] can be used to find the ϵ-optimal solution of problem P5. However, the Lambert W function was introduced in the proposed algorithm resulting in the high computing time. The bisection search method is employed to reduce the computing time of searching the optimal solution.

Specifically, using the standard optimization technique, the corresponding Lagrangian function is obtained as

$$
\begin{aligned}
L(\{P(t)\},\lambda) &= \sum_{t=0}^{T/2} g(P(t)) - \lambda\left(\sum_{t=0}^{T/2} P(t) - \left(\frac{T}{2}+1\right)P_{\text{av}}\right) \\
&= \sum_{t=0}^{T/2}\left[\ln\left(\eta \ln\left(1+\frac{P(t)}{N(t)}\right)\right) - \lambda P(t) + \lambda P_{\text{av}}\right].
\end{aligned}
\tag{5.24}
$$

Based on the KKT conditions, we have

$$
\frac{\partial L(\{P(t)\},\lambda)}{\partial P(t)} = \frac{1}{\eta \ln\left(1+\frac{P(t)}{N(t)}\right)} \frac{\eta}{P(t)+N(t)} - \lambda = 0, \tag{5.25}
$$

which can be rewritten by

$$
\ln\left(1+\frac{P(t)}{N(t)}\right)(P(t)+N(t)) = \frac{1}{\lambda} \tag{5.26}
$$

Let $f(P(t)) = \ln\left(1+\frac{P(t)}{N(t)}\right)(P(t)+N(t))$, which is a monotonically increasing function of $P(t)$ at any slot t. Let $\beta = 1/\lambda$, then (5.26) is equal to $f(P(t)) = \beta$. Due to the monotonicity of $f(P(t))$, the bisection search method can be used to find $P(t)$ satisfying $f(P(t)) = \beta$ for a given β at each slot t. In addition, for any slot t, $P(t) = f^{-1}(\beta)$ is also a monotonically increasing function of β. Thus, to satisfy the average power constraint in (5.23), the bisection search method can also be used to find the optimal β^*.

The specific steps of the bisection search method are provided in Algorithm 5.1. The search regions of $P(t)$ and β should be initialized based on their maximums and minimums. At first, it is easy to verify that the maximum and minimum of $P(t)$ at each slot t can be set as $P_{\max} = \left(\frac{T}{2}+1\right)P_{\text{av}}$ and $P_{\min} = 0$, respectively. Based on the equality $f(P(t)) = \beta$, the maximum of β can be obtained when $P(t) = P_{\max}$ and $t = 0$ in function $f(P(t))$, i.e., $\beta_{\max} = f(P_{\max})|_{t=0}$ and the minimum of β can be set as $\beta_{\min} = 0$.

Algorithm 5.1 consists of two loops to find the optimal power allocation. The outer loop is used for the bisection search of β and the inner loop is used to solve $f(P(t)) = \beta$ for a given β. In addition, the convergence of Algorithm 5.1 is ensured by the bisection search, where $\varepsilon_{\Delta P}$ and $\varepsilon_{\Delta\beta}$ are small constants to control the convergence accuracy.

Algorithm 1 Dynamic resource allocation algorithm

Initialize $\beta_{\min}$, $\beta_{\max}$, $P_{\min}$, $P_{\max}$, $\varepsilon_{\Delta\beta} = 10^{-6}$, $\varepsilon_{\Delta P} = 10^{-6}$.
while $(\beta_{\max} - \beta_{\min}) \geq \varepsilon_{\Delta\beta}$
$\beta = (\beta_{\max} + \beta_{\min})/2$;
for $t = 0$ to $\frac{T}{2}$ **do**
while $P_{\max} - P_{\min} \geq \varepsilon_{\Delta P}$ **do**
$P(t) \leftarrow (P_{\max} + P_{\min})/2$;
Calculate $f(P(t))$;
if $f(P(t)) \geq \beta$ **then**
$P_{\max} \leftarrow P(t)$;
else
$P_{\min} \leftarrow P(t)$
end if
end while
end for
if $\sum_{t=0}^{\frac{T}{2}} P(t) \geq \left(\frac{T}{2}+1\right) P_{\text{av}}$ **then**
$\beta_{\max} \leftarrow \beta$
else
$\beta_{\min} \leftarrow \beta$
end if
end while

5.3.9 The Greedy Algorithm

We obtain the power allocation results using Algorithm 5.1, and then the solution of virtual variable $x^*(t)$ for any slot t can be computed by using (5.21). Since the solution $x^*(t)$ is continuous, based on (5.16), it can not ensure that the packet allocation solution $\mu_k^*(t)$ is an integer for any service k and slot t. As a result, the solution $\mu_k^*(t)$ is not valid for practical purposes since the number of allocated packets must take integer value. As an alternative, an integer solution $\mu_k^*(t)$ in problem P1 can be obtained if the non-integer solution $\mu_k^*(t)$ is rounded to the nearest integer, but there is no guarantee about satisfying the problem constraints in problem P1.

According to the above analysis, if $x^*(t)$ is an integer, then the integer solution of problem P1 can be obtained by (5.16). Based on this idea, we propose a greedy algorithm to find the integer solution of the virtual variable $x(t)$ for any slot t, which is denoted as $y^*(t)$, and $y(t)$ is an integer variable corresponding to $x(t)$. Since $x^*(t)$ and $y^*(t)$ largely coincide, the non-integer solution $x^*(t)$ can be used to obtain the integer solution $y^*(t)$. Then substituting $y(t)$ for $x(t)$ in (5.21), the power allocation solution can be computed by

$$P(t) = \left(\exp\left(\frac{y(t)}{\eta}\right) - 1\right)N(t). \tag{5.27}$$

Since (5.27) establishes a one-to-one mapping between $y(t)$ and $P(t)$, the average power constraint should be considered when finding the integer solution $y^*(t)$. In particular, the proposed greedy algorithm provides a valid integer solution $y^*(t)$, derived from the non-integer solution $x^*(t)$ at any slot t, fulfills the average power constraint, achieves an objective value as close as possible to the maximum objective value in problem P5, and has low complexity.

Algorithm 2 Dynamic resource allocation algorithm

Input: $x^*(t)$ for slot $t \in [0, \frac{T}{2}]$
for $t = 0$ to $\frac{T}{2}$ **do**
$y(t) \leftarrow \lfloor x^*(t) \rfloor$;
Calculate $P(t)$ by (27);
Calculate $\tilde{P}(t)$ using $\tilde{y}(t) = y(t) + 1$ by (27);
$\Delta P(t) = \tilde{P}(t) - P(t)$;
$\Delta g(t) = g(\tilde{P}(t)) - g(P(t))$;
end for
repeat
$\mathcal{T}_A$ calculation;
$t' = \arg\max_{t \in \mathcal{T}_A} \left(\frac{\Delta g(t)}{\Delta P(t)}\right)$;
$y(t') \leftarrow \tilde{y}(t')$;
$P(t') \leftarrow \tilde{P}(t')$;
Calculate $\tilde{P}(t')$ using $\tilde{y}(t') = y(t') + 1$ by (27);
$\Delta P(t') = \tilde{P}(t') - P(t')$;
$\Delta g(t') = g(\tilde{P}(t')) - g(P(t'))$;
until $\mathcal{T}_A = \emptyset$
for $t = 0$ to $\frac{T}{2}$ **do**
$y^*(t) \leftarrow y(t)$, $P^*(t) \leftarrow P(t)$;
end for
Output: $y^*(t)$ and $P^*(t)$ for slot $t \in [0, \frac{T}{2}]$

The pseudocode of the greedy algorithm is provided in Algorithm 5.2 and its main steps are sketched as follows. In step 2, each non-integer solution $x^*(t)$ is rounded to its floor integer $y(t)$ by the floor integer function $\lfloor\cdot\rfloor$, and then the corresponding power allocation $P(t)$ is calculated based on (5.27) in step 3. This may cause that the total power is underutilized and the maximum objective value is not achieved. Thus, the remaining power will be allocated along the time to increase the objective function value in the following steps. In step 4, $y(t)$ is assumed to be added one for any slot t, and then the increased power $\Delta P(t)$ and increased objective function value $\Delta g(t)$ are calculated in step 5 and step 6, respectively. The allocation process from step 8 to step 16 is repeated to add one to the selected $y(t)$ at each process until the set $\mathcal{T}_A$ is empty. $\mathcal{T}_A$ in step 9 represents the set of active slots at which $y(t)$ can possibly be added one under the average power constraint, given by

$$\mathcal{T}_A = \left\{ t|t \in \left[0, \frac{T}{2}\right], \Delta P(t) + \sum_{t=0}^{T/2} P(t) \leq \left(\frac{T}{2} + 1\right) P_{\text{av}} \right\}. \tag{5.28}$$

In step 10, the slot t's in set $\mathcal{T}_A$ which can achieve the maximal ratio of $\Delta g(t)$ to $\Delta P(t)$ is selected, which implies that the increase of objective function value per power is maximal at slot t. Then only $y(t')$ can be added one and the corresponding power consumption $P(t')$ can be assigned in step 11 and step 12, respectively. Next, $\Delta P(t')$ and $\Delta g(t')$ can be updated from step 13 to step 15. Finally, the integer solution $y^*(t)$ and the power allocation solution $P^*(t)$ can be obtained in step 18.

According to Algorithm 5.2, $y^*(t)$ and $P^*(t)$ at slot $t \in \left[0, \frac{T}{2}\right]$ have been obtained. Based on Lemma 5.4 and (5.27), $y^*(t)$ and $P^*(t)$ at slot $t \in [0, T]$ can be calculated. And then substituting $y^*(t)$ for $x^*(t)$ in (5.16), we can obtain the packet allocation solution for any service $k \in \mathcal{K}$ at slot $t \in [0, T]$. Furthermore, the greedy algorithm with low complexity leads to a near optimal rather than an optimal solution of the problem P5, which implies that the obtained integer packet allocation solution and power allocation solution of the problem P1 are both near optimal.

5.3.10 Numerical Results and Discussions

We implement the proposed algorithm using MATLAB and present simulation results to illustrate the performance of it. In order to emphasize different service weights, without loss of generality, we set the integer weight value $\omega_k = k$ for any $k \in \mathcal{K}$. In addition, we summarize the simulation parameters in Table 5.1. A single simulation runs the algorithm when the train moves from the edge to the center of the BS coverage (5.25,000 slots).

The power allocation and link capacity along the time for the four power allocation schemes are presented in Figs. 5.3 and 5.4, respectively. The advantage and

Table 5.1 Parameters in simulation

Parameter	Description	Value
P_{av}	Average power constraint	30 W
W	System bandwidth	10 MHz
L	Packet size	240 bits
T_s	Slot duration	1 ms
α	Pathloss exponent	4
K	Number of services	6
v	Constant moving speed	100 m/s
R	Cell radius	2.5 km
d_0	Distance between BS and rail	100 m
N_0	Noise power spectral density	−157 dBm/Hz

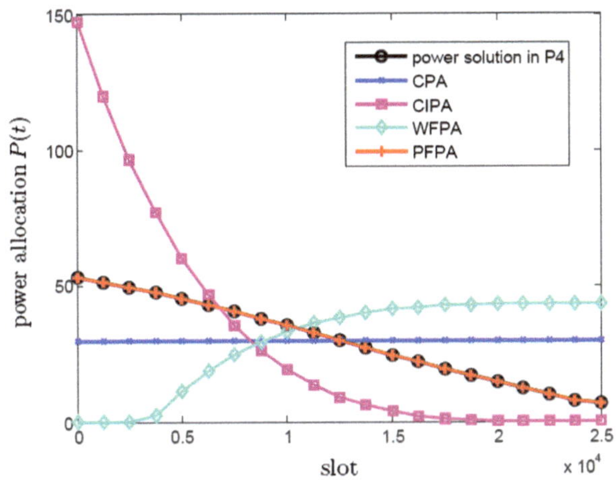

Fig. 5.3 Power allocation under different schemes

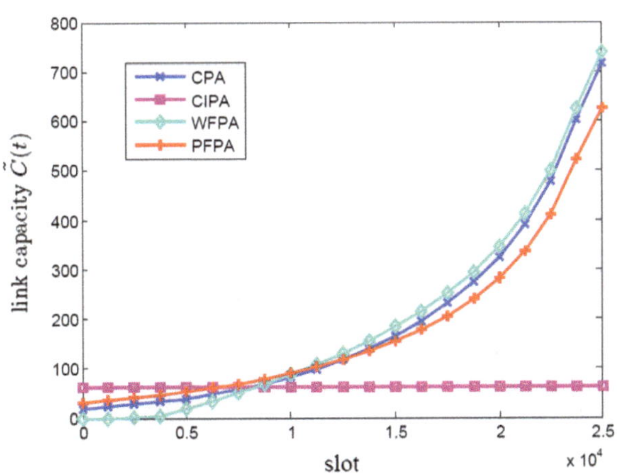

Fig. 5.4 Link capacity under different schemes

disadvantage of these schemes can be observed from these two figures. For the sake of convenience in the engineering implementation, a constant power is allocated along the time in CPA scheme while it ignores the variation of channel gain and results in the great unfairness in term of link capacity. In order to provide a stable link capacity and achieve the best fairness along the time, the CIPA scheme spends much power to compensate those bad channel states when the train is far from the BS. Similar to the traditional water-filling method, the WFPA scheme can maximum the total link capacity at the BS, whereas all the services will generally suffer from starvation when the train is near the edge of the BS coverage. In addition, the PFPA scheme can achieve a trade off between the total link capacity and the link capacity fairness along the time. Finally, from Fig. 5.3, we can observe that the power allocation solutions of PFPA problem and problem P4 are the same, which verifies Lemma 5.3 by simulations.

Figure 5.5 presents a comparison of the optimal packet solution $\mu_4^*(t)$ of problem P2 with the solution obtained by the other schemes. It can be observed that the trend of the curves in Fig. 5.5 is similar to that in Fig. 5.4. This can be explained by (5.14), which shows that the packet allocation solution $\mu_4^*(t)$ is linear with respect to the link capacity $\tilde{C}(t)$. Moreover, we can see that the optimal packet solution $\mu_4^*(t)$ of problem P2 is the same as the solution obtained by the PFPA+PAS scheme, which implies that the problem P2 can be equivalently decomposed into PFPA problem and PAS problem. The service 4 is just an example for illustrating the characteristics of the problem P2 and the same results can be obtained for the other services.

Figure 5.6 compares the power allocation solutions of three different methods solving the problem P5 and their computational complexities, where the complexity is represented by the computing time in an Intel Core 3.30 GHz computer. In this figure, the power allocation solutions are plotted in a single simulation and the computing times are obtained by averaging 100 simulations. It can be observed that the power allocation solutions of these methods are almost same, which implies that

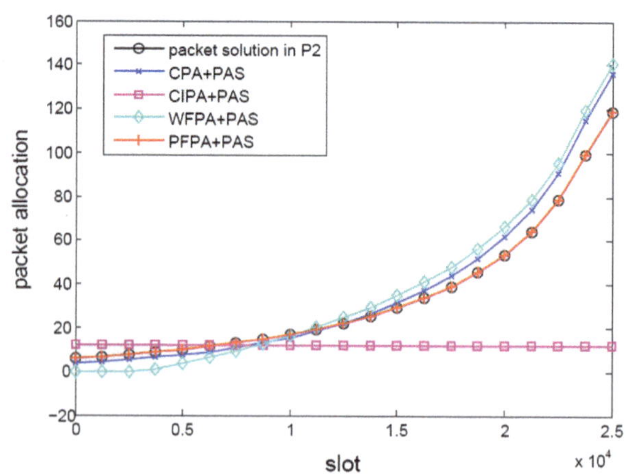

Fig. 5.5 Packet allocation solution for problem P2

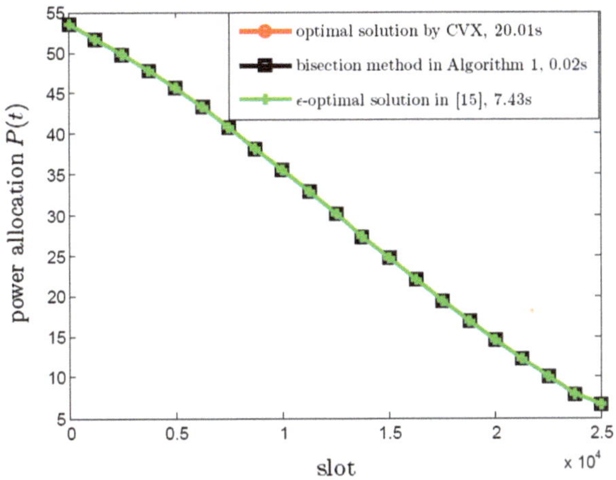

Fig. 5.6 Power allocation solution of different methods for problem P5

the optimal power allocation solution of problem P5 can be obtained by the bisection search method when $\varepsilon_{\Delta P}$ and $\varepsilon_{\Delta\beta}$ are arbitrarily small. In addition, it is worth noting that the computing time of the bisection search method is much lower than that of other two methods, which illustrates the high effectiveness of Algorithm 5.1.

The packet allocation solutions of service 2 and service 4 are described in Fig. 5.7, including the non-integer packet allocation solution of problem P2 and the integer packet allocation solution obtained by Algorithm 5.2. From this figure, we can see that more and more packets are allocated to each service when the train moves from the edge to the center of the BS coverage. The curve of the integer solution is just around that of the non-integer solution for both two services. The similar results can be obtained for other services. In addition, the number of the packets allocated to service 4 equals twice the number of the packets allocated to service 2, which can be explained by Lemma 5.1.

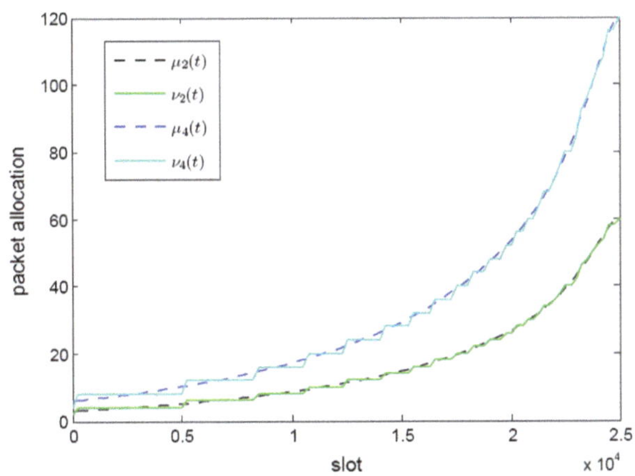

Fig. 5.7 Packet allocation solutions of service 2 and service 4

Appendix A

Proof of Lemma 5.1 Introduce a dual variable $\lambda \geq 0$ for the first constraint of (5.12). The Lagrangian function of the optimization problem (5.12) is

$$\begin{aligned} L(\{\mu_k\}, \lambda) &= \sum_{k \in \mathcal{K}} \omega_k \ln(\mu_k) - \lambda \left(\sum_{k \in \mathcal{K}} \mu_k - \tilde{C} \right) \\ &= \sum_{k \in \mathcal{K}} (\omega_k \ln(\mu_k) - \lambda \mu_k) + \lambda \tilde{C}. \end{aligned} \tag{5.29}$$

The dual function $g(\lambda)$ can now be stated as

$$g(\lambda) = \begin{cases} \max & L(\mu_k, \lambda) \\ \text{s.t.} & \mu_k \geq 0, \forall k \in \mathcal{K}. \end{cases} \tag{5.30}$$

From the solution of (5.30), the resource allocation vector $\boldsymbol{\mu}$ can be determined by solving K decomposed problems with an explicit solution $\mu_k = \frac{\omega_k}{\lambda}$.The dual problem of (5.12) can be expressed as

$$\min_{\lambda \geq 0} g(\lambda) = \sum_{k \in \mathcal{K}} \left(\omega_k \ln\left(\frac{\omega_k}{\lambda}\right) - \omega_k \right) + \lambda \tilde{C}, \tag{5.31}$$

whose optimal solution is $\lambda_* = \frac{\sum_k \omega_k}{\tilde{C}}$. Since the problem (5.12) is convex and satisfies the Slater's condition, strong duality holds between (5.12) and its dual problem (5.31). The solution of (5.12) can be obtained by the solution of (5.31) and is given by

$$\mu_k = \frac{\omega_k}{\lambda^*} = \frac{\omega_k \tilde{C}}{\sum\limits_k \omega_k}, \forall k \in \mathcal{K}. \tag{5.32}$$

In addition, we can obtain that

$$\frac{\mu_i^*}{\omega_i} = \frac{\mu_j^*}{\omega_j} = \frac{1}{\lambda^*}, \forall i, j \in \mathcal{K} \tag{5.33}$$

This completes the proof of Lemma 5.1.

Appendix B

Proof of Lemma 5.2 Based on the Definition 3 in [35], an optimal solution vector $\mathbf{x}^* = [x^*(1), \ldots, x^*(T)]^T$ is said to be proportionally fair along the time if and only if, for any feasible solution vector $\mathbf{x}' = [x'(1), \ldots, x'(T)]^T$, we have

$$\sum_{t=0}^{T} \frac{x'(t) - x^*(t)}{x^*(t)} \leq 0. \tag{5.34}$$

Firstly, we will prove that the solution of problem P3 satisfies (5.34) for any feasible vector $\mathbf{x}'$. For the ease of exposition, let $U(\mathbf{x})$ denote the objective function of (5.18) in problem P3. Since the strictly concave increasing property of $U(\mathbf{x})$, the following condition holds at $\mathbf{x} = \mathbf{x}^*$

$$\sum_{t=0}^{T} \left.\frac{\partial U(\mathbf{x})}{\partial x(t)}\right|_{x(t)=x^*(t)} (x'(t) - x^*(t)) = \sum_{t=0}^{T} \frac{x'(t) - x^*(t)}{x^*(t)} \leq 0. \tag{5.35}$$

This is can be explained by the fact that movement along any direction $(\mathbf{x} - \mathbf{x}^*)$ at the optimal vector $\mathbf{x}^*$ can not improve the objective function. Thus, the optimal solution vector $\mathbf{x}^*$ is proportionally fair.

Secondly, due to $\mu_k(t) = \omega_k x(t)$ in (5.16), for any service k, we have

$$\sum_{t=0}^{T} \frac{\mu_k'(t) - \mu_k^*(t)}{\mu_k^*(t)} = \sum_{t=0}^{T} \frac{\omega_k x'(t) - \omega_k x^*(t)}{\omega_k x^*(t)} = \sum_{t=0}^{T} \frac{x'(t) - x^*(t)}{x^*(t)} \leq 0, \tag{5.36}$$

where $\mu_k'(t)$ is the feasible solution corresponding to $x'(t)$ and $\mu_k^*(t)$ is the optimal solution corresponding to $x^*(t)$. Thus, the optimal solution of problem P3 provides proportionally fair resource allocation along the time for each service.

Appendix C

Proof of Lemma 5.3 First, substituting $\tilde{C}(t)$ into PFPA problem yields

$$\begin{aligned} &\text{maximize} && \sum_{t=0}^{T} \ln\left(\xi \ln\left(1 + \tfrac{P(t)}{N(t)}\right)\right) \\ &\text{subject to} && \tfrac{1}{T+1}\sum_{t=0}^{T} P(t) \leq P_{\text{av}}, \\ &\text{variables} && P(\tau) \geq 0, \tau \in [0, T], \end{aligned} \tag{5.37}$$

where $\xi = T_s W/L \ln 2$. For the first constraint in (5.37), when the optimal solution is achieved, the equality holds. In this sense, the constraint in problem P4 and PFPA problem is same. To compare these two problems, the only difference is that η in problem P4 and ξ in PFPA problem. Thus, if we can prove that the optimal solutions of these two problems are independent of η and ξ respectively, then the optimal solution of problem P4 is equivalent to that of the PFPA problem.

Consider the Lagrangian function of PFPA problem

$$\begin{aligned} L(\{P(t)\},\lambda) &= \sum_{t=0}^{T} \ln\left(\eta\left(1+\frac{P(t)}{N(t)}\right)\right) - \lambda\left(\sum_{t=0}^{T} P(t) - (T+1)P_{\mathrm{av}}\right) \\ &= \sum_{t=0}^{T} \left[\ln\left(\eta\left(1+\frac{P(t)}{N(t)}\right)\right) - \lambda P(t) + \lambda P_{\mathrm{av}}\right]. \end{aligned} \tag{5.38}$$

Since the PFPA problem is convex, by applying the KKT conditions, we have

$$\frac{\partial L(\{P(t)\},\lambda)}{\partial P(t)} = \frac{1}{\ln\left(1+\frac{P(t)}{N(t)}\right)(P(t)+N(t))} - \lambda = 0, \tag{5.39}$$

and

$$\lambda\left(\sum_{t=0}^{T} P(t) - (T+1)P_{\mathrm{av}}\right) = 0. \tag{5.40}$$

Thus, the optimal solution of PFPA problem can be obtained by solving (5.39) and (5.40), which is independent of η.

Similarly, we can show that the optimal solution of problem P4 is independent of ξ. Therefore, the optimal solution of problem P4 is the same as that of the PFPA problem, which completes the proof.

Appendix D

Proof of Lemma 5.4 We proof this lemma by contradiction. Without loss of generality, we assume $P^*(t_1) \neq P^*(T-t_1)$ for certain $t_1 \in [0,T]$. Construct another solution vector $\mathbf{P}'$ by replacing the elements $P^*(t_1)$, $P^*(T-t_1)$ in $\mathbf{P}^*$ with $P'(t_1)$, $P'(T-t_1)$ and keeping all other elements unchanged, where $P'(t_1) = P'(T-t_1) = \frac{P^*(t_1)+P^*(T-t_1)}{2}$. Notice that the following equality holds

$$P'(t_1) = P'(T-t_1) = P^*(t_1) + P^*(T-t_1), \tag{5.41}$$

which implies the solution vector $\mathbf{P}'$ satisfies the first constraint in (5.22) and a feasible solution vector.

For any $t \in [0,T]$, since $d(t) = d(T-t)$ and $N(t) = WN_0 d^{\alpha}(t)$, we have $N(t) = N(T-t)$. Since $\ln(\ln(\cdot))$ is a concave function, based on Jensen's inequality, we can obtain

$$
\begin{aligned}
&\ln\left(\eta \ln\left(1+\frac{P^*(t_1)}{N(t_1)}\right)\right)+\ln\left(\eta \ln\left(1+\frac{P^*(T-t_1)}{N(T-t_1)}\right)\right)\\
&=\ln\left(\eta \ln\left(1+\frac{P^*(t_1)}{N(t_1)}\right)\right)+\ln\left(\eta \ln\left(1+\frac{P^*(T-t_1)}{N(t_1)}\right)\right)\\
&\leq \ln\left(\eta \ln\left(1+\frac{P'(t_1)}{N(t_1)}\right)\right)+\ln\left(\eta \ln\left(1+\frac{P'(T-t_1)}{N(t_1)}\right)\right)\\
&=\ln\left(\eta \ln\left(1+\frac{P'(t_1)}{N(t_1)}\right)\right)+\ln\left(\eta \ln\left(1+\frac{P'(T-t_1)}{N(T-t_1)}\right)\right)
\end{aligned} \tag{5.42}
$$

which illustrates the solution vector $\mathbf{P}^*$ is not optimal, which contradicts with the assumption. Therefore, for any $t \in [0, T]$, there must be the case $P^*(t) = P^*(T-t)$ in the optimal solution vector $\mathbf{P}^*$.

5.4 Dynamic Resource Management

With the rapid development of high-speed railway (HSR) system, the resource allocation problem in HSR wireless network becomes one of the key issues to improve the efficiency of resource utilization. In this section, we investigate the downlink resource allocation problem for multimedia services delivery in HSR MIMO-OFDM system with a cellular/infostation integrated network architecture. Taking the train trajectory and network stability into account, we formulate the problem as a stochastic network optimization programming, which aims at maximizing the overall system utility while keeping the system stable under the total transmission power constraint. To address the NP-hard mixed-integer programming, the original problem is firstly transformed into a queue stability problem, and then decomposed into two separate subproblems by the drift-plus-penalty approach. Finally, based on the stochastic optimization technique, a dynamic resource allocation algorithm is proposed and its efficient is illustrated by theoretical analysis and numerical simulations.

5.4.1 *System Model*

We consider a HSR downlink MIMO-OFDM system with a cellular/infostation integrated network architecture. Similar to [27], H infostations with small coverage areas are deployed along the rail track to provide efficient data transmission, while the cellular network with seamless coverage is used for supporting the control channels over the region. For instance, we have $H = 3$ in Fig. 5.8. Each infestation with power P watts covers a segment of the rail line based on its wireless transmission range. A central controller (CC) is deployed which allocates the network radio resources

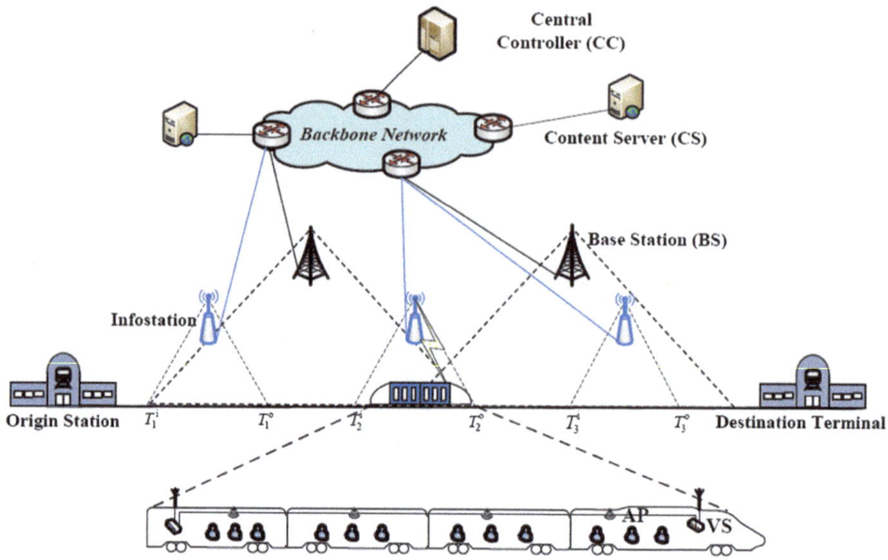

Fig. 5.8 System model

based on the train trajectory and wireless channel condition for the whole network. The CC is connected with the base stations (BSs) and content servers (CSs) via the backbone network, and the CSs are connected to the BSs via wireline links. The bandwidth of the links from the CSs to infostations is supposed to be sufficiently large. The infostations can communicate with the vehicle station (VS), which is connected to the corresponding vehicle antennas with high-quality links. In order to ensure reliable communication between infostation and the train, two VSs are installed on the top of the first and the last carriages, respectively. They can work independently or cooperatively depending on the specific circumstances. Meanwhile, the VS is further connected to the access points (APs) inside the train. Assume that the data transmission rate from the VS to passenger devices is sufficiently high, hence the data packet can be received successfully if it has been delivered to the VS. When the train is moving, the passengers send service requests from the VS to the CS through the cellular network. The requested data packets are then delivered from the CS to the VS via the infostations, and the VS will eventually forward the data to the passenger devices. For simplicity, we assume that the buffer space of VS is unlimited.

In addition, multiple-input multiple-output (MIMO) antennas are applied to the cellular/infostation integrated HSR network. Assume that infostation and VS are equipped with N_t transmit antennas and N_r receive antennas, respectively. We denote the N_t dimensional transmitted signal vector as x, and the Nr-dimensional received signal vector as y. The corresponding $N_r \times N_t$ hannel matrix and the Nr-dimensional noise vector are denoted as H and n, respectively. Then, the received signal is given by

$$\mathbf{y} = \mathbf{Hx} + \mathbf{n}. \tag{5.43}$$

For a MIMO link without interference, the infostation can deliver multiple data streams to the VS using spatial multiplexing. Specifically, for a $N_r \times N_t$ MIMO link, $I = \text{rank}(\mathbf{H})$ data streams are multiplexed by using a precoding matrix $\mathbf{V}$ at the infostation and are reconstructed with a decoding matrix $\mathbf{U}$ at the VS. The matrices $\mathbf{V}$ and $\mathbf{U}$ are obtained from the singular value decomposition (SVD) of the channel matrix H [41]

$$\mathbf{H} = \mathbf{U\Lambda V}^{\text{H}} \tag{5.44}$$

where $\mathbf{U}$ is a $N_r \times N_t$ unitary matrix, $\mathbf{V}$ is a $N_r \times N_t$ unitary matrices, $\mathbf{\Lambda}$ is a $N_r \times N_t$ rectangular matrix with nonnegative main diagonal elements $\{g_1, g_2, \ldots, g_I\}$ and all other elements equal to zero, and superscript $\mathbf{H}$ means conjugate transpose. By applying the above precoding and decoding matrices, the MIMO channel is transformed into I parallel single input single-output (SISO) channels that do not interfere with each other. Therefore, the channel gain of each SISO channel is $g_i(i = 1, 2, \ldots, I)$. It is assumed that the knowledge of channel state information (CSI) is available at CC.

Note that each high-speed train moves on a predetermined rail line with highly stable time schedule [0, T] from the origin station to its destination terminal. Thus, the information of train trajectory and network resources can be obtained by the CC in advance with high accuracy. We consider a slotted system, where time is divided into slots of equal length. Let T_h^i and T_h^o represent time instants for the train to come into and go out of the transmission range of the hth $(h \in \{1, 2, \ldots H\})$ infostation, respectively. Accordingly, we have $0 \leq T_1^i$ and $T_H^o \leq T$. Taking account of the intermittent network connectivity, we have $T_h^o \leq T_{h+1}^i$ for $1 \leq h \leq H - 1$. Moreover, the transmission period and idle period are defined as the time when the train is in and out of the coverage area of an infostation, respectively.

We consider the scenario where the data packets requested from the users arrive at their associated CSs according to a stationary process. The CSs maintains a transmission queue for each of its intended users. Assume that there are K users in the HSR wireless networks and the user set is denoted by $\mathcal{K} \triangleq \{1, 2, \ldots, k\}$. Let $Q_k(t)$ represent the current queue backlog for user $k(k \in \mathcal{K})$ at the beginning of time slot $t(t \in \{0, 1, \ldots, T\})$ in the buffer of CSs. The corresponding data arrival rate and service rate for user k during time slot t are denoted as $A_k(t)$ and $R_k(t)$, respectively. The packet arrival process for each user is assumed to be independent and identically distributed (i.i.d.) across slots. Then, the system queues evolve according to the following stochastic difference equation

$$Q_k(t+1) = \max\{Q_k(t) - R_k(t)\} + A_k(t), \quad \forall k \in \mathcal{K} \tag{5.45}$$

Since the total buffer size of CSs is finite in practice, it is thus important to consider the system stability. Therefore, the downlink resource allocation scheme should guarantee that all queues in the system are strongly stable. For convenience, we introduce the following definition.

Definition 1 A discrete-time queue Q_k is strongly stable if $\limsup_{t\to\infty}\frac{1}{t}\sum_{\tau=0}^{t-1}\mathbb{E}\left[Q_k(\tau)\right]<\infty$. The system is stable if all queues in the system are strongly stable.

The above definition implies that the system is stable when the average backlog of each queue is bounded. According to [42], we can know that to guarantee the stability of the system, the average data arrival rate of each queue should be no larger than the corresponding average service rate. That is $\bar{A}_k \leq \bar{R}_k, \forall k \in \mathcal{K}$, where $\bar{A}_k = \lim_{t\to\infty}\frac{1}{t}\sum_{\tau=0}^{t-1}\mathbb{E}\{A_k(\tau)\}$ and $\bar{R}_k = \lim_{t\to\infty}\frac{1}{t}\sum_{t=0}^{T-1}\mathbb{E}\{R_k(\tau)\}$.

5.4.2 Problem Formulation

In this section, we will formulate a discrete-time stochastic optimization problem which aims to maximize the overall system utility while stabilizing all transmission queues under the total power constraint. Consider a HSR downlink MIMO-OFDM system, we assume a frequency selective fading channel model, where different sub-carriers will experience different channel gains and the channel gain is independent and identically distributed (i.i.d.) in different slots. There are N sub-carriers that can be assigned to K users from I antennas during T time slots, and the corresponding sub-carrier set and antenna set are denoted by $\mathcal{N} \triangleq \{1,2,\ldots,N\}$, and $\mathcal{I} \triangleq \{1,2,\ldots,I\}$, respectively. Moreover, it is assumed that each sub-carrier is assigned to one user from one antenna at each time slot. We use a binary variable $\rho_{kni}(t) \in \{0,1\}$ to represent the situation of resource allocation, indicating whether the nth sub-carrier at tth time slot is allocated to the kth user from the ith antenna or not. For the HSR communication scenario, the instant data rate for the user k is given by

$$R_k(t) = \sum_{n=1}^{N}\sum_{i=1}^{I}\rho_{kni}(t)B\log\left(1+\frac{p_{kni}(t)g_{kni}(t)^2}{N_0B+\mathrm{ICI}_n(t)}\right), \tag{5.46}$$

where B denotes the bandwidth of the sub-carrier, N_0 represents the single-sided noise power spectral density (PSD) of each sub-carrier, $p_{kni}(t)$ and $g_{kni}(t)$ are the transmission power and channel gain for the kth user on the nth sub-carrier from the ith antenna at the tth time slot, respectively, and $\mathrm{ICI}_n(t)$ is the inter-carrier-interference experienced on the nth sub-carrier at the tth slot.

Notice that the ICI power $\mathrm{ICI}_n(t)$ caused by Doppler shift is not coordinated among different sub-carriers, but its average impact is considered. By doing so, the wireless system can adapt to the varying channel timely and easily. The $\mathrm{ICI}_n(t)$ can be expressed as

$$\mathrm{ICI_n}(t) \approx \frac{(f_d T_s)^2}{2} \sum_{l=1, l \neq n}^{N} \frac{p_l(t)}{(l-n)^2}, \tag{5.47}$$

where $p_l(t) = \sum_{k=1}^{K} \sum_{i=1}^{I} p_{kli}(t)$ is the transmission power of the lth sub-carrier at the tth time slot, and T_s is the OFDM symbol duration. Moreover, $f_d = \upsilon f_c / c$ represents maximum Doppler shift with the moving speed v, the carrier frequency f_c and the velocity of light c. A tight universal upper bound on the ICI power [43] can be employed to ease the challenge

$$\mathrm{ICI_{upperbound}} \leq \frac{1}{12} (2\pi f_d T_s \sqrt{P})^2. \tag{5.48}$$

Our objective is to maximize a utility function over the trip of a train subject to network stability and the total transmission power. The utility function is usually chosen as a concave, nondecreasing function of the service rates and should reflect a certain fairness criterion. Thus, the stochastic network optimization problem can be formulated as

$$\begin{aligned}
(\mathrm{P1}) \quad \max_{\{\rho_{kni}(t), P_k(t)\}} \quad & \phi(\overline{\mathrm{R}}) \\
s.t. \quad & \overline{R}_k \geq \overline{A}_k, \ \forall k \in \mathcal{K} \\
& P_k(t) = 0, \ T_h^o \leq t \leq T_{h+1}^i, \ h \in \{1, 2, \ldots, H\}, \forall k \in \mathcal{K} \\
& 0 \leq \sum_{k \in \mathcal{K}} P_k(t) \leq P, \ T_h^i \leq t \leq T_h^o, h \in \{1, 2, \ldots H\}, \forall k \in \mathcal{K} \\
& \sum_{k \in \mathcal{K}} \sum_{i \in \mathcal{I}} \rho_{kni}(t) = 1, \ \forall n \in \mathcal{N}, \mathrm{t} \in \{0, 1, \ldots, T\} \\
& \rho_{kni}(t) \in \{0, 1\}, \ \forall k \in \mathcal{K}, \ \forall n \in \mathcal{N}, \ \forall i \in \mathcal{I}, \ t \in \{0, 1, \ldots T\}
\end{aligned} \tag{5.49}$$

where the proportional fair utility function $\phi(\overline{\mathrm{R}}) = \sum_{k \in \mathcal{K}} \log(\overline{R}_k)$, the vector $\overline{\mathrm{R}} = (\overline{R}_k, k \in \mathcal{K})$, $\overline{R}_k = \lim_{t \to \infty} \frac{1}{t} \sum_{\tau=0}^{t-1} \mathbb{E}\{\mathrm{R}_k(\tau)\}$, and $P_k(t) = \sum_{n \in \mathcal{N}} \sum_{i \in \mathcal{I}} p_{kni}(t)$. The first constraint is to maintain stability of the system while the second constraint implies that users in the train can only be serviced during the transmission period. The third constraint shows that the total power that can be allocated to the users within the h th infestation coverage is limited by the infostation power P. The fourth and fifth constraints imply that the nth sub-carrier can only be allocated to one user from one antenna at the t th time slot.

The optimization problem *P1* is a mixed-integer nonlinear programming (MINLP) problem, which is in general NP-hard since it combines the difficulty of optimizing over integer variables with the handling of nonlinear functions. Therefore, it is difficult to directly find an optimal solution $\overline{\mathrm{R}}^*$ for this problem. Besides, since solving the above problem requires CSI for all time slots, which is not possible in practice, an efficient online algorithm should be considered. To solve the above dynamic optimization problem, we can use the stochastic optimization framework and propose a dynamic resource allocation policy that operates arbitrarily closely to the optimal point $\overline{\mathrm{R}}^{*}$, which will be discussed in the following section.

5.4.3 Dynamic Resource Management Schemes

We propose an efficient dynamic resource allocation scheme that achieves near optimal performance by applying the stochastic network optimization approach [42]. The main idea of the aforementioned framework is to transform the original problem that involves maximizing a function of time averages into an equivalent problem that involves maximizing a single time average. Then, the drift-plus-penalty approach can be applied. This transformation is achieved through the use of auxiliary variables $\gamma_k(t)$ for each user k at time slot t and corresponding virtual queues $W_k(t)$ with buffer evolution

$$W_k(t+1) = \max\{W_k(t) - R_k(t) + \gamma_k(t), 0\}, \ \forall k \in \mathcal{K}. \tag{5.50}$$

Then, the problem P1 can be transformed to the following problem:

$$\begin{aligned}
(\text{P2}) \quad \max_{\{\rho_{kni}(t), P_k(t)\}} \quad & \overline{\phi(\gamma)} \\
s.t. \quad & \overline{\gamma}_k \leq \overline{R}_k, \ \forall k \in \mathcal{K} \\
& 0 \leq \overline{\gamma}_k \leq \gamma_{\max}, \ \forall k \in \mathcal{K} \\
& \overline{R}_k \geq \overline{A}_k, \ \forall k \in \mathcal{K} \\
& P_k(t) = 0, \ T_h^o \leq t \leq T_{h+1}^i, h \in \{1, 2, \ldots, H\}, \ \forall k \in \mathcal{K} \\
& 0 \leq \sum_{k \in \mathcal{K}} P_k(t) \leq P, \ T_h^i \leq t \leq T_h^o, h \in \{1, 2, \ldots H\}, \forall k \in \mathcal{K} \\
& \sum_{k \in \mathcal{K}} \sum_{i \in \mathcal{I}} \rho_{kni}(t) = 1, \ \forall n \in \mathcal{N}, \mathrm{t} \in \{0, 1, \ldots, T\} \\
& \rho_{kni}(t) \in \{0, 1\}, \ \forall k \in \mathcal{K}, \ \forall n \in \mathcal{N}, \ \forall i \in \mathcal{I}, t \in \{0, 1, \ldots, T\}.
\end{aligned} \tag{5.51}$$

where $\overline{\phi(\gamma)} = \lim_{t \to \infty} \frac{1}{t} \sum_{\tau=0}^{t-1} \mathbb{E}\{\phi(\gamma(\tau))\}$ and $\gamma(t) = (\gamma(t) \forall k \in \mathcal{K})$. Notice that the first constraint correspond to stability of the virtual queues given in (5.9), since $\overline{\gamma}_k$ and $\overline{R}_k$ are the time-averaged arrival rate and the time-averaged service rate for the virtual queue $W_k(t)$ respectively. We can find that the optimal utility value is the same for both problems P1 and P2.

5.4.4 Lyapunov Drift-Plus-Penalty Approach

To solve the problem P2 effectively, the drift-plus-penalty (DPP) is obtained as follows. Let $\mathbf{\Theta}(t) = [\mathbf{Q}(t), \mathbf{W}(t)]$, where $\mathbf{Q}$(t) and $\mathbf{W}$(t) represent the queue vector and virtual queue vector, respectively. We define the following quadratic Lyapunov function

$$L(\mathbf{\Theta}(t)) \triangleq \frac{1}{2}\left(\sum_{k\in\mathcal{K}} Q_k(t)^2 + \sum_{k\in\mathcal{K}} W_k(t)^2\right). \tag{5.52}$$

Then, the conditional Lyapunov drift for slot t is defined as

$$\Delta(\mathbf{\Theta}(t)) \triangleq \mathbb{E}\{L(\mathbf{\Theta}(t)) - L(\mathbf{\Theta}(t))|\mathbf{\Theta}(t)\}. \tag{5.53}$$

From the Eqs. (5.45), (5.50), and (5.52), we have

$$\begin{aligned}
&L(\mathbf{\Theta}(t+1)) - L(\mathbf{\Theta}(t)) \\
&= \frac{1}{2}\sum_{k\in\mathcal{K}} (Q_k(t+1)^2 - Q_k(t)^2) + \frac{1}{2}\sum_{k\in\mathcal{K}} (W_k(t+1)^2 - W_k(t)^2) \\
&= \frac{1}{2}\sum_{k\in\mathcal{K}} \left((\max\{Q_k(t) - R_k(t), 0\} + A_k(t))^2 - Q_k(t)^2\right) \\
&\quad + \frac{1}{2}\sum_{k\in\mathcal{K}} \left((\max\{W_k(t) - R_k(t) + \gamma_k(t), 0\})^2 - W_k(t))^2\right) \\
&\le \frac{1}{2}\sum_{k\in\mathcal{K}} (A_k(t)^2 + \gamma_k(t)^2 + 2R_k(t)^2) + \sum_{k\in\mathcal{K}} Q_k(t)(A_k(t) - R_k(t)) \\
&\quad + \sum_{k\in\mathcal{K}} W_k(t)(\gamma_k(t) - R_k(t)).
\end{aligned} \tag{5.54}$$

Plugging above equations directly into (5.11) yields

$$\begin{aligned}
\Delta(\Theta(t) &= \mathbb{E}\left\{\frac{1}{2}\sum_{k\in\mathcal{K}} (Q_k(t+1)^2 - Q_k(t)^2 + W_k(t+1)^2 - W_k(t)^2)|\Theta(t)\right\} \\
&\le \left\{\sum_{k\in\mathcal{K}} \frac{1}{2}(A_k(t)^2 - \gamma_k(t)^2 + 2R_k(t)^2)|\mathbf{\Theta}(t)\right\} \\
&\quad + \mathbb{E}\left\{\sum_{k\in\mathcal{K}} Q_k(t)(A_k(t) - R_k(t))|\mathbf{\Theta}(t)\right\} + \mathbb{E}\left\{\sum_{k\in\mathcal{K}} W_k(t)(\gamma_k(t) - R_k(t))|\mathbf{\Theta}(t)\right\} \\
&\le D + \mathbb{E}\left\{\sum_{k\in\mathcal{K}} Q_k(t)(A_k(t) - R_k(t))|\mathbf{\Theta}(t)\right\} + \mathbb{E}\left\{\sum_{k\in\mathcal{K}} W_k(t)(\gamma_k(t) - R_k(t))|\mathbf{\Theta}(t)\right\},
\end{aligned} \tag{5.55}$$

where D is a finite constant that satisfies

$$D = \frac{1}{2}\sum_{k\in\mathcal{K}} (A_{\max}^2 + \gamma_{\max}^2 + 2R_{\max}^2) \geq \mathbb{E}\left\{\sum_{k\in\mathcal{K}} \frac{1}{2}(A_k(t)^2 + \gamma_k(t)^2 + 2R_k(t)^2)|\boldsymbol{\Theta}(t)\right\}.$$

Adding the penalty term $-\beta\mathbb{E}\{\phi(\gamma(t))|\boldsymbol{\Theta}(t)\}$ on both sides of the inequality (5.13), we have

$$\begin{aligned}
&\Delta(\boldsymbol{\Theta}(t)) - \beta\mathbb{E}\{\phi(\gamma(t))|\boldsymbol{\Theta}(t)\} \leq D - \beta\mathbb{E}\{\phi(\gamma(t))|\boldsymbol{\Theta}(t)\} \\
&+ \mathbb{E}\left\{\sum_{k\in\mathcal{K}} Q_k(t)(A_k(t) - R_k(t))|\boldsymbol{\Theta}(t)\right\} \\
&+ \mathbb{E}\left\{\sum_{k\in\mathcal{K}} W_k(t)(\gamma_k(t) - R_k(t))|\boldsymbol{\Theta}(t)\right\},
\end{aligned} \tag{5.56}$$

where $\beta > 0$ is a control parameter of the DPP policy that affects the utility-backlog trade off. The DPP policy acquires information about current queue lengths and the current channel state at every slot t and chooses a control decision to minimize the right hand side of the above inequality. Thus, at each time slot t, the resulting DPP policy is given by the maximization of the following expression

$$\underbrace{\left[\beta\phi(\gamma(t)) - \sum_{k\in\mathcal{K}} \gamma_k(t)W_k(t)\right]}_{\text{utility maximization subproblem}} + \underbrace{\left[\sum_{k\in\mathcal{K}} (Q_k(t) + W_k(t))R_k(t)\right]}_{\text{resource allocation subproblem}} \tag{5.57}$$

Then, the transformed problem is decomposed into two separate optimization problems which will be discussed in the following subsection. And, we will discuss them in the following subsection.

5.4.5 Dynamic Resource Management Algorithm

Based on above approach, we propose a dynamic resource allocation framework as follows. Consider the scenario where queue backlog information and global CSI are available at the CC. At the beginning of time slot $t (t = 0, 1, \ldots, T)$, two cases are considered: slot t is in an idle period (i.e., $T_h^o \leq t \leq T_{h+1}^i$) or in a transmission period (i.e., $T_h^i \leq t \leq T_h^o$). If slot t is in an idle period, we only need to update the actual queue backlog information since no data can be delivered; If t is in a transmission period, the proposed dynamic resource allocation scheme consists of three steps given as follows

Utility Maximization The first step is to determinate the values of auxiliary variables $\gamma_k(t)$ $(k \in \mathcal{K})$ by solving the following problem:

$$\begin{aligned} \max_{\gamma(t)} \ & \beta\phi(\gamma(t)) - \sum_{k\in\mathcal{K}} \gamma(t)W_k(t) \\ s.t. \ & 0 \leq \gamma(t) \leq \gamma_{\max}, \quad \forall k \in \mathcal{K}. \end{aligned} \tag{5.58}$$

where $W_k(t)$ is the corresponding virtual queue backlog known at the CC, vector $\gamma(t) = (\gamma_k(t), k \in \mathcal{K})$ and $\beta, \gamma_{\max} > 0$ are system parameters. These decisions can push the system to approach the maximum of the network utility function, and the optimum solution can be obtained by CVX (http://cvxr.com/cvx).

Resource Allocation The next step is to solve a weighted sum rate maximization (WSRM) problem as follows:

$$\begin{aligned} \max_{\{\rho_{kni}(t), P_k(t)\}} \ & \sum_{k\in\mathcal{K}} (Q_k(t) + W_k(t))R_k(t) \\ s.t. \ & 0 \leq \sum_{k\in\mathcal{K}} P_k(t) \leq P, T_h^i \leq t \leq T_h^o, h \in \{1, 2, \ldots H\}, \forall k \in \mathcal{K}, \\ & \sum_{k\in\mathcal{K}} \sum_{i\in\mathcal{I}} \rho_{kni}(t) = 1, \quad \forall n \in \mathcal{N}, t \in \{1, 2, \ldots T\}, \\ & \rho_{kni}(t) \in \{0, 1\}, \forall k \in \mathcal{K}, \forall n \in \mathcal{N}, \forall i \in \mathcal{I}, t \in \{0, 1, \ldots, T\}. \end{aligned} \tag{5.59}$$

The weight of each user rate is sum of the corresponding actual and virtual queue backlogs. This resource allocation problem is a MINLP problem which is in general very difficult to solve, and it will be discussed in the following.

Queue Updates Finally, we update all the actual queues $Q_k(t+1)$ and virtual queues $W_k(t+1)$ according to (5.45) and (5.50), respectively.

Based on the above framework, we propose a dynamic resource allocation scheme as shown in Algorithm 5.3. The CC first initializes all system parameters before the trip begins. At the beginning of each time slot, we first judge if the time slot is in an idle period or not. If it is in an idle period, we only need to update the actual queue backlog information since no data can be delivered. Otherwise, each CS collects the CSI from its subscribed users, and passes the CSI and queue backlog information to the CC. Next, the CC numerically computes the values of auxiliary variables according to (5.58) in step 6 and selects the desired users and the corresponding antenna allocations, sub-carrier allocations and power allocations by searching for the optimal solution to (5.59) in step 7. Then, the CC updates the virtual queue backlogs and sends the resource allocation decision to each CS in steps 8 and 9, respectively. Finally, at the end of the time slot, CSs update their actual queues according to (5.45). This dynamic resource allocation process will be repeated until the end of the trip.

Algorithm 3 Dynamic resource allocation algorithm

1: Initializeβ, $\gamma_{\max}$, and $W_k(0), \forall k \in \mathcal{K}$
2: **for** each $t \in [0, T]$**do**
3: **if**$T_h^i \leq t \leq T_h^o, h \in \{1, 2, \ldots H\}$**then**
4: Each CS collects the CSI from its intended users.
5: The CC obtains global CSI and queue backlog information from CSs.
6: The CC finds$\gamma_k^*(t), \forall k \in \mathcal{K}$by solving (58).
7: The CC finds$\rho_{kni}^*(t)$ and $P_k^*(t)$by solving (59).
8: The CC updates $W_k(t+1), \forall k \in \mathcal{K}$according to (50).
9: The CC sends out the resource allocation decision to each CS.
10: The CSs update $Q_k(t+1), \forall k \in \mathcal{K}$according to (45) at the end of the time slot.
11: **else**
12: The CSs update $Q_k(t+1), \forall k \in \mathcal{K}$according to (45), with$R_k(t) = 0$.
13: **end if**
14: **end for**

By applying the stochastic network optimization framework, the following theorem can be obtained.

Theorem 1 *Assume the data arrival rate and the service rate are upper bounded by* Amax *and* Rmax, *respectively. For given constants* $\beta, \gamma_{\max}$ *and a concave and entry-wise nondecreasing utility function* $\phi(\cdot)$, *if there exist at least one feasible resource allocation policy, then we have*:

$$\lim_{t \to \infty} \inf \phi(\bar{\mathbf{R}}(t)) \geq \phi^{\text{opt}} - \mathcal{O}(1/\beta) \tag{5.60}$$

$$\lim_{t \to \infty} \sup \frac{1}{t} \sum_{\tau=0}^{t-1} \sum_{k \in \mathcal{K}} \mathbb{E}\{Q_k(\tau)\} \leq D/\epsilon + \mathcal{O}(\beta), \tag{5.61}$$

where D is defined in (5.55), $\epsilon < 0$ *is a parameter, and* ϕ^{opt} *is the maximum utility associated with the problem P1.*

Proof The proof of Theorem 5.1 is provided in Appendix E.

According to this theorem, the proposed algorithm can achieve a utility which is arbitrarily close to ϕ^{opt} by increasing β, while the actual queue backlog of each user grows linearly with β. Therefore, we can obtain a utility-backlog trade off of $[\mathcal{O}(1/\beta), \mathcal{O}(\beta)]$.

5.4.6 Dual Optimization Framework

To implement the proposed dynamic resource allocation scheme, in each time slot of transmission period we need to solve (5.59) in step 2, which is a non-convex problem with exponential complexity. Fortunately, (5.59) is separable across the sub-carriers, and is tied together only by the power constraint. Thus, it is useful to approach the problem using duality principles. Let $\alpha_k(t) = Q_k(t) + W_k(t)$, then the WSRM problem for the high-speed railway downlink MIMO-OFDM system can be reformulated as

$$\begin{aligned}
\max_{\{\rho_{kni}(t),p_{kni}(t)\}} \quad & \sum_{k=1}^{K} \alpha_k(t) \sum_{n=1}^{N} \sum_{i=1}^{I} \rho_{kni}(t) B \log_2\left(1 + \frac{p_{kni}(t) g_{kni}(t)^2}{N_0 B + \mathrm{ICI}_n(t)}\right) \\
s.t. \quad & 0 \leq \sum_{k\in\mathcal{K}} \sum_{n\in\mathcal{N}} \sum_{i\in\mathcal{I}} p_{kni}(t) \leq P_{\text{total}}, t \in \{1,2,\ldots\} \\
& \sum_{k\in\mathcal{K}} \sum_{i\in\mathcal{I}} \rho_{kni}(t) = 1, \quad \forall n \in \mathcal{N}, t \in \{1,2,\ldots\} \\
& \rho_{kni}(t) \in \{0,1\}, \quad \forall k \in \mathcal{K}, \quad \forall n \in \mathcal{N}, \forall i \in \mathcal{I}, t \in \{1,2,\ldots\}.
\end{aligned} \tag{5.62}$$

Denote the domain $\mathcal{D}$ as the set of all nonnegative $p_{kni}(t)$ for all $k \in \mathcal{K}, i \in \mathcal{I}$ and $n \in \mathcal{N}$ at time slot t. For each sub-carrier n, as it can only be allocated to one user from one antenna at each time slot, only one $p_{kni}(t)$ is positive for all $k \in \mathcal{K}$ and $i \in \mathcal{I}$ at time slot t. Introduce dual variable λ to the first constraint, then the Lagrangian of WSRM problem in (5.62) is defined over domain $\mathcal{D}$ as

$$\begin{aligned}
& L(p_{kni}(t), R_{kni}(t), \lambda) \\
&= \sum_{k=1}^{K} \alpha_k(t) \sum_{n=1}^{N} \sum_{i=1}^{I} B \log_2(1 + p_{kni}(t)\gamma_{kni}(t)) - \lambda \sum_{k\in\mathcal{K}} \sum_{n\in\mathcal{N}} \sum_{i\in\mathcal{I}} p_{kni}(t) - P_{\text{total}} \\
&= \sum_{k=1}^{K} \alpha_k(t) \sum_{n=1}^{N} \sum_{i=1}^{I} R_{kni}(t) - \lambda \left(\sum_{k=1}^{K} \sum_{n=1}^{N} \sum_{i=1}^{I} p_{kni}(t) - P_{\text{total}} \right),
\end{aligned} \tag{5.63}$$

where $\gamma_{kni}(t) = \frac{g_{kni}(t)^2}{N_0 B + \mathrm{ICI}_n(t)}$ denotes the instantaneous channel-to-noise ratio (CNR), and $R_{kni}(t) = B \log_2(1 + p_{kni}(t)\gamma_{kni}(t))$. Then, the Largrange dual function is

$$g(\lambda) = \max_{\{p_{kni(t)} \in \mathcal{D}\}} L(p_{kni}(t), R_{kni}(t), \lambda), \tag{5.64}$$

and the dual problem is given by

$$d^* = \min_{\lambda \geq 0,} g(\lambda). \tag{5.65}$$

From (5.63), the maximization of L can be decomposed into the following N separable optimization problems $\tilde{g}_n(\lambda)$, then the Largrange dual function becomes

$$\begin{aligned} g(\lambda) &= \sum_{n=1}^{N} \max_{\{p_{kni}(t) \in D\}} \left\{ \sum_{k=1}^{K} \sum_{i=1}^{I} \alpha_k(t) R_{kni}(t) - \lambda \sum_{k=1}^{K} \sum_{i=1}^{I} p_{kni}(t) \right\} + \lambda P_{\text{total}} \\ &= \sum_{n=1}^{N} \tilde{g}_n(\lambda) + \lambda P_{\text{total}}, \end{aligned} \tag{5.66}$$

where $\tilde{g}_n(\lambda) = \max_{\{p_{kni}(t) \in D\}} \left\{ \sum_{k=1}^{K} \sum_{i=1}^{I} \alpha_k(t) R_{kni}(t) - \lambda \sum_{k=1}^{K} \sum_{i=1}^{I} p_{kni}(t) \right\}$

Assume that user k is active on sub-carrier n from antenna I at time slot t. With a fixed λ, $\tilde{g}_n(\lambda)$ is a concave function of $p_{kni}(t)$. Therefore, by taking the derivative of $\tilde{g}_n(\lambda)$ with respect to $p_{kni}(t)$, we can obtain a closed-form expression for the optimal powers as follows:

$$p_{kni}(t) = \left[\frac{\alpha_k(t) B}{\lambda \ln 2} - \frac{1}{\gamma_{kni}(t)} \right]^+, \tag{5.67}$$

where $[x]^+ = \max\{0, x\}$, and it is a "multi-level water-filling" power allocation with cutoff CNR $\lambda \ln 2/\alpha_k(t)B$, below which we do not transmit any power, and above which we transmit more power when the CNR $\gamma_{kni}(t)$ is higher. Then, by searching over all K possible user assignments and I possible antenna assignments for sub-carrier n, $\tilde{g}_n(\lambda)$ can be obtained as

$$\tilde{g}_n(\lambda) = \max_{k,i} \left\{ \alpha_k(t) B \log_2 \left(1 + \left[\frac{\alpha_k(t) B}{\lambda \ln 2} - \frac{1}{\gamma_{kni}(t)} \right]^+ \gamma_{kni}(t) \right) - \lambda \left[\frac{\alpha_k(t) B}{\lambda \ln 2} - \frac{1}{\gamma_{kni}(t)} \right]^+ \right\} \tag{5.68}$$

Once above equation is solved for all n, the overall Lagrange dual function $g(\lambda)$ is derived from (5.66). Finally, it is required to find $\lambda^* \geq 0$ that minimizes $g(\lambda)$. The update of λ can be done by using a simple bisection method until the sum power converges. Moreover, if the converged sum power is equal to the total power constraint, the duality gap is zero, and thus solving the dual problem implies that the primal problem is also solved. In our numerical results, the power constraints are met almost exactly, resulting in relative optimality gaps that are practically zero ($<10^{-4}$). Thus, our approach can, for all practical purposes, be considered as an optimal solution to the problem.

Once λ^* is determined, we plug it back into the optimal power allocation function (5.67) and arrive at the following simple user assignment and power allocation for each sub-carrier n given as

$$k_n^*(t) = \arg \max_{k \in \mathcal{K}, i \in \mathcal{I}} \{\alpha_k(t) B \log(1 + p_{kni}^*(t)\gamma_{kni}(t)) - \lambda^* p_{kni}^*(t)\} \tag{5.69}$$

$$p_{kni}^*(t) = p_{kni}^*(t) 1(k = k_n^*(t)), \tag{5.70}$$

where 1(x) is the indicator function, which evaluates to 1 if x is true and 0 if false.

5.4.7 Simulation Results

In this section, we evaluate the performance of the proposed dynamic resource allocation scheme for HSR downlink MIMO-OFDM system. Consider a cellular/infostation integrated network where the transmission range of each infostation is approximately 500 m and the distance between the two infostations is 1 km. Each train moves at a constant speed when it travels from the origin station to its destination terminal. The transmission power at each infostation is $P = 5$ W. All wireless links experience Rician fading, then the channel gain $g_{kni}(t)$ is generated by Rician distribution with Rice factor 1. The duration of a time slot is 1 ms. The data packet size is 100 bits. Moreover, the related simulation parameters are set to be $T_s = 1.33 \times 10^{-4}$ s, $f_c = 2.6$ GHz, $c = 3 \times 10^8$ m/s, $B = 7.5$ KHz, $N_0 = 2 \times 10^{-7}$ W/Hz [44]. We consider Poisson arrival processes with different average data rates for users, where the average arrival data rate for a user is randomly selected from the set $\{3, 4, 5, 6\}$ (packets per time slot). The base stations store 50 packets for each user before transmission starts. Other parameters are $\gamma_{\max} = 100$ packets/s and $\beta = 100$.

Figure 5.9 shows the relationship between average utility and the moving speed for different numbers of sub-carriers and antennas, where $K = 8$ and $H = 10$. We

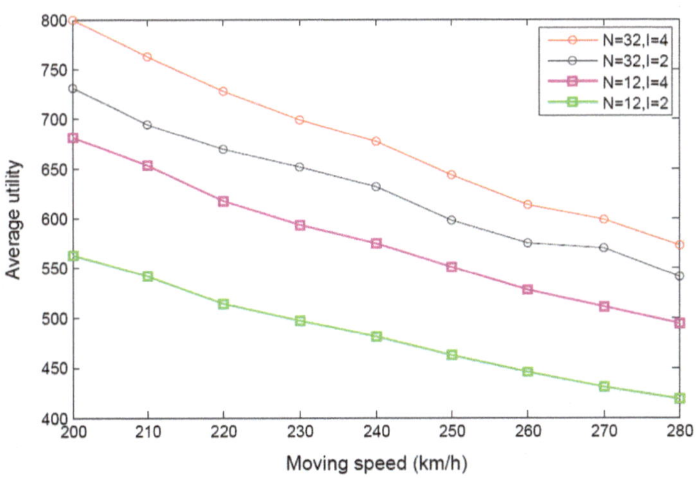

Fig. 5.9 Average utility versus moving speed with $K = 8$ and $H = 10$

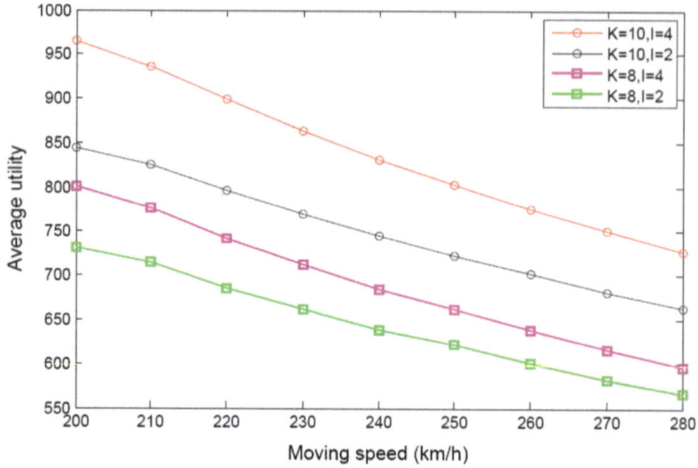

Fig. 5.10 Average utility versus moving speed with $N = 32$ and $H = 10$

can see that the average utility decreases gradually as the moving speed increases from 200 to 280 km/h. That is because Doppler shift becomes more and more serious, which results in an increase in ICI power. In addition, since more bits can be transferred by more antennas during the same time, the average utility grows with the increase of the number of antennas.

Figure 5.10 illustrates the relationship between average utility and the moving speed for different numbers of users and antennas, with $N = 32$ and $H = 10$. It can be seen that the variation trend of average utility and the moving speed is similar to that of Fig. 5.9, and the average utility grows with the increase of number of antennas from 2 to 4, too. Moreover, the average utility rises up as the number of users increases from 8 to 10. It can be explained that if the number of sub-carriers is enough, the larger the number of users, the more the multiuser diversity gain.

Figure 5.11 describes the relationship between average utility and the moving speed for different numbers of infostations and antennas, with $K = 8$ and $N = 32$. Besides the similar changing trend between average utility and the moving speed, we can find that the average utility increases as the number of infostations grows from 8 to 10. The reason is that more packets will be transferred during a longer transmission time when the numbers of users and sub-carriers are constant.

When the moving speed of the train ranges from 200 to 280 km/h, we carry out our resource allocation algorithm with or without ICI consideration, and then we can observe the actual average utility shown in Fig. 5.12. In the simulations, we use the same numbers of infostations $H = 10$ and different numbers of sub-carriers, users, and antennas. From the figure, we can clearly see that the average utility decreases as the moving speed goes up for all curves and the curves with ICI consideration get the large average utility, i.e., the proposed algorithm with ICI consideration has a better performance in terms of average utility.

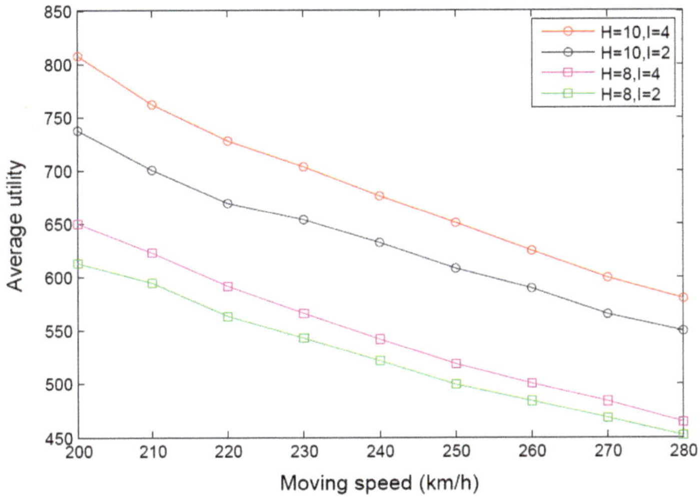

Fig. 5.11 Average utility versus moving speed with $K = 8$ and $N = 32$

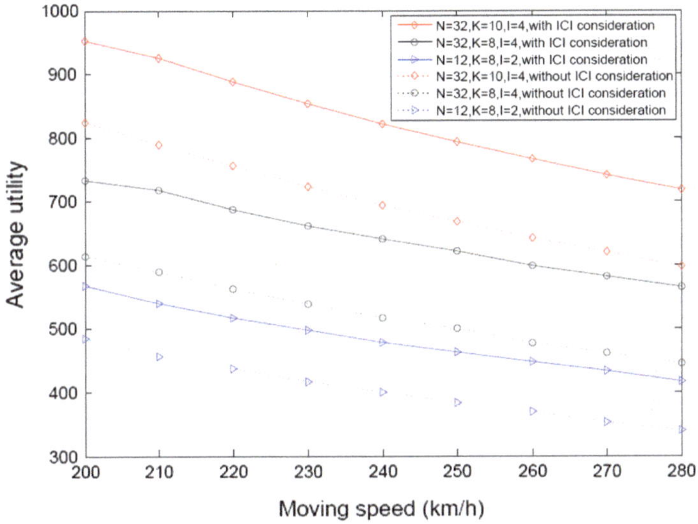

Fig. 5.12 Average utility versus moving speed with $H = 10$

Appendix E

Proof of Theorem 5.1 First, we prove inequality (5.60). Let $\tilde{\gamma}(t) = \tilde{\gamma}_k(t), k \in \mathcal{K})$ is any vector in the feasible region, $\tilde{A}_k$ is any arrival rate, and $\tilde{R}_k$ is derived from any feasible resource allocation policy. From (5.56), we have

$$\begin{aligned}\Delta(\mathbf{\Theta}(t)) - \beta\mathbb{E}\{\phi(\gamma(t))|\mathbf{\Theta}(t)\} &\leq D - \beta\mathbb{E}\{\phi(\tilde{\gamma})\} \\ &+ \mathbb{E}\left\{\sum_{k\in\mathcal{K}} Q_k(t)(\tilde{A}_k(t) - \tilde{R}_k(t))|\mathbf{\Theta}(t)\right\} \\ &+ \mathbb{E}\left\{\sum_{k\in\mathcal{K}} W_k(t)(\tilde{\gamma}_k(t) - \tilde{R}_k(t))|\mathbf{\Theta}(t)\right\}.\end{aligned} \tag{5.71}$$

According to [42], if a feasible solution of the original problem exists, for any $\delta > 0$, there is a feasible resource allocation policy and a vector $\tilde{\boldsymbol{\gamma}}(t)$ such that

$$\begin{aligned}&-\mathbb{E}\{\phi(\tilde{\gamma})\} \leq -\phi^{\text{opt}} + \delta,\\ &\mathbb{E}\{\tilde{A}_k(t) - \tilde{R}_k(t)\} \leq \delta,\\ &\mathbb{E}\{\tilde{\gamma}_k(t) - \tilde{R}_k(t)\} \leq \delta.\end{aligned} \tag{5.72}$$

Taking $\delta \rightarrow 0$ and plugging the above into (5.71) gives

$$\Delta(\mathbf{\Theta}(t)) - \beta\mathbb{E}\{\phi(\gamma(t))|\mathbf{\Theta}(t)\} \leq D - \beta\phi^{\text{opt}}. \tag{5.73}$$

By using iterated expectations and telescoping sums in the above inequality, for all $t > 0$, we have

$$\lim_{t\rightarrow\infty}\frac{1}{t}\sum_{\tau=0}^{t-1}\mathbb{E}\{\phi(\gamma(t))|\mathbf{\Theta}(t)\} \geq \phi^{\text{opt}} - D/\beta - \mathbb{E}\{L(\mathbf{\Theta}(0))\}/\beta t. \tag{5.74}$$

According to Jensen's inequality for the concave function $\phi(\cdot)$ and taking a lim inf of both sides, we get

$$\lim\inf_{t\rightarrow\infty}\phi(\bar{\gamma}(t)) \geq \phi^{\text{opt}} - D/\beta. \tag{5.75}$$

On the other hand, rearranging (5.73) yields

$$\Delta(\mathbf{\Theta}(t)) \leq D + \beta(\mathbb{E}\{\phi(\gamma(t))|\mathbf{\Theta}(t)\} - \phi^{\text{opt}}). \tag{5.76}$$

By using the Lyapunov Drift Theorem in [42], we find that all queues are mean rate stable, which means

$$\lim\sup(\bar{\gamma}_k(t)) - \overline{R}_k(t)) \leq 0 \tag{5.77}$$

Then, using this along with the continuity and entry-wise nondecreasing properties of $\phi(\cdot)$, we have

$$\lim_{t\to\infty}\inf \phi(\overline{\mathbf{R}}(t)) \geq \lim_{t\to\infty}\inf \phi(\bar{\gamma}(t)), \tag{5.78}$$

Plugging the above inequality into (5.75), we get

$$\lim_{t\to\infty}\inf \phi(\overline{\mathbf{R}}(t)) \geq \phi^{\text{opt}} - D/\beta = \phi^{\text{opt}} - \mathcal{O}(1/\beta). \tag{5.79}$$

Next, we prove inequality (5.61). We assume there exists a feasible scheduling scheme π, which leads to $0 \leq \mathbb{E}\{R_k^{\pi}(t)\} \leq \gamma_{\max}, \mathbb{E}\{A_k(t) - R_k^{\pi}(t)\} \leq -\epsilon$, and $\phi\left\{\mathbb{E}(R_k^{\pi}(t)\}\right) = \phi_{\epsilon}$. From (5.71) and (5.78), we have

$$\Delta(\mathbf{\Theta}(t)) \leq D + \beta(\mathbb{E}\{\phi(\gamma(t))|\mathbf{\Theta}(t)\} - \phi_{\epsilon}) - \epsilon \sum_{k\in\mathcal{K}} Q_k(t). \tag{5.80}$$

By taking iterated expectations, summing the telescoping series, and rearranging terms, we get

$$\frac{1}{t}\sum_{\tau=0}^{t-1}\sum_{k\in\mathcal{K}} \mathbb{E}\{Q_k(\tau)\} \geq \frac{D + \beta\left(\frac{1}{t}\sum\limits_{\tau=0}^{t-1}\mathbb{E}\{\phi(\gamma(\tau))\} - \phi_{\epsilon}\right)}{\epsilon} - \frac{E\{L(\mathbf{\Theta}(0))\}}{\epsilon t}. \tag{5.81}$$

Taking a lim sup of (5.81) as $t \to \infty$, and using $\lim_{t\to\infty}\frac{1}{t}\sum_{\tau=0}^{t-1}\mathbb{E}\{\phi(\gamma(\tau))\} \leq \phi^{\text{opt}}$, we obtain

$$\lim_{t\to\infty}\sup \frac{1}{t}\sum_{\tau=0}^{t-1}\sum_{k\in\mathcal{K}} \mathbb{E}\{Q_k(\tau)\} \geq \frac{D + \beta(\phi^{\text{opt}} - \phi_{\epsilon})}{\epsilon} = D/\epsilon + \mathcal{O}(\beta). \tag{5.82}$$

This completes the proof. ■

5.5 Challenges and Open Issues

As we have seen from the previous sections, RRM services as an effective method to optimize the system resource utilization and provide QoS guarantees. Besides the existing research efforts, there are still some challenges and open research issues on RRM design, which will be discussed as follows. In addition, we also recommend the readers refer to [45] for more challenges and opportunities related to radio communications that railways will meet in both the near and far future.

5.5.1 Location-Aware Resource Management

Train positioning technique is one of the main techniques in HSR communication systems. How to use the position information to enhance the HSR communication performance has become a research trend. However, only a limited number of studies have considered the train position information to facilitate the system design, such as position-based channel modeling, position-assisted handover scheme, position-based limited feedback scheme, and location information-assisted opportunistic beamforming. In HSR scenarios, the channel condition mainly depends on the signal transmission distance, which results in the different channel conditions along the rail. Therefore, further research is needed to exploit the train position information to facilitate RRM design, by taking use of the future channel or signal prediction. Moreover, the seamless handover scheme can be further strengthened with the help of train position information and signal prediction.

Accurate position and speed measurements of the high-speed train are critical for the location-aware RRM in HSR communications. However, there exist measurement errors in the train positioning techniques, such as GPS and odometer. Thus, studies on the effect of location uncertainty are needed to assess performance in practical HSR scenarios. Also, models that capture the variability of train location information, as well as methods that are robust to such inaccuracies are required.

5.5.2 Cross-Layer Based Resource Management

To satisfy different requirements stemming from various layers, an appropriate cross-layer model will be highly beneficial to improve the RRM performance in HSR wireless communications. These models should account for the parameters at the PHY layer (e.g., channel, power, and modulation order), MAC layer (e.g., scheduling, queuing and automatic repeat request), network layer (e.g., routing and packet forwarding) and APP layer with different QoS requirements, and then optimally determine the resource management actions. Such cross-layer models are able to manage the inherent trade offs at different layers in a comprehensive manner. More models and designs of both physical layer and higher layer for high mobility wireless communications have been discussed in [46]. However, note that the incorporation of the constraints from different layers can further complicate the feasibility of RRM problems. In light of practicality, it is necessary to devise the low-complexity cross-layer techniques so that the close-to-optimal solutions can be identified while not severely compromising the overall system performance.

Considering the cross-layer model, most available studies in HSR wireless communications only focus on one single type of RRM, such as independent resource allocation and handover scheme. To achieve higher system performance, it is necessary and challenging to jointly optimize RRM schemes. Further studies are needed to develop joint RRM schemes, such as jointly optimizing admission control, power control and resource allocation. Additionally, the tradeoff between the complexity and performance deserves future research.

5.5.3 Energy-Efficient Resource Management

Providing multimedia services in HSR wireless communications will become a reality in the very near future, such as on-demand media services and social network services. Some network architectures have been presented for multimedia service transmission over HSR communications, such as a mobile proxy architecture. However, most of the existing studies only focus on RRM design for both voice and data services while multimedia services receive little attention. Thus, the effective RRM schemes for multimedia services delivery are critical with the purpose of high efficiency and green train communications. Proper RRM design can lead to significant improvements in energy efficiency due to the strong dependence of power consumption on the distance between the train and base station. For instance, solely for a fairly constant rate, much power will be consumed to compensate for the fading effect when the train is far from the base station. From the perspective of energy efficiency, it is intuitive to transmit more data when the signal is strong and less data when the signal is weak. Therefore, energy-efficient media delivery problem should be further analyzed, and the related schemes such as power control and resource allocation are developed. Furthermore, the QoS requirements should be also considered into RRM design for media delivery, such as playback delay and quality level.

5.5.4 Robust Resource Management

Reliability requirements of HSR wireless communications pose great challenges to RRM design. Generally, the HSR services especially critical core services have strict reliability requirements, such as very low bit error rate and packet loss probability. The reliable transmission of critical core services directly affects the safe operation of the train. Thus, how to design a robust RRM scheme for reliability assurance is an important issue. However, most existing RRM schemes for HSR wireless communications are built on the perfect CSI assumption, while few investigations have been performed on the impact of channel estimation error in terms of resource allocation and beamforming design. Indeed, the channel estimation error that may result from the time-varying wireless channel, has a direct effect on the communication reliability, such as the bit error rate performance. Further research is, therefore, needed to address a range of issues related to communication robustness, and improve the reliability of service transmission. A typical example is the robust beamforming design problem for MIMO-based HSR wireless communications with imperfect CSI, where the worst-case performance optimization will be exploited and the robust algorithms with low complexity are more attractive for practical implementation.

5.5.5 *Resource Management for 5G Communications*

The research for the fifth generation (5G) communications is now on its way. The European Union, the United States, Korea, China, and Japan have developed their organizations for 5G development. The HSR scenario has been recognized as one of the typical scenarios for 5G. The related system designs such as network architecture and transmission technique have received much attention. Meanwhile, the advanced RRM for 5G HSR communications is urgently needed to further improve overall system performance and face the increasing demands.

To meet the emerging massive capacity demands in 5G communications, a control and data signaling decoupled architecture bas been presented for railway wireless communications [47], in which the relatively important control plane is kept on high-quality lower frequency bands to handle mobility, while the corresponding user plane is moved to higher frequency bands to gain broader spectra. Since the control plane and user plane are physically separated, the RRM schemes for this novel architecture, such as handover and spectrum allocation schemes, are required to further investigated. One of the 5G key techniques, massive MIMO technique, has been involved into HSR wireless communications [48]. Wireless coverage based on massive MIMO for railway stations and train cars is proposed to fulfill the requirement of high-data-rate and high spectrum efficiency. Further investigations on channel modeling and system-level modeling for HSR communications are still needed. Additionally, the bandwidth shortage has motivated the exploration of the underutilized millimeter wave (mm-wave) frequency spectrum for future HSR broadband mobile communications [49]. However, mm-wave communications suffer from huge propagation loss, which indicates that it would be beneficial to investigate the use of power control to improve system performance.

5.6 Summary

Radio resource management is a powerful tool that enables high-resource utilization and results in improved QoS performance. However, compared with common cellular communications, some characteristics in HSR wireless communications, such as high mobility, unique channel conditions and heterogeneous QoS requirements, impose some challenges to the RRM design .This leads to significant attention on the study of RRM under HSR scenarios. In this chapter, we provide a literature survey on RRM schemes for HSR wireless communications, with an in-depth discussion on admission control, power control, and resource allocation. Then, we study joint power control and resource allocation problem, and dynamic resource management problem, respectively. Finally, challenges and open issues on RRM design of HSR wireless communications are outlined.

References

1. Ai B, Cheng X, Kurner T, Zhong Z, Guan K, He R, Xiong L, Matolak D, Michelson D, Briso-Rodriguez C (2014) Challenges toward wireless communications for high-speed railway. IEEE Trans Intell Transp Syst 15(5):2143–2158
2. Zhong ZD, Ai B, Liu QY, Lin SY (2009) Basis theory of railway digital mobile communication system. Tsinghua University Press and Beijing Jiaotong University Press, Beijing
3. Guan K, Zhong ZD, Ai B (2011) Assessment of LTE-R using high speed railway channel model. In: Proceedings of international conference on communications and mobile computing, pp 461–464
4. Barbu G (2010) E-Train, broadband communication with moving trains. International Union of Railways (UIC), Technical Report
5. Fokum D, Frost V (2010) A survey on methods for broadband. Surv Tutor 12(2):171–185
6. Liu L, Tao C, Qiu J, Chen H, Li Y, Dong W, Yuan Y (2012) Position-based modeling for wireless channel on high-speed railway under a viaduct at 2.35 GHz. IEEE J Sel Areas Commun 30(4):834–845
7. Ai B, He R, Zhong Z, Guan K, Chen B, Liu P, Li Y (2012) Radio wave propagation scene partitioning for high-speed rails. Int J Antennas Propag 2012:7. Article ID 815232
8. Zander J (2001) Radio resource management for wireless networks. Artech House, Boston, MA
9. Ahmed MH (2005) Call admission control in wireless networks. Surv Tutor 7(1):49–68
10. Zhao YS, Li X, Ji H (2011) Spring model-based call admission control mechanism for high-speed railway environment. In: Proceedings of international ICST conference on communications and networking in China, pp 1196–1200
11. Kim Y, Ko H, Pack S, Lee W, Shen X (2013) Mobility-aware call admission control algorithm with handoff queue in mobile hotspots. IEEE Trans Veh Technol 62(8):3903–3912
12. Lattanzi F, Giambene G, Acar G, Evans B (2010) Admission control and handover management for high speed trains in vehicular geostationary satellite networks with terrestrial gap-filling. Int J Satell Commun Netw 28(1):1–27
13. Zhou YZ, Ai B (2014) Access control schemes for high-speed train communications. In: Proceedings International Workshop on High Mobility Wireless Communications (HMWC), pp 33–37
14. Xu Q, Li X, Ji H, Yao LP (2013) A cross-layer admission control scheme for high-speed railway communication system. In: Proceedings of International Conference on Communications (ICC), pp 4936–4940
15. Xu QS, Li X, Ji H, Yao LP (2013) Radio admission control scheme for high-speed railway scenario: an optimal accepted service number approach. In: National doctoral academic forum on information and communications technology, pp 1–6
16. Zhao Y, Li X, Ji H (2012) Radio admission control scheme for high-speed railway communication with MIMO antennas. In: Proceedings of IEEE international conference on communications (ICC), pp 5005–5009
17. Xu SF, Zhu G, Shen C, Lei Y, Zhong Z (2014) Analysis and optimization of resource control in high-speed railway wireless networks. Math Probl Eng 2014: 13. Article ID 781654

18. Gong M, Zhang C, Lu JH, Lin XK (2008) Dynamic resouce allocation in high speed mobile OFDMA system. In: Proceedings of IEEE international conference on communications (ICC), pp 3335–3339
19. Zhao Y, Li X, Li Y, Ji H (2013) Resource allocation for high-speed railway downlink MIMO-OFDM system using quantum-behaved particle swarm optimization. In: Proceedings of IEEE international conference on communications (ICC), pp 2343–2347
20. Qiu JH, Lin ZH, Hardjawana W, Vucetic B, Tao C., Tan ZH (2014) Resource allocation for OFDMA system under high-speed railway condition. In: Proceedings of IEEE wireless communications and networking conference (WCNC), pp 2725–2729
21. Xu SY, Kwak KS (2013) Effective uplink interference avoidance scheme for high-speed railway communications. In: Proceedings of IEEE vehicular technology conference (VTC fall)
22. Zhang J, Jin F, Tan Z, Wang H, Huang Q, Hanzo L (2014) Performance analysis of high-speed railway communication systems subjected to co-channel interference and channel estimation errors. IET Commun 8(7):1151–1157
23. Zhou YZ, Ai B (2014) Quality of service improvement for high-speed railway communications. China Commun 11(11):156–167
24. Xu QS, Ji H, Li X (2014) Goodput performance improvement in high-speed railway communication systems: a link adaptation approach. In: Proceedings of IEEE international conference on communications (ICC), pp 1059–1064
25. Zhu HL (2012) Radio resource allocation for OFDMA systems in high speed environments. IEEE J Sel Areas Commun 30(4):748–759
26. Lei L, Lu J, Jiang Y, Shen X, Li Y, Zhong Z, Lin C (2016) Stochastic delay analysis for train control services in next-generation high-speed railway communications system. IEEE Trans Intell Transp Syst 17(1):48–64
27. Liang H, Zhuang WH (2012) Efficient on-demand data service delivery to high-speed trains in cellular/ infostation integrated networks. IEEE J Sel Areas Commun 30(4):780–791
28. Xu SF, Zhu G, Shen C, Ai B (2014) A QoS-aware scheduling algorithm for high-speed railway communication system. In: Proceedings of IEEE International Conference on Communications (ICC), pp 2855–2860
29. Chen TY, Shan HG, Wang X (2014) Optimal scheduling for wireless on-demand data packet delivery to high-speed trains. IEEE Trans Veh Technol 64(9):4101–4112
30. Xu SF, Zhu G, Shen C, Lei Y (2013) Delay-aware fair scheduling in relay-assisted high-speed railwaynetworks. In: Proceedings of international conference on communications and networking in China, pp 17–21
31. Xu SF, Zhu G, Shen C, Lei Y, Zhong Z (2014) Delay-aware online service scheduling in high-speedrailway communication systems. Math Probl Eng 2014:10. Article ID 248174
32. Yang X, Liu H (2014) A resource allocation scheme for VoIP traffic under high speed train scenario. In: Proceedings of international workshop on high mobility wireless communications (HMWC), pp 23–26
33. Lin S, Zhong Z, Cai L, Luo Y (2012) Finite state markov modelling for high speed railway wireless communication channel. In: Proceedings of IEEE global communications conference (GLOBECOM), pp 5421–5426
34. Dong YQ, Fan PY, Letaief KB (2013) High speed railway wireless communications: efficiency versus fairness. IEEE Trans Veh Technol 63(2):925–930
35. Dong YQ, Fan PY (2013) Bounds on the average transmission rate in high speed railway wireless communications. In: Proceedings of international workshop on high mobility wireless communications(HMWC), pp 142–145
36. Xu SF, Zhu G, Shen C, Lei Y, Zhong Z (2014) Utility-based resource allocation in high-speed railway wireless networks. EURASIP J Wirel Commun Netw 2014(1):1–14
37. Thai CD, Berbineau M, Sun F (2013) Exploiting the forecast channels for power allocation in high-speed railway communications. In: Proceedings of international conference on ITS telecommunications, pp 402–406

38. Zhang C, Fan PY, Xiong K, Fan PZ (2015) Optimal power allocation with delay constraint for signal transmission from a moving train to base stations in high-speed railway scenarios. IEEE Trans Veh Technol 64(12):5775–5788
39. Ma X, Liu J, Jiang H (2012) Energy-efficient mobile data uploading from high-speed trains. Mob Netw Appl 17(1):143–151
40. Xu SF, Zhu G, Shen C, Li SC, Zhong ZD Delay-aware dynamic resource management for high-speed railway wireless communications. http://arxiv.org/abs/1601.01526
41. Tse D, Viswanath P (2005) Fundamentals of wireless communication. Cambridge University Press, New York
42. Neely MJ (2010) Stochastic network optimization with application to communication and queueing systems. Morgan and Claypool, San Rafael
43. Li Y, Cimini L (2006) Bounds on the inter channel interference of OFDM in time varying impairments. IEEE Trans Commun 49(3):401–404
44. Zhao Y, Li X, Li Y, Ji H (2013) Resource allocation for high-speed railway downlink MIMO-OFDM system using quantum-behaved particle swarm optimization. IEEE ICC, pp 2343–2347
45. Moreno J, Riera JM, Haro LD, Rodriguez C (2015) A survey on future railway radio communications services: challenges and opportunities. IEEE Commun Mag 53(10):62–68
46. Wu J, Fan P (2016) A survey on high mobility wireless communications: challenges, opportunities and solutions. IEEE Access 4:450–476
47. Yan L, Fang X, Wang CX (2015) Position based limited feedback scheme for railway MU-MIMO systems. IEEE Trans Veh Technol. doi:10.1109/TVT.2015.2511021
48. Liu ZY, Wu J, Fan P (2014) On the uplink capacity of high speed railway communications with massive MIMO systems. In: Proceedings of IEEE vehicular technology conference (VTC Spring) 2014
49. Dat P, Kanno A, Yamamoto N, Kawanishi T (2015) WDM RoF-MMW and linearly located distributed antenna system for future high-speed railway communications. IEEE Commun Mag 53(10):86–94

Chapter 6
LTE-R Network

6.1 LTE-R Network Services

Recently, high-speed railway (HSR) has been developed rapidly all over the world, which puts forward requirements for a reliable and efficient wireless communication system between the moving train and the ground. According to International Union of Railways (UIC) E-Train Project [1], the train–ground wireless communication services for HSR system mainly include the following: (1) *train control services*, which are specific data and voice transmissions dedicated to the train crew with respect to the train control, train operator or other correspondents; (2) *train monitoring services*, which are data transmission in provenience from the train automatic monitoring and diagnosis systems; and (3) *passenger services* from/to Internet (all multimedia services accessible through Internet connection). The first and second categories are special services for train needs and provided by train operators mainly using Global System for Mobile Communications—Railway (GSM-R) [2], which is an international wireless communications standard for railway communication and applications. The third category is commercial services for passengers and currently provided by mobile network operators using cellular network standards, e.g., GSM/General Packet Radio Service (GPRS), Universal Mobile Telecommunications System (UMTS) and 3rd Generation Partnership Project (3GPP) Long Term Evolution (LTE).

Among the cellular network standards, LTE/LTE-Advanced represents the latest progress. It aims at providing a unified architecture to real-time and non-real-time services and providing users with high data transfer rate, low latency and optimized packet wireless access technology. Although LTE is designed to support up to 350 km/h or even up to 500 km/h mobility speed, network performance is only optimized for 0–15 km/h and high performance is possible only when the mobility speed is under 120 km/h [3]. This means that the quality of service (QoS) provided to the passengers on high-speed trains may be far from satisfactory. On the other hand, although GSM-R is specifically standardized for communication between

Z.-D. Zhong et. al., *Dedicated Mobile Communications for High-speed Railway*,
Advances in High-speed Rail Technology, DOI 10.1007/978-3-662-54860-8_6

train and railway regulation control centers, it is built on the GSM technology, which is a 2nd Generation (2G) cellular standard and much less efficient compared with the 4th Generation (4G) LTE standard. Therefore, it is important to design the next generation HSR communications system based on LTE technology while addressing the specific challenges of HSR environment, such as high mobility speeds (from 120 km/h for regional trains to 350 km/h for high-speed trains) and stringent QoS requirement of some railway-specific signaling, so that the above mentioned three types of communication services can be well supported by a unified network. Such a wireless communications system is commonly referred to as LTE-Railway (LTE-R). LTE-R system has the significant characteristics of wide bandwidth, low delay, all-IP network architecture and so on, which is applicable to the next generation railway wireless communications network technology mechanism.

Current railway applications are focused on ensuring essential radio communications such as train driver and dispatcher communications, driver-to-driver operational communications, and trackside maintenance team communications. Railway applications may be classified into on-train applications, trackside applications, station applications, and depot, control center, and office applications [4]. In addition, there are voice and data applications which can be further categorized into critical core services and noncritical communications services.

Critical core services are usually referred to as mission-critical services, and include critical railway communications, train operational voice services and operational data applications. Additional communication services include passenger experience services and business process support services such as voice and data train crew communications and train support applications [4]. Figure 6.1 shows the different categories of railway group services.

The railway communication system of the future must address both critical and noncritical applications. The main constraints for noncritical applications include coverage, network capacity, and cost requirements, while constraints for critical services are mainly related to require more for reliability, availability, and prioritization. Clearly, there is a mismatch between core services and additional services

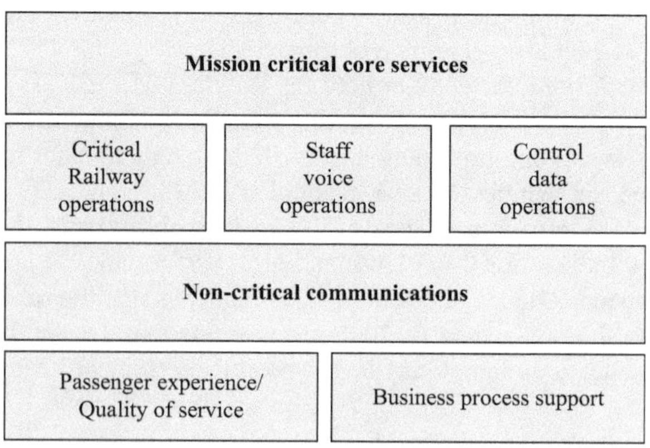

Fig. 6.1 The different categories of railway group services

QoS requirements. Mission-critical services demand assurances for low delay and high reliability, availability, and safety. The additional services are constrained only by available bandwidth.

The mismatch between critical and noncritical applications can play a key role in the deployment strategy for train radio infrastructure. Railway operators have the option of deciding between deploying a private radio system for railway purposes only, or deploying a hybrid solution with a public telecommunication company. Figure 6.2 offers a comprehensive list of the future services necessary in the railway environment.

It is clear that the significant improvement in data rates that users experience with LTE-R has the potential to meet the requirements for noncritical railway services. However, as we stated in the previous section, LTE's potential to address the challenge of supporting critical railway applications has not yet been analyzed. New requirements and functionalities could arise from new services and applications in railway environments and at the very least, future train radio systems will have to fulfill current requirements around RAMS and QoS [5].

Railway services demand-specific functionalities for train radio systems. Through GSM-R, the GSM standard was enhanced with advanced speech call item (ASCI) functionalities to meet railway needs, and LTE must be similarly enhanced in order to take its place. Table 6.1 shows the LTE features and mechanisms necessary to implement the railway functionalities.

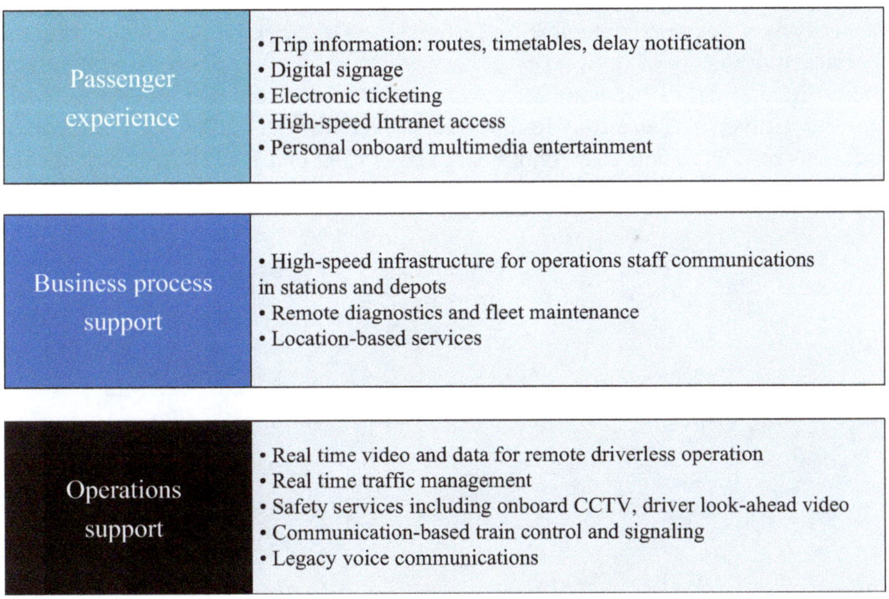

Fig. 6.2 The future services necessary in the railway environment

Table 6.1 LTE-R and GSM-R features and mechanisms necessary

LTE-R	GSM-R
LTE IMS-based VoIP (VoLTE) + IMS-based push-to-talk over cellular (PoC) *(This will be enhanced with the 3GPP Release 12 GCSE_LTE)*	Voice group call service (VGCS)
VoLTE + PoC: IP multicast of voice and video services *(This will be enhanced with the 3GPP Release 12 GCSE_LTE)*	Voice broadcast calls (VBS)
Access class barring mechanisms + policy control rules +QoS mechanisms	Priority and preemption (eMLPP)
Session Initiation Protocol (SIP) addressing	Functional addressing (FN)
Localization services in LTE (Release 10)	Location-dependent addressing (LDA, eLDA)
Emergency and critical safety voice services over IMS in LTE	Railway emergency calls (REC, e-REC)
Very low latency of LTE to support fast exchange of signaling (e.g., IMS-based PoC) + Access class barring	Fast call set up
IMS-based SMS service Use SG interface between MME and MSC server MME-based SMS service	Data exchange (SMS, shunting)

6.2 LTE-R Network Architecture

The 3rd Generation Partnership Project (3GPP) recently specified LTE to fulfill the increasing requirements of mobile broadband communication systems. LTE-R is a new railway wireless communication system based on the LTE, which can meet the special requirements of railway.

In Fig. 6.3, LTE-R communication system is divided into three parts: the Evolved Universal Terrestrial Radio Access Network (E-UTRAN), the Evolved Packet Core (EPC), and User Equipment (UE). EPC and E-UTRAN compose the Evolved Packet System (EPS).

Fig. 6.3 LTE-R network architecture

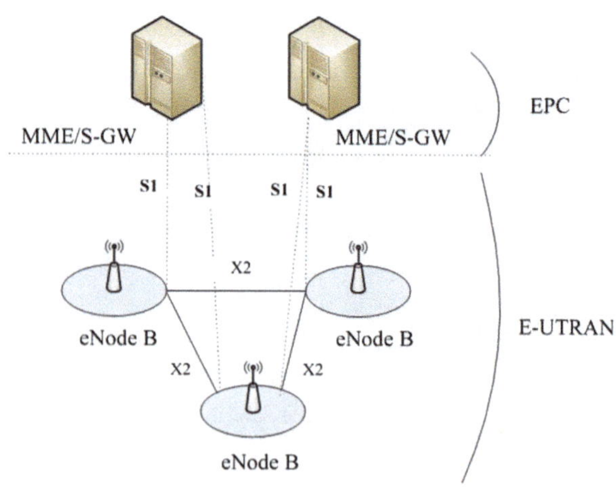

As illustrated in Fig. 6.4, LTE-R consists of two user-plane nodes, called Serving Gateway (S-GW) or Packet Data Network Gateway (P-GW) and the base station, denoted as evolved Node B (eNB), and one control-plane node, called a Mobility Management Entity (MME) [6].

The LTE-R radio-access network only consists of eNBs, connected to each other through the X2 interface, to the MME through the S1-MME interface and to the S-GW through S1-U interface. Because LTE radio-access network includes only the eNBs, the number of nodes in network architecture decreases and the network architecture is more tend to be flat.

The functionalities of network elements in LTE-R are as follows:

(1) eNB: the only one network equipment in E-UTRAN without the Base Transceiver Station(BTS) and the Base Station Control (BSC) as in GSM-R, which transmits signal to terminals or receives signals from terminals in one or more cells. It has a set of functions related to transmission and reception at physical layer of radio interface, including modulation and demodulation, channel coding and decoding. At the same time, as without BSC, it also has functions of radio resource control, wireless mobility management [7].
(2) MME: the control center of EPC, its functions include signaling processing of Non-Access Stratum (NAS), idle mode access restrictions, security key management, etc.

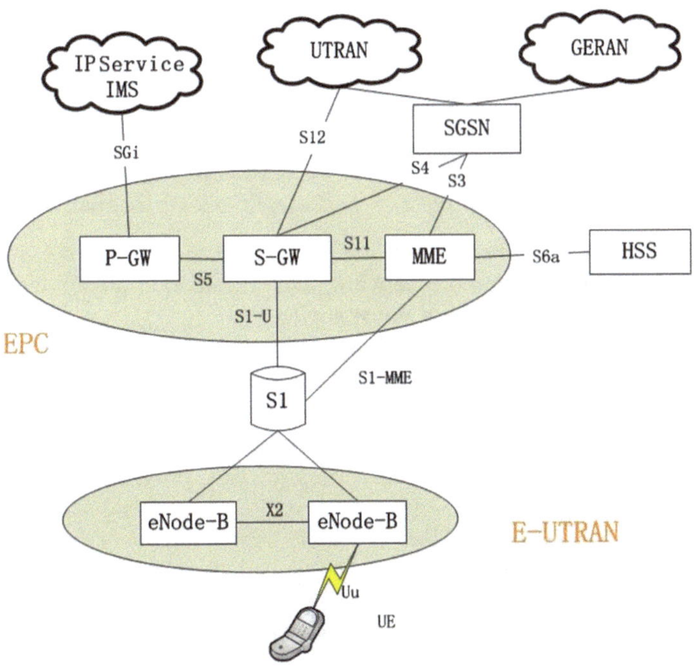

Fig. 6.4 EPC network architecture

(3) HSS: the local user server which is responsible for managing the user registration information.
(4) S-GW: S-GW is responsible for packet routing, Quality of Service (QoS) handing, handover on user-plane, etc.
(5) P-GW: P-GW provides users with a stable IP access points, IP address assignment and packet filtering, etc.
(6) Gateway GPRS Support Node (SGSN): SGSN is used for packet data transmission between the EPC and the traditional 2G/3G Radio Access Network (RAN).

There are many interfaces in LTE-R network architecture. The functions of one interface are different from those of others.

(1) S1: interface between an eNB and an EPC, providing an interconnection point between the E-UTRAN and the EPC. It is also considered as a reference point.
S1-MME is the interface between eNB and MME, which is in charge for radio-access control.
S1-U is the interface between eNB and S-GW, which is in charge for user data transmission.
(2) X2: logical interface between two eNBs. While logically representing a point-to-point link between eNBs, the physical realization need not be a point-to-point link.
(3) Uu: interface between an eNB and a UE.
(4) S3: interface between MME and SGSN for signaling interaction.
(5) S4: interface between S-GW and SGSN for inter-system handover.
(6) S5: interface between S-GW and P-GW.
(7) S6a: interface between HSS and MME for carrying authentication information.
(8) S11: interface between S-GW and MME for signaling message transmission.
(9) SGi: interface between P-GW and Public Data Network (PDN). PDN can be either external public or private IP packet network or internal IP network.

The significant difference between EPC and the core network of GSM-R is that the EPC is an all-IP mobile core network. This means that all services would be built on the PS domain, including voice services [7].

There are a few reasons why LTE should be considered the most likely future alternative to GSM-R. LTE can address all of the major shortcomings of GSM-R, and it is the latest mobile communication standard. It provides a number of capacity and capability advantages over GSM and also over the newer Universal Mobile Telecommunications System (UMTS). Moreover, EPC provides support for legacy 3GPP technologies such as GSM [8].

Compared with GSM-R, LTE-R network architecture only has two types of nodes on the user-plane. In Fig. 6.5, BSC and Wireless Network Controller (RNC) are no longer needed in LTE-R network, because their functions are incorporated into the eNB, and EPC includes all kinds of core network entities and

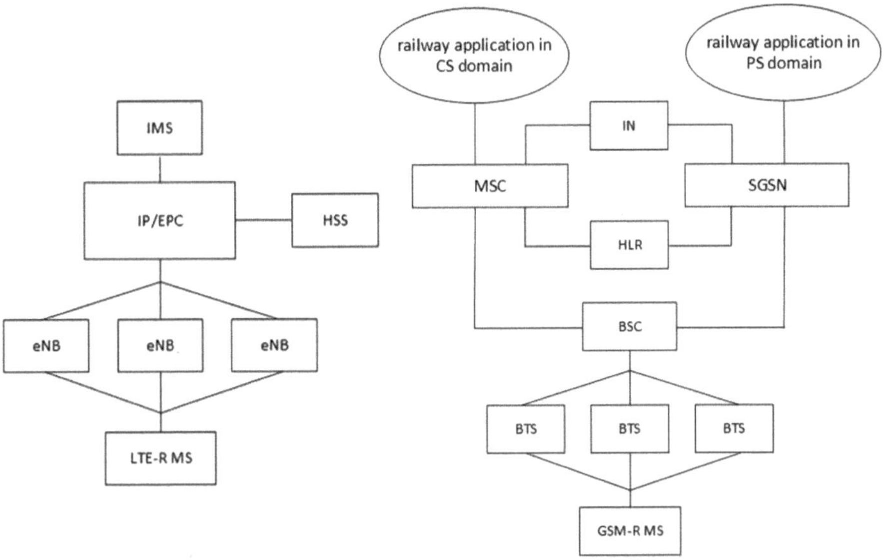

Fig. 6.5 The comparison between LTE-R network architecture and GSM-R network architecture

gateways. In LTE-R, a large amount of data services are based on PS domain and IP network architecture, and the flat structure of LTE-R reduces the number of nodes and the complexity of signal and path selection caused by multiple interfaces, improving instantaneity.

6.3 LTE-R Network Performance Evaluation

Network performance refers to measures of service quality of a network as seen by the customer.

There are many different ways to evaluate the performance of a network, as each network is different in nature and design.

6.3.1 Queueing Theory

Queueing theory is the mathematical study of waiting lines, or queues [9]. In queueing theory a model is constructed so that queue lengths and waiting time can be predicted [9]. Queueing theory is generally considered a branch of operations research because the results are often used when making business decisions about the resources needed to provide a service.

Queueing theory has its origins in research by Agner Krarup Erlang when he created models to describe the Copenhagen telephone exchange. The ideas have since seen applications including telecommunication, traffic engineering, computing and the design of factories, shops, offices, and hospitals.

Since 1909 when the first paper was published, queueing theory has been developed and applied in a wide variety of areas. In particular, queueing theory has played a fundamental role in modeling, analyzing and dimensioning traditional circuit-switched telecommunication networks. With the advance and pervasive adoption of Internet networks, it is natural to apply queueing theory to performance analysis of such networks.

6.3.2 Petri Nets

Petri Nets were developed originally by Carl Adam Petri in 1962. Since then, Petri nets have been extended and developed, and applied in a variety of areas, such as Office automation, manufacturing, programming design, computer networks, hardware structures, real-time systems, performance evaluation, operations research, embedded systems, communications, Internet, railway networks, and biological systems. The mathematical properties of Petri nets are interesting and useful. The beginner will find a good approach to learn to model systems is constructing them graphically, assisted in construction and analysis by simulation of computer software and analysis of Petri nets. Petri nets aim at construction and dynamic action of investigative model system, focusing on relationships of all state changes and transition.

Petri nets are a powerful tool for the description and the analysis of systems. Timed Petri nets in which the basic model is associated with time specifications are commonly used to evaluate the performance and reliability of complex systems. Stochastic Petri Nets (SPN) were introduced in 1980 as formalism for description of Discrete Event Dynamic Systems (DEDS) [10]. With the goal of improving the modeling power of stochastic Petri nets, Stochastic High-level Petri Nets (SHLPN) has been proposed. Although SPN models are widely used for performance and reliability evaluation of many practical systems, state-space explosion is the main problem to cope with. Model decomposition and iteration technique is effective way to solve the exponential growth of the state space.

6.3.3 Network Calculus

Network calculus is a theory dealing with queuing systems found in computer networks. Its focus is on performance guarantees. Central to the theory is the use of alternate algebras such as the min-plus algebra to transform complex network systems into analytically tractable systems. To simplify the analysis, another idea is to characterize traffic and service processes using various bounds. Since its introduction in the early 1990s, network calculus has developed along two tracks—deterministic and stochastic. This chapter is devoted to summarizing results for stochastic network calculus to provide stochastic service guarantees to the LTE-R networks.

The remainder of this chapter introduces system model including LTE-R architecture and the snetal basics. The stochastic arrival curve of train control services and the stochastic service curve for HSR fading channel are derived, respectively. Finally, numerical and simulation results are presented, compared, and discussed.

According to International Union of Railways (UIC) E-Train Project [1], the train–ground wireless communication services for HSR system mainly include: (1) *train control services*, which are specific data and voice transmissions dedicated to the train crew with respect to the train control, train operator or other correspondents; (2) *train monitoring services*, which are data transmission in provenience from the train automatic monitoring and diagnosis systems; and (3) *passenger services* from/to Internet (all multimedia services accessible through Internet connection). Among these three categories of services for HSR communications system, the first category has higher priority over the other two categories, since the communication delay of the train control services between the train and trackside infrastructure is crucial for train movement control and safety [11]. Therefore, the LTE-R system needs to provide stringent QoS guarantee for these mission-critical services. To this end, assigning dedicated radio resources to the first category of services is preferred to sharing radio resources with the other two categories of services, although higher resource efficiency can be achieved by the latter alternative due to statistical multiplexing gain. So an interesting question is how many resources should be dedicated to these mission-critical services to guarantee their QoS performance or what is the expected QoS performance given a certain amount of dedicated resources for train control services transmission? In order to answer this question, we need to evaluate and quantify the QoS performance so that useful insights can be provided for LTE-R network dimensioning and design. Although the problem of cross-layer performance modeling and analysis of cellular networks and wireless ad hoc networks has been addressed in literature [12, 13], the performance evaluation of LTE-R system is an open problem due to the following special features and requirements as compared to the LTE public communications system:

(1) Traffic model: The characteristics of train control services are different from the user services in public communications system, which have to be studied and modeled for performance evaluation.
(2) Wireless channel: The wireless channel characteristics for LTE-R system are unique due to the high mobility of the trains. The path loss varies rapidly as the train moves since it mainly depends on the distance between the train and the base station (referred to as evolved-Node B (eNodeB) in LTE system). On the other hand, the time-correlation of the fading channel becomes very small with

the increasing mobility speed. These effects together determine the instantaneous channel gains of the wireless channel.

(3) Adaptive Modulation and Coding (AMC): Due to the high mobility of the trains and the induced rapid channel variations, it is very difficult to obtain accurate instantaneous Channel State Information (CSI) at the eNodeBs considering the channel measurements inaccuracy and CSI feedback delay. This will impact the performance of the AMC scheme in LTE system.

In this chapter, we develop an analytical framework based on stochastic network calculus (snetal) taking into account the above unique characteristics to evaluate the performance of LTE-R system. The network calculus is a theory of queuing systems that has been developed as an initially deterministic framework for analysis of worst-case backlogs and delays, which are obtained by applying deterministic upper envelopes on traffic arrivals and lower envelopes on the offered service, the so-called arrival and service curves [14]. It is founded on the min-plus algebra and max-plus algebra to transform complex queuing systems into analytically tractable systems and mostly applied in the area of Internet QoS analysis. Compared with queuing theory which is largely constrained by the technical assumption of Poisson arrivals, network calculus can characterize a large variety of traffic arrival processes by their arrival curves. Although the worst-case performance bounds provided by deterministic network calculus (dnetal) were proven to be tight, the occurrence of such worst-case events is usually rare and statistical multiplexing gain can be captured when some violations of the deterministic bounds are tolerable. This has motivated considerable research for a stochastic network calculus which describes arrivals and service probabilistically while preserving the elegance and expressiveness of the original framework [15–19]. Generally speaking, existing work on snetal can be classified into two broad categories: the Moment Generating Function (MGF) approach [20] and the Complementary Cumulative Distribution Function (CCDF) approach [21]. Since it is easier to understand and simpler to implement, the MGF approach is more widely used in performance evaluation of wireless networks [22–25]. The research on CCDF approach for wireless channel focuses more on general principle and has mostly been applied for simple on–off impairment model [26, 27]. Notice that we use snetal instead of dnetal for the performance analysis of train control services mainly due to the following reasons: (1) Although the delay performance of train control services is crucial to the safety of train operation, a small amount of violation probability can be tolerated according to the related standard [28]; and (2) much tighter bound can be derived by snetal compared with dental due to the stochastic nature of HSR fading channel and train control services. In addition, statistical multiplexing gain can be exploited for passenger services, which does not exist for the train control services studied in this chapter.

This chapter focuses on train control data traffic performance analysis of HSR fading channel. More specifically, we are interested in probabilistic delay and backlog guarantees of train control services in such a system. The impact of transmission delay to railway control system and the importance to provide delay guarantee to the train control services are discussed in [29, 30]. However, the above work assumes that the transmission delay is known as a fixed value and does not discuss on how to obtain the delay value. While increasing amount of literature on cross-layer modeling and optimization of HSR communications system has been proposed in recent years, most work only addresses the problem under the *infinite backlog traffic* model, assuming there will always be data to transmit from the queues [31–34]. Moreover, the models and optimization problems are *deterministic* considering a snapshot of the system instead of its dynamic behavior as a stochastic process over time. Different from the above work, [35, 36] consider dynamic optimization of radio resource management for HSR communication system. However, the transmission mechanisms considered are quite different from that of LTE-R. In order to analyze stochastic data traffic performance of HSR communications system, the HSR fading channel has to be modeled as a link between the physical layer and higher layers. Although HSR wireless channel modeling for physical layer has been a very active area, it is too complex to be incorporated into the cross-layer models for performance analysis and optimization. On the other hand, wireless channel can be modeled as a first-order Finite-State Markov Chain (FSMC) [37], which has been widely adopted in cross-layer performance analysis. However, most FSMC models in literature consider only low to medium mobility speed and assume that the average signal-to-noise ratio (SNR) remains constant [38], which is obviously not true for HSR fading channel. An FSMC model is developed for HSR fading channel in [39], which divides the coverage area of a base station along the railway line into multiple zones, assuming that the average SNR is constant within each zone and an FSMC similar to the traditional FSMC models are formulated for each zone. However, the FSMCs for different zones are considered separately, which cannot reflect the variation of average SNR over time as a train moves along the railway line. This "one FSMC per zone" modeling methodology is also used in other literature for HSR fading channel [40, 41], which is different from [39] in that real field measurement, data is used to derive the SNR distribution.

The main contributions of this chapter lie in the following aspects:

(1) The mobility model of HSR communications system is formulated as a semi-Markov process. As such, the instantaneous data rate of wireless channel becomes a semi-Markov modulated process, which takes into account the channel variations due to both large-scale and small-scale fading effects. Moreover, the performance loss due to AMC selection with imperfect CSI is also considered. Finally, the stochastic service curve of HSR communications system is derived based on the semi-Markov modulated process.
(2) Both CCDF snetal and MGF snetal approaches are used to derive the delay and backlog bounds of train control services and the results are compared.
(3) The analytical delay and backlog bounds are validated by simulation and can be used in the design and dimensioning of LTE-R system.

6.3.4 System Model

6.3.4.1 LTE-R Architecture for Train Control System

Train control is an important part of the railway operating management system. Traditionally it connects the fixed signaling infrastructure with the trains. In modern train control systems, trains and control centers are connected by mobile communications links. Examples are European Train Control System (ETCS)/Chinese Train Control System (CTCS), which are used for main line railways in Europe/China; and Communications-Based Train Control (CBTC), which can mainly be used for urban railway lines. The current radio communication networks for ETCS/CTCS are based on GSM-R, which is envisioned to be upgraded to LTE-R in the future.

Figure 6.6 depicts a simplified view of the LTE-R communication architecture. LTE-R eNodeBs are deployed along the railway line to provide a seamless coverage over the region. Although the LTE-R specifications have not been standardized yet, it is envisioned to be mostly based on the existing LTE specifications with some adaptations for the special characteristics of HSR communications, such as the high mobility and high priority of train control services. In this chapter, the LTE-R eNodeBs can be considered as LTE eNodeBs, except that the proposed AMC scheme as described later is used in order to adapt to the HSR fading channel. The eNodeBs are connected to the core network via wireline links, while the core network provides connectivity to the train control centers. To overcome the penetration loss of train carriages, a vehicle station (VS) is fixed in the ceiling on top of the train. The data traffic dedicated to train control involves both downlink and

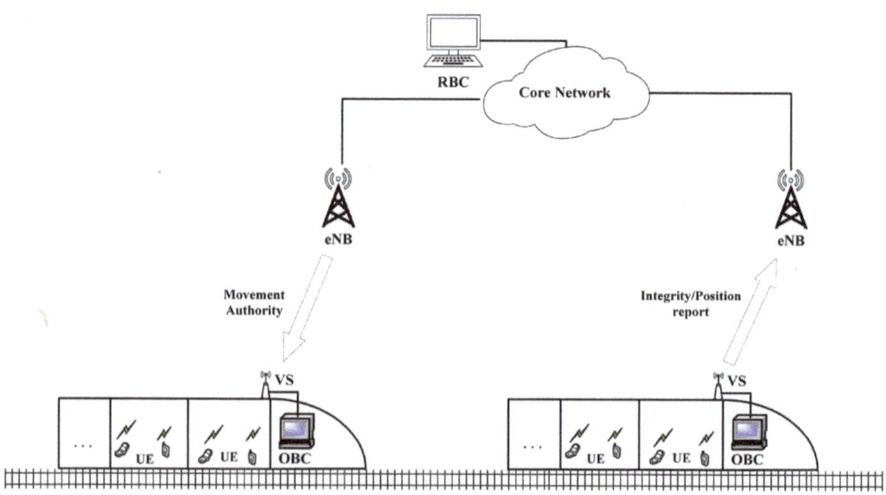

Fig. 6.6 LTE-R communications architecture for train control service

uplink wireless transmissions between the VS and eNodeB. In modern train control systems, the train movement is controlled by exchanging messages with the control center, which is referred to as radio block center (RBC) in the ETCS system. Each train features a train integrity control system and a computer (e.g., onboard controller (OBC) in ETCS) that can control train speed. It communicates via VS with eNodeBs, which are connected to the RBCs by the core network. Each train checks periodically its integrity and sends the integrity information together with the current position of the train head to the RBC, where such information is processed. The resulting information is sent to the following train, telling it either that everything is fine to go on driving (by sending a new movement authority message) or that an emergency braking is necessary immediately.

The communication delay between the VS and eNodeB of the train control services has great impact on the track utilization and speed profile of high-speed trains. The maximum track utilization will be achieved if trains are following each other with a minimum distance. Now we examine the minimum distance between trains operated under ETCS. We assume two trains (Train1 and Train2) directly follow each other with a maximum speed $v_{\max}$ and a distance d on a continuous track without stops, as shown in Fig. 6.7. At time t_1, Train1 completes its integrity check and sends a train integrity/position report to the RBC. Consider Scenario 1 where a part of Train1's carriages is lost immediately after t_1 from the main train and stop where they are. At time $t_1 + \Delta\tau$, an updated train integrity/position report is sent from Train1 to the RBC which informs the RBC that a part of its carriages is lost, where $\Delta\tau$ denotes the time between two successive integrity/position reports.

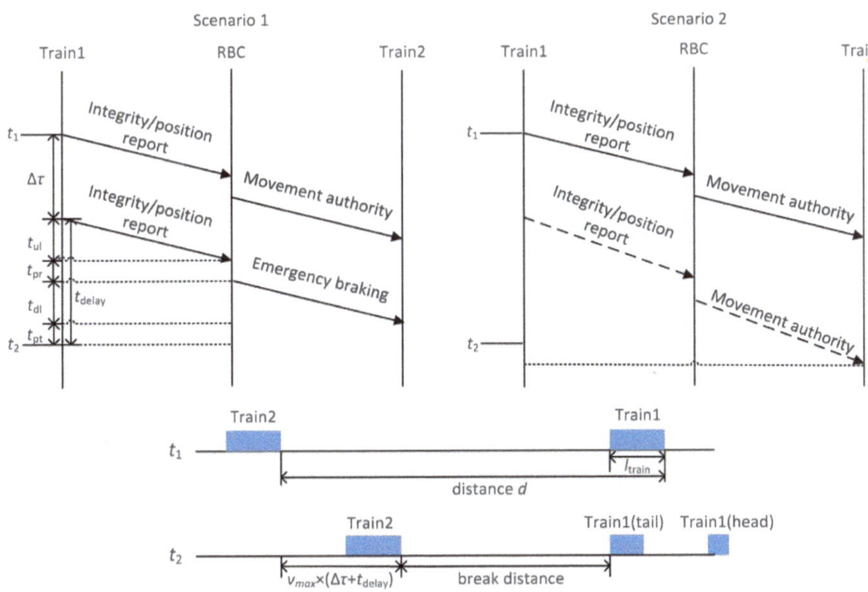

Fig. 6.7 The impact of communication delay to train distance and speed profile

After the RBC has processed this report, an emergency braking message is sent to the following Train2 which is processed there. As a result, Train2 starts to perform braking at a time no later than $t_2 = t_1 + \Delta\tau + t_{\text{delay}}$, where t_{delay} is the sum of the worst-case values of the communication delay t_{ul} of the integrity/position report to the RBC, the processing time t_{pr} at the RBC, the communication delay t_{dl} of the emergency braking message to the Train2, and the processing time t_{pt} at Train2. The distance between the head of Train2 to the stopped part of Train1 is $d - l_{\text{train}} - v_{\max}\left(\Delta\tau + t_{\text{delay}}\right)$ at time t_2, where l_{train} is the train length. Assume that the braking distance is l_{brake}. Then the minimum head-to-head distance d between the two trains should be $d = l_{\text{train}} + v_{\max}\left(\Delta\tau + t_{\text{delay}}\right) + l_{\text{brake}}$ to ensure train safety. Now consider Scenario 2 when Train1 is moving normally, but the communication delay of either the integrity/position report from Train1 to RBC or the movement authority message from RBC to Train2 exceeds the worst-case value, so consequently Train2 does not receive the second movement authority message in time. Without any information from RBC, Train2 needs to avoid accident if Scenario 1 of lost carriages as described above happens and brake at time t_2 even though it is actually safe to continue moving at the maximum speed. This means that if the communication delay exceeds the required worst-case value, the trains will perform unnecessary braking, which causes inefficiency in train operation and affects passenger comfort. Although increasing the required worst-case value of communication delay will solve this problem, the minimum distance d between trains will be increased and the track utilization decreased. Therefore, it is important to accurately evaluate the worst-case communication delay of the train control services to achieve the best tradeoff between track utilization and unnecessary braking.

6.3.4.2 Stochastic Network Calculus

This section provides a brief overview on the basic principle of stochastic network calculus and introduces the notation and basic assumptions in this chapter. First, we introduce the following min-plus convolution and deconvolution operators, denoted by $\otimes$ and Θ, respectively:

$$(f_1 \otimes f_2)(x) = \inf_{0 \le y \le x} [f_1(y) + f_2(x-y)]$$

$$(f_1 \Theta f_2)(x) = \sup_{y \ge 0} [f_1(x+y) - f_2(y)]$$

We use $\mathcal{F}$ (resp. $\bar{\mathcal{F}}$) to denote the set of nonnegative wide sense increasing (resp. decreasing) functions as follows:

$$\mathcal{F} = \{f(.) : \forall 0 \le x \le y, 0 \le f(x) \le f(y)\}$$

$$\bar{\mathcal{F}} = \{f(.) : \forall 0 \le x \le y, 0 \le f(y) \le f(x)\}$$

In this chapter, the time model is discrete starting from zero. The time indices are denoted by the symbols n, k, t, and s. The stochastic processes are all considered as stationary. The cumulative arrivals and departures of a flow at/from a system up to time n are denoted by non-decreasing processes $A(n)$ and $A^*(n)$. The doubly indexed extensions are $A(k,n) = A(n) - A(k)$ and $A^*(k,n) = A^*(n) - A^*(k)$. The delay of the flow at time n is

$$D(n) = \inf\{d \geq 0 : A(n) \leq A^*(n+d)\} \tag{6.1}$$

and the backlog of the flow at time n is

$$B(n) = A(n) - A^*(n) \tag{6.2}$$

Let $S(n)$ denote the cumulative amount of workload that can be served by the system up to time n. The departure process $A^*(n)$ is determined by $A(n)$ and $S(n)$. Specifically, for a lossless queuing system, the following equality holds according to Lindley recursion:

$$B(n) = \sup_{0 \leq k \leq n} \{A(k,n) - S(k,n)\} \tag{6.3}$$

Combining (6.2) with (6.3), we have

$$D(n) = \inf_{0 \leq y \leq x} \{A(k) + S(k,n)\} = A \otimes S(n) \tag{6.4}$$

Generally speaking, the snetal tackles the problem of performance analysis in two steps: (1) characterizing the stochastic arrival curve (SAC) for the flow arrival process $A(n)$ and stochastic service curve (SSC) for the system service process $S(n)$, respectively [42]; (2) deriving the stochastic delay and backlog bounds of the flow based on SAC and SSC. In order to achieve the above tasks, the MGF snetal and CCDF snetal take different approaches.

(1) *Step 1: Derivation of SAC and SSC*: The SAC for $A(n)$ should be its upper envelop and the SSC for $S(n)$ should be its lower envelop, i.e., for all $0 \leq k \leq n$.

$$A(k,n) - \hat{A}(n-k) \leq 0 \tag{6.5}$$

$$A \otimes \hat{S}(n) - A^*(n) \leq 0 \tag{6.6}$$

Since $A(n)$ and $S(n)$ are stochastic processes, it may be impossible to find deterministic processes $\hat{A}(n)$ and $\hat{S}(n)$ to satisfy the above inequalities as in dnetal, and will result in loose bounds even if such deterministic processes can be found.

From this point on, the MGF snetal and CCDF snetal start to take different paths in dealing with this problem.

In MGF snetal, $\hat{A}(n)$ and $\hat{S}(n)$ are considered as stochastic processes referred to as stochastic envelop processes, which can be the arrival and service processes themselves. Then, the MGFs of stochastic envelop processes are derived, where the MGF for a stochastic process $X(t)$ is defined for any θ as

$$M_X(\theta, n) = Ee^{\theta X(n)} \tag{6.7}$$

and E is the expectation of its argument. Note that another closely related concept is the effective bandwidth $\delta_X(\theta, n)$ of an arrival process $X(n)$, where

$$\delta_X(\theta, n) = 1/(\theta n) \cdot \log Ee^{\theta X(n)} = 1/(\theta n) \cdot \log M_X(\theta, n) \tag{6.8}$$

In CCDF snetal, SAC $\hat{A}(n)$ and SSC $\hat{S}(n)$ are considered as deterministic processes. However, bounding functions $f(x)$ and $g(x)$ that bound the violation probabilities of (6.5) and (6.6) are defined in the CCDF form of $P(W > x) \leq f(x)$ and $P(V > x) \leq g(x)$ for all $x \geq 0$, where W and V can be the LHS term of (6.5) and (6.6), respectively. Alternatively, W and V can also be the maximum values of the LHS term of (6.5) and (6.6) over one or both of its free variable k and n. In [8], the three versions of SACs are thus defined as traffic-amount-centric (t.a.c.), virtual-backlog-centric (v.b.c.), and maximum (virtual)-backlog-centric (m.b.c.), while the two versions of SSC are defined as weak stochastic service curve and stochastic service curve. In this chapter, we use the v.b.c. stochastic arrival curves and weak stochastic service curve, and their formal definitions are given below. For ease of understanding, we will use notations $\alpha(n)$ and $\beta(n)$ instead of $\hat{A}(n)$ and $\hat{S}(n)$ to represent SAC and SSC in CCDF snetal, respectively, where they are deterministic processes.

Definition 6.1 A flow $A(n)$ is said to have a v.b.c. stochastic arrival curve (SAC) $\alpha \in \mathcal{F}$ with bounding function $ff \in \bar{\mathcal{F}}$, denoted by $A \sim_{ta} \langle f, \alpha \rangle$, if, for all $x \geq 0$ and $n \geq 0$

$$P\left\{ \sup_{0 \leq k \leq n} \{A(k, n) - \alpha(n - k)\} > x \right\} \leq f(x) \tag{6.9}$$

Definition 6.2 A system S is said to provide a weak stochastic service curve $\beta \in \mathcal{F}$ with bounding function $g \in \bar{\mathcal{F}}$, denoted by $S \sim_{ws} \langle g, \beta \rangle$, if, for all $x \geq 0$ and $n \geq 0$

$$P\{A \otimes \beta(n) - A^*(n) > x\} \leq g(x) \tag{6.10}$$

(2) *Step 2: Derivation of Backlog and Delay Bounds:* The objective is to find the error functions $\varepsilon_b(x)$ and $\varepsilon_d(x)$ for backlog and delay, respectively, such that $P\{B(n) > x\} \leq \varepsilon_b(x)$ and $P\{D(n) > x\} \leq \varepsilon_d(x)$.

Specifically, the backlog satisfies

$$P\{B(n) > x\} = P\{A(n) - A^*(n) > x\} \leq P\left\{ \sup_{0 \leq k \leq n} \{\hat{A}(k,n) - \hat{S}(k,n)\} > x \right\} \tag{6.11}$$

where the first equality follows (6.2), while the second inequality can be derived by combining (6.5) and (6.6) with (6.2).

Moreover, the definition of delay in (6.1) implies that for any $x \geq 0$, if $D(n) > x$, $A(0,n) > A^*(0,n+x)$ is true [14]. Therefore, we have

$$P\{D(n) > x\} \leq P\{A(n) > A^*(n+x)\} \leq P\left\{ \sup_{0 \leq k \leq n} \{\hat{A}(k,n) - \hat{S}(k,n+x)\} > 0 \right\} \tag{6.12}$$

where the second inequality follows taking (6.5) and (6.6) into its LHS term.

In MGF snetal, the second inequalities of (6.11) and (6.12) are used to derive the stochastic backlog and delay bounds, i.e., $\varepsilon_b(x)$ and $\varepsilon_d(x)$ using Boole's inequality and Chernoff bound. The results are given in Theorem 6.1 [20, 23]. Note that the second inequalities of both (6.11) and (6.12) become equalities if the stochastic envelop processes $\hat{A}(n)$ and $\hat{S}(n)$ are the arrival and service processes $A(n)$ and $S(n)$ themselves [8, 13, 16].

Theorem 6.1 *Given the stochastic arrival envelop process $\hat{A}(n)$ with MGF $M_A(\theta,n)$ and stochastic service envelop process $\hat{S}(n)$ with MGF $\overline{M}_S(\theta,n) = M_S(-\theta,n)$. If $\hat{A}(n)$ is independent of $\hat{S}(n)$, then an upper backlog bound and an upper delay bound, each with at most violation probability $\varepsilon \in (0,1]$, are given by*

$$x_b(\varepsilon) = \inf_{\theta > 0} \left[\frac{1}{\theta} \left(\ln \sum_{k=0}^{\infty} M_A(\theta,k)\overline{M}_S(\theta,k) - \ln \varepsilon \right) \right] \tag{6.13}$$

$$x_d(\varepsilon) = \inf_{\theta > 0} \left\{ \inf \left[\tau : \frac{1}{\theta} \left(\ln \sum_{k=0}^{\infty} M_A(\theta,k-\tau)\overline{M}_S(\theta,k) - \ln \varepsilon \right) \leq 0 \right] \right\} \tag{6.14}$$

Note that $x_b(\varepsilon)$ is the inverse function of $\varepsilon_b(x)$, i.e., $x_b(\varepsilon) = x$, if and only if $\varepsilon_b(x) = \varepsilon$.

In CCDF snetal, the first equality of (6.11) and first inequality of (6.12) are used to calculate $\varepsilon_b(x)$ and $\varepsilon_d(x)$, respectively. For example, by adding and subtracting $A \otimes \beta(n)$ to $A(n) - A^*(n)$, we derive

$$\begin{aligned} & A(n) - A^*(n) \\ & = \sup\{A(k,n) - \alpha(n-k) + \alpha(n-k) - \beta(n-k)\} + [A \otimes \beta(n) - A^*(n)] \\ & \leq \sup_{0 \leq k \leq n} \{A(k,n) - \alpha(n-k)\} + \sup_{0 \leq k \leq n} \{\alpha(k) - \beta(k)\} + [A \otimes \beta(n) - A^*(n)] \\ & \leq \sup_{0 \leq k \leq n} \{A(k,n) - \alpha(n-k)\} + [A \otimes \beta(n) - A^*(n)] + \sup_{n \geq 0}\{\alpha(n) - \beta(n)\} \end{aligned} \tag{6.15}$$

Since the CCDFs of random variables (r.v.s) $X = \sup_{0 \leq k \leq n}\{A(k,n) - \alpha(n-k)\}$ and $Y = A \otimes \beta(n) - A^*(n)$ are bounded by the bounding functions $f(x)$ and $g(x)$ by the definitions of v.b.c. stochastic arrival curve and weak stochastic service curve, we have $P\{B(n) > x\} = P(A(n) - A^*(n) > x)$ *is bounded by* $\varepsilon_b(x) = f \otimes g(x + \inf_{k \geq 0}[\beta(k) - \alpha(k)])$ *according to probability theory given in Lemma 6.2.*

Similarly, the stochastic delay bound $\varepsilon_d(x)$ can also be derived for $P\{D(n) > x\} \leq P\{A(n) - A^*(n+x) > 0\}$. The results are summarized in the following theorem.

Theorem 6.2 *Consider a system S with input A. If the input has a v.b.c. stochastic arrival curve c with bounding function* $f \in \bar{\mathcal{F}}$, *(i.e.,* $A \sim_{vb} \langle f, \alpha \rangle$*), the server provides to the input a weak stochastic service curve* $\beta \in \mathcal{F}$ *with bounding function* $g \in \bar{\mathcal{F}}$, *i.e., (* $S \sim_{ws} \langle g, \beta \rangle$*), then the backlog B(n) and delay D(n) are guaranteed such that, for all* $x \geq 0$ *and* $n \geq 0$

$$P\{B(n) > x\} \leq f \otimes g(x + \inf_{k \geq 0}[\beta(k) - \alpha(k)]) \tag{6.16}$$

$$P\{D(n) > x\} \leq f \otimes g(\inf_{k \geq 0}[\beta(k) - \alpha(k-x)]) \tag{6.17}$$

According to Lemma 6.2 of Appendix B, if X and Y are independent r.v.s, the backlog and delay bounds can be further improved. However, since both X *and* Y defined above depend on the arrival process $A(n)$, they are not independent even if the arrival and service processes are independent. In order to further improve the performance bounds, the concept of a stochastic strict server is introduced in [21] which characterizes the service process $S(n)$ by a deterministic ideal service process with strict service curve $\hat{\beta}(n)$ and an impairment process $I(n)$ according to the following definition.

Definition 6.3 A system $S(n)$ is said to be a stochastic strict server providing strict service curve $\hat{\beta}(n) \in \bar{\mathcal{F}}$ with impairment process $I(n)$ if, during any backlog period $(k, n]$, the actual service $S(k,n)$ provided by the system satisfies

$$S(k,n) \geq \hat{\beta}(n-k) - I(k,n) \tag{18}$$

If $I(n)$ has a v.b.c. stochastic arrival curve $\xi(n)$ with bounding function $g(x)$, it can be proved that the service process satisfies $A \otimes \beta(n) - A^*(n) \leq \sup_{0 \leq k \leq n}\{I(k,n) - \xi(n-k)\}$, where $\beta(n) = \hat{\beta}(n) - \xi(n)$. Since $P\{\sup_{0 \leq k \leq n}\{I(k,n) - \xi(n-k)\} > x\} \leq g(x)$ by the definition of v.b.c. stochastic arrival curve and also because $I(n)$ is independent from $A(n)$.

Theorem 6.3 *Consider a system S with input A. If the input has a v.b.c. stochastic arrival curve* $\alpha \in \mathcal{F}$ *with bounding function* $f \in \bar{\mathcal{F}}$, *(i.e.,* $A \sim_{vb} \langle f, \alpha \rangle$*). Also suppose the server is a stochastic strict server providing strict service curve* $\hat{\beta}$ *with impairment process* $I \sim_{vb} \langle g, \xi \rangle$. *If A and I are independent, the backlog B(n) and delay D(n) are guaranteed such that, for all* $x \geq 0$

$$P\{B(n) > x\} \leq 1 - \bar{f} * \bar{g}(x + \inf_{k \geq 0}[\beta(k) - \alpha(k)]) \tag{6.19}$$

$$P\{D(n) > x\} \leq 1 - \bar{f} * \bar{g}(\inf_{k \geq 0}[\beta(k) - \alpha(k - x)]) \tag{6.20}$$

where, $\beta(n) = \hat{\beta}(n) - \xi(n)$, $\bar{f}(x) = 1 - \min[f(x), 1]$, $\bar{g}(x) = 1 - \min[g(x), 1]$

6.3.5 Stochastic Arrival Curve for Train Control Service

As discussed in Section II, Position Report (PR) messages are transmitted in uplink direction from OBC (train) to RBC (ground) and Movement Authority (MA) messages are transmitted in downlink direction from RBC (ground) to OBC (train) periodically. Therefore, we use a periodic traffic source $A(n)$ to model each type of traffic with different parameters. The source generates σ units of workload at times $\{n = U\tau + c\tau,\ c = 0,1,\ldots\}$ where τ is the period of the source and U is the initial start time which is uniformly distributed in the interval [0, 1].

For all $n \geq 0$ and $\theta \geq 0$ it is known that the MGF of $A(n)$ is

$$M_A(\theta, n) = e^{\theta\sigma\lfloor \frac{n}{\tau} \rfloor}\left[1 + \left(\frac{n}{\tau} - \left\lfloor \frac{n}{\tau} \right\rfloor\right)\left(e^{\theta\sigma} - 1\right)\right] \tag{6.21}$$

while the effective bandwidth of $A(n)$ is

$$\delta_A(\theta, n) = \frac{\sigma}{n}\left\lfloor \frac{n}{\tau} \right\rfloor + \frac{1}{\theta n}\log\left[1 + \left(\frac{n}{\tau} - \left\lfloor \frac{n}{\tau} \right\rfloor\right)\left(e^{\theta\sigma} - 1\right)\right] \tag{6.22}$$

Now we derive the v.b.c. stochastic arrival curve of the periodic source $A(n)$ according to the following theorem.

Theorem 6.4 *A flow* $A(n)$ *with effective bandwidth* $\delta_A(\theta, n)$ *has stationary increments, then it has a v.b.c. stochastic arrival curve* $A \sim_{vb} \langle \alpha, f \rangle$, *where*

$$\alpha(n) = [\delta_A(\theta, n) + \theta_1] \times n \tag{6.23}$$

$$f(x) = \frac{e^{-\theta\theta_1}}{1 - e^{-\theta\theta_1}} e^{-\theta x} \tag{6.24}$$

for any $\theta_1 > 0$ and $\theta > 0$.

6.3.6 Stochastic Service Curve for HSR Fading Channel

6.3.6.1 Mobility Model

We divide the communication region of a serving eNodeB along the railway line into multiple zones, $\mathbf{Z} = \{1, 2, \ldots, Z\}$, as shown in Fig. 6.3, where in each spatial zone z, $z \in \mathbf{Z}$, the average received Signal-to-Interference-and-Noise Ratio (SINR) over the wireless channel between the serving eNodeB and the VS on the train is approximately the same, denoted by $\overline{\gamma}_z^{DL}$ for downlink and $\overline{\gamma}_z^{UL}$ for uplink. Let d_z denote the length of zone z and c_z denote the average distance between the serving eNodeB and a train in zone z. The average received SINR is determined by

$$\overline{\gamma}_z^{DL} = \frac{P_{\mathrm{eNB}} \times PL(c_z)}{N_0 W + I_z^{DL}} \tag{6.25}$$

$$\overline{\gamma}_z^{UL} = \frac{P_{\mathrm{VS}} \times PL(c_z)}{N_0 W + I_z^{UL}} \tag{6.26}$$

where P_{eNB} and P_{VS} are the transmit power of eNodeB and VS, respectively. $PL(c_z)$ is the path loss between the eNodeB and the VS given their distance c_z. N_0 is the noise spectral density and W is the system bandwidth. I_z^{DL} and I_z^{UL} denote the average received interference power in uplink and downlink for zone z, respectively.

Since the eNodeBs are deployed along the railway line, we only consider interference from the two neighboring cells to the left and right of the considered cell. We consider the worst-case scenario where both neighboring cells are active and cause interference to the considered cell as shown in Fig. 6.8. Let cr_z and cl_z denote the average distances between a train in zone z and its right and left neighbor eNodeB, respectively. For the downlink, we have

$$I^{DL} = P_{\mathrm{eNB}} \times PL(cl_z) + P_{\mathrm{eNB}} \times PL(cr_z) \tag{6.27}$$

For the uplink, we do not know the exact locations of the trains in the neighboring cells. However, we consider that the stationary probability π_z of there is a train in zone z, $\forall z \in \mathbf{Z}$ is the same for all the cells, which will be determined by our mobility model below. Therefore, we calculate the expected path loss between a train in a neighboring cell to the serving eNodeB given π_z, and the uplink interference power can be derived as

$$I^{UL} = P_{\mathrm{VS}} \times \underbrace{\sum_{z=1}^{Z} (\pi_z \times PL(cl_z))}_{\substack{\text{expected\_path\_loss\_in} \\ \text{right\_neighboring\_cell}}} + P_{\mathrm{VS}} \times \underbrace{\sum_{z=1}^{Z} (\pi_z \times PL(cr_z))}_{\substack{\text{expected\_path\_loss\_in} \\ \text{left\_neighboring\_cell}}} \tag{6.28}$$

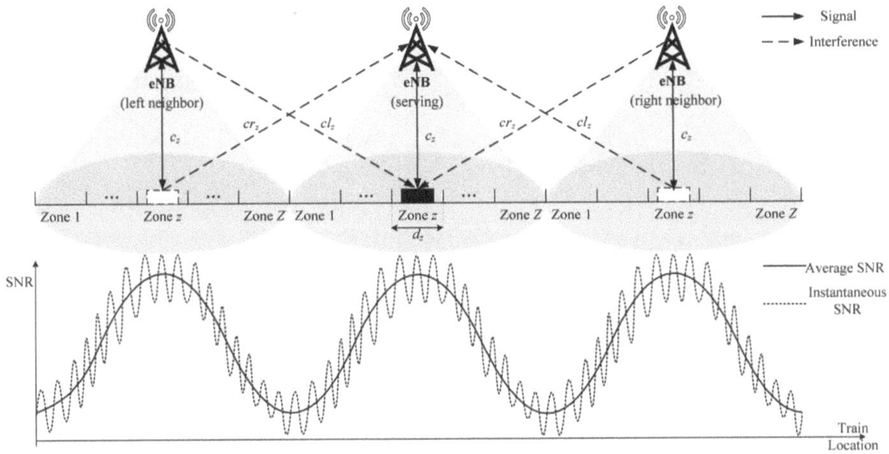

Fig. 6.8 HSR fading channel

Note that the first term is the uplink interference power from the right neighboring cell because the distance between a train in zone z of right neighboring cell and the serving eNodeB equals the distance cl_z between a train in zone z of the considered cell and the left neighbor eNodeB. For similar reason, the second term is the uplink interference power from the left neighboring cell. We will use $\overline{\gamma}_z$ to represent either uplink or downlink average SINR in the rest of the chapter.

The movement of trains is modeled by a stochastic process $\{\mathbf{Z}_t, t = 0, 1, \ldots\}$ with discrete state space Z, in which each state corresponds to one spatial zone. A discrete and integer time scale is adopted: t and t + 1 correspond to the beginning of two consecutive time slots, where the duration of a time slot ΔT = 1 ms in LTE system. Within the duration of a time slot, a train either moves to the next zone, or remains in the current zone. If a train leaves the current eNodeB and connects to a new eNodeB, it is regarded to move from state Z back to state 1 in the stochastic process, representing a new round of communication. Let the duration for which the trains stay in zone z be a random variable (r.v.) t_z, which is determined by the length of the partition zone d_z and the speed of trains vt representing the distance the trains move during a time

The stochastic process $\{\mathbf{Z}_t, t = 0, 1, \ldots\}$ representing the movement of trains as described above is a semi-Markov process associated with a Markov renewal process $\{(X(n), T(n)), n = 0, 1, \ldots\}$ with some semi-Markov kernel $Q(z, y, t)$.

Specifically, $X(n) \in \mathbf{Z}$ is the n-th state visited by the semi-Markov process and $T(n)$ is the time of this visit such that

$$\mathbf{Z}_t = X(n) \text{whenever}\, T(n) \leq t < T(n+1) \tag{6.29}$$

Moreover, the semi-Markov kernel gives

$$Q(z, y, t) = P[X(n+1) = y, T(n+1) - T(n) \leq t | X(n) = z] \tag{6.30}$$

for all $z, y \in \mathbf{Z}$ and $t \geq 0$. The process $\{X(n), n = 0, 1, \ldots\}$ is a Markov chain with transition matrix $\mathbf{P}$, where each element $P(z, y)$ equals

$$P(z, y) = Q(z, y, +\infty) = \begin{cases} 1 & \text{if } z < Z, y = z + 1 \text{ or } z = Z, y = 1 \\ 0 & \text{otherwise} \end{cases} \tag{6.31}$$

Note that from (6.31) we can derive the stationary probabilities $\{\pi_z, z \in \mathbf{Z}\}$ if a train in zone z, which are used in (6.28) to derive the uplink interference from neighboring cells.

The above semi-Markov model of the train mobility process does not require that the speed of trains to be a deterministic value as assumed in this chapter. Specifically, if $T(n)$ is geometrically distributed, the stochastic process $\{\mathbf{Z}_t, t = 0, 1, \ldots\}$ reduces to a Markov chain [36]. In the discrete-time Markov chain $\{X(n), n = 0, 1, \ldots\}$, n and n + 1 correspond to the beginning of two consecutive time units, where the duration $T(n)$ of a time unit n equals the duration for which the train stay in zone $z \in \mathbf{Z}$ if $X(n) = z$, i.e., t_z time slots or t_z ms. We will use s and t for the index of 1 ms time slots and k and n for the index of t_z ms time units in the rest of the chapter.

6.3.6.2 Data Rate Process

Within any spatial zone z, the instantaneous received SINR $\gamma_{z,t}$ over the wireless channel between the eNodeB and the train is also affected by small-scale fading apart from large-scale fading, which makes $\gamma_{z,t}$ deviate from its average value $\overline{\gamma}_z$, as shown in Fig. 6.8. Due to the large fading rate $f_D \Delta T$ induced by the high mobility speed of the trains, $\gamma_{z,t}$ can be regarded as i.i.d. random variables over different time slots t [37]. Since the high-speed trains typically run on the viaduct (such as in Chinese HSR), so the line of sight (LoS) path typically exists in the multipath environment. Thus, the multipath fast fading can be described using a Rician channel model [32, 37]. The instantaneous received SINR $\gamma_{z,t}$ in the downlink can be derived as

$$\gamma_{z,t}^{DL} = \frac{P_{\text{eNB}} \times PL(c_z) \times |h_t|^2}{N_0 W + P_{\text{eNB}} \times PL(cr_z) \times |ir_t|^2 + P_{\text{eNB}} \times PL(cl_z) \times |il_t|^2} \tag{6.32}$$

where $|h_t|$, $|ir(t)|$ and $|il(t)|$ are Rice distributed random variables whose square represent the small-scale fading gain of received signal power, received interference power from the right and left neighboring cells at time slot t, respectively. The instantaneous received SINR in the uplink can be derived Similarly.

Since the average received signal power $P_{\mathrm{eNB}} \times PL(cr_z)$ or $P_{\mathrm{VS}} \times PL(cl_z)$ is usually much stronger than the average received interference power I_z^{DL} or I_z^{UL}, we ignore the effect of fast fading on the received interference power and approximate the denominator of (6.32) by $N_0 + I_z^{DL}$. Therefore, we have $\gamma_{z,t} = \overline{\gamma}_z |h_t|^2$. Using the more general and simpler Nakagami$m$ fading model to approximate the Rician fading model, the probability distribution function (PDF) of the SINR $\gamma_{z,t}$ can be presented by [38]

$$f(\gamma_{z,t}) = \left(\frac{m}{\overline{\gamma}_z}\right)^m \frac{\gamma_{z,t}^{m-1}}{\Gamma(m)} \exp\left(-\frac{m\gamma_{z,t}}{\overline{\gamma}_z}\right) \tag{6.33}$$

where $\Gamma(\cdot)$ is the Gamma function, $m = \frac{(K+1)^2}{2K+1}$ is the fading parameter, and K is the Rice factor.

The instantaneous data rate $r_{z,t}$ within spatial zone z can be determined from the instantaneous received SNR $\gamma_{z,t}$. The simplest method is using the Shannon formula, where the instantaneous data rate within spatial zone z is a random variable $r_{z,t} = C \log(1 + \gamma_{z,t})$. However, this method can only provide an approximate data rate which is not accurate. In this chapter, we consider the practical scenario where the instantaneous data rate within spatial zone z is determined by the adaptive modulation and coding (AMC) scheme. The SINR values are divided into L nonoverlapping consecutive regions. Ideally, perfect channel state information (CSI) is available at the eNodeB, based on which the optimum modulation and coding scheme (MCS) can be selected. For any $l \in \{1, \ldots, L\}$, the l-th MCS is selected if the instantaneous SINR value γ falls within the l-th region $[\Gamma_l, \Gamma_{l+1})$. Obviously, $\Gamma_0 = 0$ and $\Gamma_{L+1} = \infty$. However, due to the rapidly varying channel condition induced by the high mobility of HST, the CSI at the eNodeB may be highly inaccurate. Therefore, the performance of the CSI-based AMC scheme may be seriously degraded. As an alternative, we propose to perform AMC based on the average received SINR instead, since the average received SINR is mainly impacted by the large-scale fading effect and varies on a much slower time scale than the instantaneous SINR. Based on the above assumption,

a fixed MCS scheme is selected for each zone z according to its average SINR $\overline{\gamma}_z$, i.e., the l-th MCS is selected for zone z if $\overline{\gamma}_z$ falls within the l-th region $[\Gamma_{l-1}, \Gamma_l)$. The selected MCS determines the ideal transmission capability r_z^{ideal} of the wireless channel in zone z. However, as the channel condition is also impacted by the small-scale fading effect, transmission errors may be incurred when the channel is in deep fade. To simplify performance analysis, we will rely on the following approximate block error rate (BLER) expression over Additive White Gaussian Noise (AWGN) channel [46]:

$$\mathrm{BLER}_l(\gamma) = \begin{cases} 1 & \text{if } 0 < \gamma < \gamma_{pl} \\ a_l \exp(-g_l \gamma) & \text{if } \gamma > \gamma_{pl} \end{cases} \tag{6.34}$$

Parameters a_l, g_l, and γ_{pl} are MCS-dependent, and are obtained by fitting and comparing curves by (6.34) to the simulated BLER according to the Monte Carlo simulations with parameters given by 3G LTE specification [46]. We select $L = 6$ MCSs from the 32 MCSs in LTE and the parameters are given in Table 6.2. In LTE system, a terminal can be allocated in the downlink or uplink with a minimum of 1 Resource Block (RB) during 1 subframe (1 ms), where an RB occupies 12 subcarriers (12 × 15 kHz = 180 kHz) in frequency domain. Therefore, the data rate r_z^{ideal} in Table 6.2 is the number of bits that can be transmitted on one RB within 1 ms time slot. We consider an infinite-persistent ARQ protocol in the link layer, where an erroneous block is retransmitted until it is received correctly at the receiving end. Depending on the transmission outcome in each time slot, an acknowledgment (ACK) or a negative acknowledgement (NACK) is replied by the receiver to the transmitter for each transmitted packet. We assume that the ACK/NACK packets are available at the end of the transmission time slot, and the feedback channel carrying ACK/NACK packets is a reliable one. Based on the above assumptions, the instantaneous data rate of zone z given MCS index l is a random variable [48]

$$r_{z,t} = r_z^{\text{ideal}}\left(1 - \text{BLER}_l\left(\gamma_{z,t}\right)\right) \tag{6.35}$$

The *data rate process* of a HSR wireless communication channel as described above can be modeled by a semi-Markov Modulated Process (SMMP). The modulation is done via a discrete-time homogeneous semi-Markov process (SMP) $\{\mathbf{Z}_t, t = 0, 1, \ldots\}$ on the states $\{1, 2, \ldots, Z\}$. Let $\{r_{z,t}, t = 0, 1, \ldots\}, z = 1, \ldots, Z$, be Z sequences of i.i.d. random variables, representing the instantaneous data rate at time slot t when the SMP $\mathbf{Z}_t$ is at state z. The *data rate process* $r_t = r_{\mathbf{Z}_t,t}$ is then an SMMP with the modulating process $\mathbf{Z}_t$.

6.3.6.3 Stochastic Service Curve

Define the *service process* $S_t \equiv \sum_{s=1}^{t} r_s$ as the cumulative amount of service provided by the wireless channel by time t. If the *data rate process* r_t as defined above

Table 6.2 AMC parameters for LTE

	Mode 1	Mode 2	Mode 3	Mode 4	Mode 5	Mode 6
Modulation order	2	2	4	4	6	6
Rate r_z^{ideal} (bits/ms/180 kHz)	56	120	208	280	408	552
a_l	4.194	5.521	8.013	16.7	12.7	15.12
g_l	3.133	1.521	0.947	0.6359	0.2964	0.1211
γ_{pl}(dB)	−3.395	0.505	3.419	6.462	9.332	13.508
Γ_l (dB)	−0.37	3.09	5.63	8.31	11.23	15.31

is a Markov Modulated Process (MMP), the stochastic service curve of S_t can be derived. However, as r_t is an SMMP, we construct an *equivalent data rate process* which is an MMP. The modulation is done via a discrete-time homogeneous Markov process $\{X(n), n = 0, 1, \ldots\}$ on the states $\{1, 2, \ldots, Z\}$. Let $\{r_z(n) := \sum_{t=1}^{t_z} r_{z,t}, n = 0, 1, \ldots\}, z = 1, \ldots, Z$, be Z sequences of i.i.d. random variables, representing the total achievable data rate during the n-th state visited by the SMP $\{\mathbf{Z}_t, t = 0, 1, \ldots\}$, when the n-th state is state z. The *equivalent data rate process* $r(n) = r_{X(n)}(n)$ is then an MMP with the modulating process $X(n)$.We define the *equivalent service process* $S(n)$ as

$$S(n) := \sum_{k=1}^{n} r(k) = S\left(\sum_{k=1}^{n} t_{X(k)}\right) \tag{6.36}$$

(1) *MGF Snetal:* Define $\phi_{S,z}(\theta) := E\left[\mathrm{e}^{-\theta r_z(n)}\right] = \left(E\left[\mathrm{e}^{-\theta r_{z,1}}\right]\right)^{t_z}$ as the MGF of $r_z(n)$ and let $\phi_S(\theta)$ be the diagonal matrix $\mathrm{diag}\{\phi_{S,1}(\theta), \ldots, \phi_{S,Z}(\theta)\}$. For all $n \geq 0$ and all $\theta \geq 0$, the MGF of the *equivalent service process* $S(n)$ can be derived as [42]

$$\overline{M}_S(\theta, n) = \boldsymbol{\pi}(\phi_S(-\theta)\mathbf{P})^{n-1}\phi_S(-\theta)\mathbf{1} \tag{6.37}$$

where $\boldsymbol{\pi}$ is a row vector of the stationary state distribution of the modulating process $X(n)$, $\mathbf{P}$ is the transition matrix of $X(n)$ given in (6.31), and (6.1) is a column vector of ones.

(2) *CCDF Snetal:* Now we use two stochastic processes to characterize the equivalent service process $S(n)$, i.e., an ideal deterministic service process $\hat{S}(n) = \hat{r} \times n$ and an impairment process $I(n)$, where $S(n) = \hat{S}(n) - I(n)$ with $\hat{S}(0) = I(0) = 0$ by convention. Therefore, $I(n) = \sum_{k=1}^{n} (\hat{r} - r(k))$ according to (6.35). We can see that the impairment process is also the cumulative process of an MMP $i(n) = i_{X(n)}(n)$ with modulating process $X(n)$, where $i_z(n) = \hat{r} - r_z(n), n = 0, 1, \ldots, z = 1, \ldots, Z$ are Z sequences of i.i.d. random variables, representing the amount of impaired services during the n-th state visited by the SMP $\{\mathbf{Z}_t, t = 0, 1, \ldots\}$, when the n-th state is state z. Now, the *equivalent service process* $S(n)$ is a stochastic strict server with strict service curve $\hat{\beta}(n) = \hat{S}(n)$ and impairment process $I(n)$ by Definition 6.3.

Define $\phi_{I,z}(\theta) := E\left[\mathrm{e}^{-\theta i_z(n)}\right] = \mathrm{e}^{\theta\hat{r}}\left(E\left[\mathrm{e}^{-\theta r_{z,1}}\right]\right)^{t_z}$ as the MGF of $i_z(n)$ and let $\phi_I(\theta)$ be the diagonal matrix $\phi_{I,z}(\theta) := E\left[\mathrm{e}^{-\theta i_z(n)}\right] = \mathrm{e}^{\theta\hat{r}}\left(E\left[\mathrm{e}^{-\theta r_{z,1}}\right]\right)^{t_z}$. Let $sp(\phi_I(\theta)\mathbf{P})$ be the spectral radius of the matrix $\phi(\theta)\mathbf{P}$, where the transition matrix $\mathbf{P}$ is given in (6.31).

For all $n \geq 0$ and all $\theta \geq 0$, the effective bandwidth of the impairment process $I(n)$ can be derived as [49]

$$\delta_I(\theta, n) = \frac{1}{\theta n} \log\left(\boldsymbol{\pi}(\phi_I(\theta)\mathbf{P})^{n-1}\phi_I(\theta)\mathbf{1}\right) \tag{6.38}$$

The impairment process *I*(*n*) can be characterized by v.b.c. stochastic arrival curve according to the following Lemma, whose proof is similar to that of Theorem 6.4 and omitted here.

Lemma 6.1 If impairment process *I*(*n*) with effective bandwidth $\delta_I(\theta, n)$ has stationary increments, then it has a v.b.c. stochastic arrival curve $A \sim_{vb} \langle \xi, g \rangle$, where

$$\xi(n) = [\delta_I(\theta, n) + \theta_1] \times n \tag{6.39}$$

$$g(x) = \frac{e^{-\theta\theta_1}}{1 - e^{-\theta\theta_1}} e^{-\theta x} \tag{6.40}$$

for any $\theta_1 > 0$ and $\theta > 0$.

Given the stochastic strict server and v.b.c. stochastic arrival curve of the impairment process, Theorem 6.3 can be applied to derive the stochastic backlog and delay bounds using independence case analysis. Alternatively, we can first characterize the *equivalent service process S*(*n*) using weak stochastic service curve according to the following theorem.

Theorem 6.5 *The equivalent service process S(n) provides a weak stochastic service curve, i.e.,* $S \sim \langle g, \beta \rangle$*, where*

$$\beta(n) = [\hat{r} - \delta_I(\theta, n) - \theta_1]^+ n \tag{6.41}$$

$$g(x) = \frac{e^{-\theta\theta_1}}{1 - e^{-\theta\theta_1}} e^{-\theta x} \tag{6.42}$$

for $\forall\theta > 0$ *and* $\theta_1 > 0$.

Given the weak stochastic service curve of the *equivalent service process S*(*n*), Theorem 6.2 can be applied to derive the stochastic backlog and delay bounds

6.3.7 Performance Evaluation

6.3.7.1 Derivation of Delay Bound

Given the SAC of train control services in Sect. 6.3.3 and the SSC provided by the HSR fading channel in Sect. 6.3.4, the stochastic delay bound of the flow can be determined using the following three methods.

(1) MGF method: Theorem 6.1 is used to derive the delay bound where the MGFs of the arrival process and service process are derived from (6.21) and (6.37), respectively;

(2) CCDF method: For ease of notation, we denote $m = \inf_{k \geq 0}[\beta(k) - \alpha(k-x)]$

(a) Method 1: With the v.b.c. stochastic arrival curve of arrival process given in Theorem 6.4 and the weak stochastic service curve of service process given in Theorem 6.5, the delay bound can be derived by Theorem 6.2. Taking $f(x) = g(x) = \frac{e^{-\theta\theta_1}}{1-e^{-\theta\theta_1}} e^{-\theta x}$ from (6.24) and (6.42) into (6.17), we have

$$P\{D(n) > x\} \leq \frac{2e^{-\theta\theta_1}}{1-e^{-\theta\theta_1}} e^{\frac{-\theta m}{2}} \tag{6.43}$$

(b) Method 2: With the v.b.c. stochastic arrival curve of arrival process given in Theorem 6.4, and the stochastic strict server of service process with v.b.c. stochastic arrival curve of impairment process given in Lemma 6.1, the delay bound can be derived by Theorem 6.3. When taking $f(x) = g(x) = \frac{e^{-\theta\theta_1}}{1-e^{-\theta\theta_1}} e^{-\theta x}$ into (6.20),

we have

$$P\{D(n) > x\} \leq 1 - \frac{e^{-\theta\theta_1}}{1-e^{-\theta\theta_1}}(1 - e^{-\theta m}) + \left(\frac{e^{-\theta\theta_1}}{1-e^{-\theta\theta_1}}\right)^2 \theta m e^{-\theta m} \tag{6.44}$$

Note that $m = \inf_{k \geq 0}[(\hat{r} - \delta_I(\theta,k) - \delta_A(\theta,k-x) - 2\theta_1)k + (\delta_A(\theta,k-x) + \theta_1)x]$ according to (6.41) and (6.23). θ and θ_1 are free parameters to optimize the performance of the delay bound so that $P\{D(n) > x\}$ can be as small as possible

6.3.7.2 System Parameter

The system parameters are given in Table 6.3. We use the Winner Phase II model D2a sub-scenario to calculate the path loss $PL(d)$ in (6.25)–(6.28), which is a measurement-based physical layer channel model for links between the trackside base station and the roof-top antenna of a train

$$PL(d) = \begin{cases} 44.2 + 21.5\log(d) + L & d < d_{bp} \\ 44.2 + 40\log(d/d_{bp}) + L_{bp} + L & d \geq d_{bp} \end{cases} \tag{6.45}$$

where d is the distance from the roof-top antenna of the train to the eNodeB, which could be either c_z, cr_z or cl_z in (6.14)–(6.17). The distance from the eNodeB to the track is set to 50 m in order to calculate the above distances. L and L_{bp} are constants in dB. $L = 20\log(f_c/(5\times10^9))$ is the carrier frequency loss, where f_c is the carrier frequency in Hz. L_{bp} = 21.5log(d_{bp}), where d_{bp} is the break point of the path loss

Table 6.3 System parameters

Parameter	Value
Transmit power of eNodeB P_{eNB}	43 dBm
Transmit power of VS P_{VS}	23 dBm
Bandwidth W	3 MHz
Noise spectral density N_0	−174 dBm/Hz
Carrier frequency f_c	1.9 GHz
Train velocity $v_{max}/\Delta T$	100 m/s
Inter-site distance	3 km
length of a zone d_z	5 m

curve. d_{bp} equals to $4h_{eNB}h_{VS}f_c/c$, where $h_{eNB} = 45$ m and $h_{VS} = 5$ m are the eNodeB and VS antenna heights in meter compared to the ground, respectively, and c is velocity of light in vacuum.

The system bandwidth is 3 MHz containing 15 RBs. However, we assume that only one RB is dedicated to the train control services in the following numerical experiments. We set the velocity of trains to be 100 m/s if not specified otherwise in the following numerical experiments, so that $v_{max} = 0.1$ m/time slot. Moreover, the inter-site distance between two neighboring eNodeBs is set to be 3 km. Let the length of a zone z be $d_z = 5$ m, and we have the duration for which a train stays in zone z is $t_z = d_z/v_{max} = 50$ time slots or ms for any $z \in \mathbf{Z}$. Moreover, the number of zones $Z = 600$. Note that smaller zone size d_z will result in smaller time unit duration t_z.

As our delay bound is in terms of time unit, shorter length of a time unit shall result in more precise measurement of the delay bound. Therefore, we should set d_z to be as small as possible.

However, smaller d_z will lead to larger number of zones Z within a cell and thus larger state space of the semi-Markov process, which results in larger computational complexity I analysis. Moreover, if d_z is too small, a train may move from zone z to $z + i$ with $i > 1$ during a time slot, which further complicates the analysis. Therefore, the zone size d_z should be set to a proper value according to the train speed and coverage region of a cell considering the above tradeoff.

6.3.7.3 Numerical and Simulation Results

In this section, the delay performance of train control services over HSR fading channel is evaluated through both simulation and numerical results. Our simulation program is built on the MATLAB platform. The BS and the VS each has a buffer, where the arrived packets wait for transmission. At the start of each 50 ms time unit when the train is in zone z, the instantaneous SNR values $\gamma_{z,t}$ of $t_z = 50$ i.i.d. Rician fading channels with mean SNR $\bar{\gamma}_z$ are generated, each of which represents the instantaneous SNR of the HSR fading channel during 1 ms time slot. Then, the instantaneous data rate of each Rician fading channel is derived using both the AMC method by (6.35) and the Shannon method by the Shannon formula, respectively. The sum of the $t_z = 50$ instantaneous data rates represents the total amount of data that can be transmitted by the HSR fading channel during the period when the train is in zone z. The sojourn time of each data unit in buffer is recorded when it is transmitted. With this, the delay performance is obtained. The results are collected over Z simulations, each of which runs for 10^6 time units, where in the z-th simulation $(z \in \{1, \ldots, Z\})$ the train is assumed to be in zone z when the simulation starts. In the following experiments, we focus on the downlink transmission, since the analytical principles for deriving the stochastic delay bounds of uplink and downlink transmissions are the same. Figures 6.9 and 6.10 compare the analytical bounds by MGF and CCDF snetals and the simulation results under different violation probabilities, where the burst size and period of the periodical source are set to $\sigma = 4000$ bits and $\tau = 120$ time units (6 s), respectively. Figure 6.9 uses the proposed AMC method to obtain instantaneous data rate, while Fig. 6.10 uses the Shannon method. As expected, the estimated delay bound with Shannon method is smaller than that with the AMC method, which means that the Shannon method will provide results that are more optimistic than that can be actually achieved. It can be observed that in both figures, the analytical bounds provided by the MGF method are the tightest while those provided by the CCDF method 1 are the loosest. This is because the MGF method only uses the Boole's inequality and Chernoff bound when deriving the analytical bound by (6.46), while both CCDF methods use the above inequalities twice (when obtaining the stochastic arrival curves for periodical source and impairment process by (6.51)). Moreover, CCDF method 2 provides tighter bound than CCDF method 1 in Fig. 6.9 and the same bound with CCDF method 1 in

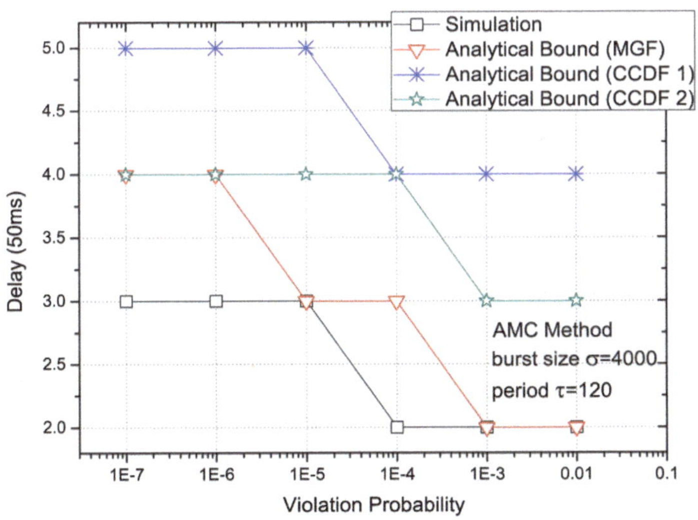

Fig. 6.9 Comparison of simulation results and analytical bounds under different violation probabilities with AMC method (burst size $\sigma = 4000$ bits, period $\tau = 120$ time units (6 s))

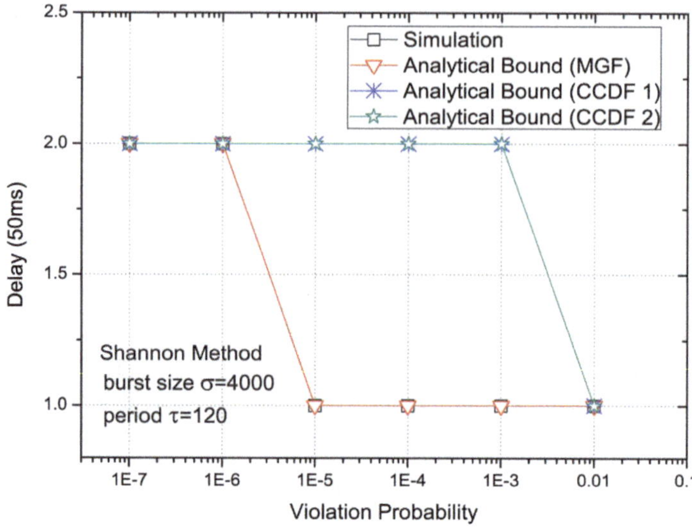

Fig. 6.10 Comparison of simulation results and analytical bounds under different violation probabilities with Shannon method (burst size $\sigma = 4000$ bits, period $\tau = 120$ time units (6 s))

Fig. 6.10, because the second bound in Lemma 6.2 is generally better than the first bound. Note that the MGF snetal is easier to implement than the CCDF snetal, because that in the MGF snetal, only one free parameter θ needs to be optimized in order to derive the performance bound as in (6.14). In the CCDF snetal, on the other hand, two free parameters θ and θ_1 need to be optimized as in (6.43) and (6.44). Therefore, in Figs. 6.11 and 6.12, we only use MGF snetal to derive the analytical bounds.

Figure 6.11 compares the analytical bounds by MGF snetal and the simulation results under different burst sizes with AMC method and Shannon method, where the period of the periodical source is set to $\tau = 120$ time units (6 s) and the violation probability is set to 1e-7, respectively. Figure 6.6 shows that the delay bound increases with the increasing burst size, and the increasing rate of Shannon method is slower than that of the AMC method. This is because that the period of burst arrival is $\tau = 120$ time units and the largest delay experienced by the data in the buffer (when the burst size is 14000 bits with AMC method) is larger than 7 time units with probability no larger than 1e-7. Therefore, we can safely conclude that a burst is fully transmitted before the next one arrives, which means that the largest backlog equals the burst size and thus the delay bound depends on the burst size and the instantaneous data rate at every zone. From Fig. 6.11 we can also observe that the MGF method can provide a relatively tight bound with both AMC method and Shannon method.

Figure 6.12 compares the analytical bounds by MGF snetal and the simulation results under different periods with AMC method and Shannon method, where the burst size is set to $\sigma = 14000$ bits and the violation probability is set to $1e - 7$,

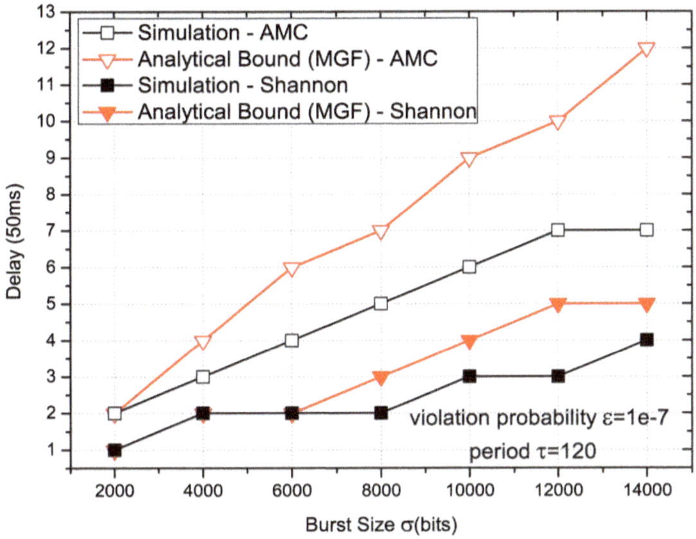

Fig. 6.11 Comparison of simulation results and analytical bounds for periodical source under different burst sizes with AMC method and Shannon method (violation probability $\varepsilon = 1e^{-7}$, period $\tau = 120$ time units (6 s))

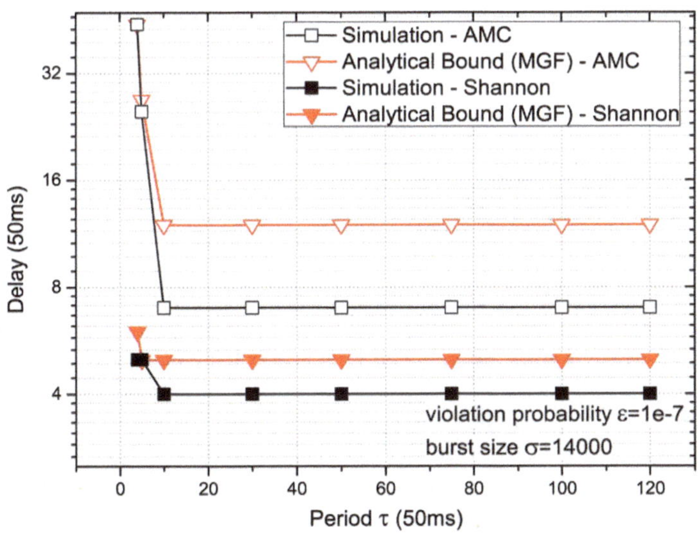

Fig. 6.12 Comparison of simulation results and analytical bounds under different periods with AMC method and Shannon method (violation probability $\varepsilon = 1e^{-7}$, burst size $\sigma = 14000$ bits)

respectively. It can be observed that the delay bound remains to be the same when the period reduces from 120 time units to 10 time units for both the AMC method and Shannon method. However, when the period reduces from 10 time units to 4 time units, the delay bound grows quickly for the AMC method (from 7 to 44 time units in the simulation results) and grows a little for the Shannon method (from 4 to 5 time units in the simulation results). This observation is because when the period is larger than 10 time units, a burst is almost always fully transmitted before the next one arrives, as explained in Fig. 6.6. However, when the period becomes smaller than 10 time units, a burst may not be fully transmitted before the next one arrives, so the remaining data in the previous burst will be backlogged to the next period for transmission. For the AMC method, since the delay bound corresponding to violation probability 1e − 7 is 7 time units (in the simulation result) when the period is 10 time units, the delay bound increase quickly when the period reduces to 4 and 5 time units. When the period further reduces to be smaller than 4 time units, the MGF snetal fails to derive the delay bound since the term $M_A(\theta, k - \tau)\bar{M}_S(\theta, k)$ in (6.14) increases with increasing k and the sum of this term over $k = \{0, 1, \ldots\}$ becomes infinity. This is because the traffic intensity becomes too large so the queuing system is not stable anymore and the backlog accumulates to infinity over time. On the other hand, for the Shannon method, since the delay bound corresponding to violation probability 1e − 7 is 4 time units (in the simulation result) when the period is 10 time units, the bursts are fully transmitted before the next one arrives even when the period is reduced to 4 time units. Therefore, the delay bound only increases a little in this case. From both Figs. 6.11 and 6.12, it can be seen that the analytical bounds for Shannon method is much tighter than those for AMC method. The reason for this difference is due to the mathematical principle of snetal. In snetal, some general purpose methods are commonly used in order to derive the performance bounds, such as the Chernoff's bound and Boole's inequality. The derived bounds may be tight or loose depending on the specific distribution of the arrival and service processes. Therefore, the usage of Shannon method or AMC method leads to different distributions of the service process, which results in the different tightness of the derived bounds

In the above numerical experiments, we set the velocity of trains to be 100 m/s. Note that the train speed may affect the delay bound, since the channel variation due to changing path loss shall be faster with higher train speed. In order to examine its impact on the delay bound, we vary the train speed from 50 to 200 m/s in a step of 50 m/s. Moreover, in order to facilitate comparison, we set the zone size d_z correspondingly so that the duration of a time unit t_z remains to be 50 ms irrespective of the train speed. Figure 6.13 shows the analytical bound and simulation results under different train speeds with AMC method when the burst size $\sigma = 14000$ bits, period $\tau = 4$ time units and violation probability $\varepsilon = 1e^{-7}$. It can be seen that the delay bound is improved with increasing train speed. Although not shown in Fig. 6.13, the impact of train speed reduces to almost zero when we reduce the burst

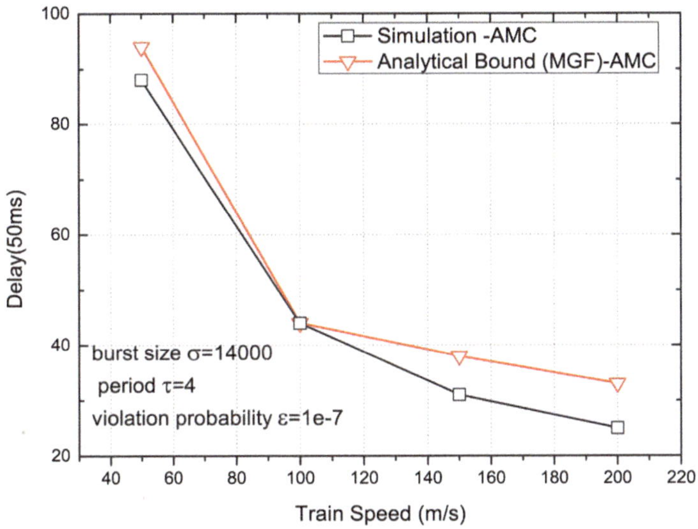

Fig. 6.13 Impact of train speed on delay bound (violation probability $\varepsilon = 1e^{-7}$, burst size $\sigma = 14000$ bits, period $\tau = 4$ time units, AMC method)

size or increase the period. This is because the delay bound improves when the train speed is higher because the train will travel longer distance and thus the HSR fading channel will experience larger channel state variation during the transmission of a message. When the burst size is reduced or the period is increased, it can be seen from Figs. 6.6 and 6.7 that the message transmission delay is significantly reduced compared to that corresponds to the parameter setting in Fig. 6.13, which results in reduced impact of train speed on the delay bound.

We would like to remark that the length of the MA message in the current ETCS/CTCS system is typically around 1600 bits and the length of the PR message is 192 bits, and the arrival periods of both messages are typically around 6 s (120 time units) based on our measurement data in practical system. Moreover, it is required in the ETCS/CTCS system that the maximum end-to-end transfer delay should be ≤ 0.5 s (10 time units) under 99% probability [28]. Therefore, we can conclude that the LTE system can provide satisfactory performance guarantee for the train control services using one RB based on our analytical and simulation results above.

Generally speaking, the analytical principles and delay bounds derived for uplink and downlink transmissions are the same for both FDD-LTE and TD-LTE. However, for the total end-to-end transfer delay from the time when an integrity/position report is transmitted by Train1 until a movement authority message is received by Train2 as illustrated in Fig. 6.7, TD-LTE may cause several milliseconds of more delay than FDD-LTE. This is because in TD-LTE, a 10 ms frame is divided into 10 1 ms subframes, which are reserved for downlink or uplink transmissions according to different configurations. Therefore, a message buffered at the train or base station needs to wait for the proper type of subframe for transmission, which may cause some further delay.

References

1. Barbu G (2010) E-Train—broadband communication with moving trains technical report—technology state of the art, Int. Union Railways, Paris, France, Technical Report
2. Coraiola A, Antscher M (2000) GSM-R network for the high speed line Rome-Naples. Signal Draht 92(5):42–45
3. Requirements for further advancements for evolved universal terrestrial radio access (E-UTRA) (LTE-Advanced) (Release 11), 3GPP TR specification 36.913, Technical Specification Group Radio Access Network, Sept 2012
4. Rail safety and standards board, assessing bandwidth demand for future communications needs on GB Railways, RSSB research programme, operations and management, Aug 2010
5. Masur K-D, Mandoc D (2009) LTE/SAE—The future railway mobile radio system? long-term visions on railway mobile radio technologies, international union of railways (UIC), Technical Report, v1.1
6. Larmo A, Lindström M, Meyer M, Pelletier G, Torsner J, Wiemann H (2009) The LTE link-layer design. IEEE Commun Mag 47(4):52–59
7. Tingting G, Bin S (2010) A high-speed railway mobile communication system based on [8] LTE. Electron Inf Eng (ICEIE) 1:414–417
8. Sniady A, Soler J (2014) LTE for railways: impact on performance of ETCS railway signaling. IEEE Veh Technol Mag 9(2):69–77
9. Sundarapandian V (2009) 7 queueing theory. Probability, statistics and queueing theory. PHI learning. ISBN 8120338448
10. Gianfranco Balbo: Introduction to stochastic petri nets
11. Ai B et al (2014) Challenges toward wireless communications for high-speed railway, IEEE Trans Intell Trans Syst 15(5):2143–2158
12. Abboud K, Zhuang W (2014) Stochastic analysis of a single-hop communication link in vehicular ad hoc networks. IEEE Trans Intell Transp Syst 15(5):2297–2307
13. Lu N et al (2013) Vehicles meet infrastructure: toward capacity-cost tradeoffs for vehicular access networks. IEEE Trans Intell Transp Syst 14(3):1266–1277
14. Cruz R (1991) A calculus for network delay—I: network elements in isolation. IEEE Trans Inf Theory 37(1):114–131
15. Jiang Y, Liu Y (2008) Stochastic network calculus. Springer, London, U.K
16. Burchard A, Liebeherr J, Patek S (2006) A min-plus calculus for end-toend statistical service guarantees. IEEE Trans Inf Theory 52(9):4105–4114
17. Ciucu F, Burchard A, Liebeherr J (2005) A network service curve approach for the stochastic analysis of networks. In: Proceedings of the ACM SIGMETRICS, pp 279–290
18. Li C, Burchard A, Liebeherr J (2007) A network calculus with effective bandwidth. IEEE/ACM Trans Netw 15(6):1442–1453
19. Fidler M (2010) Survey of deterministic and stochastic service curve models in the network calculus. IEEE Commun Surv Tuts 12(1):59–86. (1st Quart.)
20. Fidler M An end-to-end probabilistic network calculus with moment Generating functions. In: Proceedings of the 14th IEEE IWQoS, 2006, pp. 261–270
21. Jiang Y (2006) A basic stochastic network calculus. In: Proceedings of the ACM SIGCOMM, pp 123–134

22. Fidler M (2006) A network calculus approach to probabilistic quality of service analysis of fading channels. In: Proceedings of the IEEE GLOBECOM, pp 1–6
23. Zheng K, Liu F, Lei L, Lin C, Jiang Y (2013) Stochastic performance analysis of a wireless finite-state Markov channel. IEEE Trans Wirel Commun 12(2):782–793
24. Al-Zubaidy H, Liebeherr J, Burchard A (2013) A (min, ×) network calculus for multi-hop fading channels. In: Proceedings of the IEEE INFOCOM, pp 1833–1841
25. Wu D (2003) Providing quality-of-service guarantees in wireless networks, Ph.D. dissertation, Dept. Elect. Comput. Eng., Carnegie Mellon Univ., Pittsburgh, PA, USA
26. Jiang Y, Emstad PJ (2005) Analysis of stochastic service guarantees in communication networks: a server model. In: Proceedings of the IWQoS, pp 233–245
27. Gao Y, Jiang Y (2012) Analysis on the capacity of a cognitive radio network under delay constraints. IEICE Trans 95-B(4):1180–1189
28. EEIG ERTMS User group, ETCS/GSM-R quality of service—Operational analysis [04E117], Int. Union Railways, Paris, France, Technical Report, 2005
29. Zimmermann A, Hommel G (2005) Towards modeling and evaluation of ETCS real-time communication and operation. J Syst Softw 77(1):47–54
30. Xun J, Ning B, Li K, Zhang W (2013) The impact of end-to-end communication delay on railway traffic flow using cellular automata model. Trans Res C Emerg Technol 35:127–140
31. Dong Y, Fan P, Letaief KB (2014) High-speed railway wireless communications: Efficiency versus fairness. IEEE Trans Veh Technol 63(2):925–930
32. Karimi O, Liu J, Wang C (2012) Seamless wireless connectivity for multimedia services in high speed trains. IEEE J Sel Areas Commun 30(4):729–739
33. Zhao Y, Li X, Li Y, Ji H (2013) Resource allocation for high-speed railway downlink MIMO-OFDM system using quantum-behaved particle swarm optimization. In: Proceedings of the IEEE ICC, pp 2343–2347
34. Xu Q, Li X, Ji H, Yao L (2013) A cross-layer admission control scheme for high-speed railway communication system. In: Proceedings of the IEEE ICC, pp 6343–6347
35. Zhu L, Yu F, Ning B, Tang T (2012) Handoff performance improvements in MIMO-enabled communication-based train control systems. IEEE Trans Intell Transp Syst 13(2):582–593
36. Liang H, Zhuang W (2012) Efficient on-demand data service delivery to high-speed trains in cellular/infostation integrated networks. IEEE J Sel Areas Commun 30(4):780–791
37. Sadeghi P, Kennedy R, Rapajic P, Shams R (2008) Finite-state Markov modeling of fading channels—a survey of principles and applications. IEEE Signal Process Mag 25(5):57–80
38. Zhang R, Cai L (2010) Joint AMC and packet fragmentation for error control over fading channels. IEEE Trans Veh Technol 59(6):3070–3080
39. Lin S, Zhong Z, Cai L, Luo Y (2012) Finite state Markov modeling for high speed railway wireless communication channel. In: Proceedings of the IEEE GLOBECOM, pp 5421–5426
40. Wang H, Yu F, Zhu L, Tang T, Ning B (2014) Finite-state Markov modeling for wireless channels in tunnel communication-based train control systems. IEEE Trans Intell Transp Syst 15(3):1083–1090
41. Li X, Shen C, Bo A, Zhu G (2013) Finite-state Markov modeling of fading channels: a field measurement in high-speed railways. In: Proceedings of the IEEE/CIC ICCC, pp 577–582
42. Mao S, Panwar SS (2006) A survey of envelope processes and their applications in quality of service provisioning. IEEE Commun Surv Tuts 8(3):2–20. (3rd Quart.)
43. Luan T, Ling X, Shen X (2012) MAC in motion: Impact of mobility on the MAC of drive-thru internet. IEEE Trans Mobile Comput 11(2):305–319
44. He R et al (2013) Measurements and analysis of propagation channels in highspeed railway viaducts. IEEE Trans Wireless Commun 12(2):794–805
45. Stber GL (2001) Principles of mobile communication, 2nd edn. Kluwer, Norwell
46. Liu Q, Zhou S, Giannakis G (2005) Queuing with adaptive modulation and coding over wireless links: Cross-layer analysis and design. IEEE Trans Wirel Commun 4(3):1142–1153

47. Evolved universal terrestrial radio access (E-UTRA); Physical channels and modulation, 3GPP TR Specification 36.211, Technical Specification Group Radio Access Network, June 2008
48. Zheng H, Viswanathan H (2005) Optimizing the ARQ performance in downlink packet data systems with scheduling. IEEE Trans Wirel Commun 4(2):495–506
49. Chang C-S (2000) Performance guarantees in communication networks. Springer, London, U.K

Chapter 7
Security of Dedicated Mobile Communications for Railway

7.1 Security Threats of Mobile Communications for Railway

7.1.1 Security Threats

In railway mobile communication networks, there are many vulnerabilities such as the open air interface, some communication protocol defects, and so on, which will result in security threats to the network. Threats like eavesdropping, impersonation attacks, information tampering, denial of service attacks, retransmission attacks, theft, and loss will seriously affect the safety of railway mobile communication networks [1–3].

Eavesdropping attacks: In railway mobile communication system, all communication messages are transmitted via wireless channel whose openness makes them easy to be eavesdropped by some special wireless devices. Eavesdroppers can obtain the important information transmitted via air interface, such as the user's identity information, location information, data information, and control signaling. Moreover, even if eavesdroppers cannot acquire real messages, they can speculate the purpose and content of communication through the message transmission flows after obtaining the addresses of message sender and receiver.

Impersonation attacks: The information exchange between user and network center (including the transmission of authentication information) is not entirely dependent on any fixed physical channels such as optical fibers and cables, but passes through a completely open wireless channel. The impersonation attack means that the eavesdroppers of wireless channel can fake the user's identity and then pass through the authentication of the network center and access to the legal network after obtaining the user's identity information. A typical case of impersonation attack is that the attacker counterfeits network control center to obtain the user's identity after defrauding the user's access or sending illegal short messages by short message service.

Z.-D. Zhong et. al., *Dedicated Mobile Communications for High-speed Railway*, Advances in High-speed Rail Technology, DOI 10.1007/978-3-662-54860-8_7

Information tampering: Information tampering mainly means that the attackers modify the information intercepted through the transfer party, including replacing or deleting part of the information even the entire content, and then sends the modified result to the previous designated receiver. However, in the wireless communication system, the network center or the wireless base station sometimes need to act as an intermediate party to forward the communication information between two wireless base stations. These intermediate parties may become the source of information tampering.

Denial of service attacks: Denial of service attack means that one of the parties involved denies its participation in the service to evade responsibility when the service is achieved. Specific performances are as follows.

(1) Sender repudiation. The sender is unwilling to take responsibility for the service of sending message.
(2) Receiver repudiation. The receiver is unwilling to take responsibility for the service of receiving message.

Retransmission attacks: Attackers will send the intercepted information to the original receiver when the information expires. They aim to use the expired information under a changed situation to achieve the same purpose as that could be achieved by valid information. For instance, if the attacker sends the intercepted scheduling command information to a train when the information expires, the train may execute the wrong control command, even cause accidents.

Theft and loss: Mobile terminals are becoming more and more intelligent and mobile terminal applications are carrying lots of user's privacy and security information. Some users may also be able to store the company's confidential business information in their mobile devices like mobile phones. These devices are generally small and easy to be stolen or lost, and once getting lost, the important information will be disclosed. Therefore, this problem is also worthy of attention.

7.1.2 Security Issues in GSM-R

As an open wireless transmission system, GSM-R is vulnerable to external damage and attacks. So its security and reliability are key performances of the railway mobile communication system. Since GSM-R network is based on GSM platform, the security threats in GSM system also remain in the GSM-R system. Due to the particularity of GSM-R network, there are also other types of security threats in GSM-R. Figure 7.1 shows the main security threats in GSM-R [4].

The following section is going to analyze the potential security threats of GSM-R network in terms of four points, namely radio interface, network terminal, mobile terminal, and other security issues.

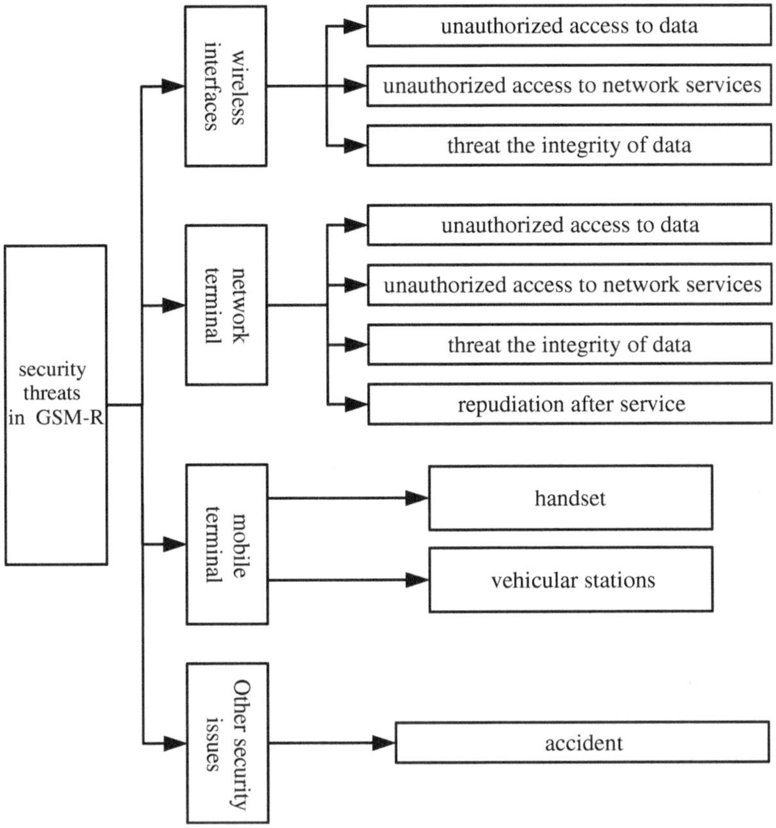

Fig. 7.1 Main security threats in GSM-R

1. **Security threats in radio interface**

In GSM-R network, mobile terminals and base stations communicate via the radio interface. Since the radio interface is open, to deceive the network side and achieve the purpose of impersonate legitimate users, any user with appropriate wireless devices can obtain the messages by eavesdropping on the wireless channel, even modify, insert, delete, or retransmit the messages transmitted through radio interface. Apparently, insecurity factors of the radio interface in GSM-R network mainly contain wireless eavesdropping, spoofing identity, and tampering with data. These insecurity factors can be divided into three categories of attacks: unauthorized access to data, unauthorized access to network services, and data integrity threat.

Unauthorized access to data attack: The main purpose of the unauthorized access to data is to obtain the user data or signaling data transmitted in GSM-R radio interface. The attacker can acquire the user's communications content, network management information, and other information conducive to active attacks by eavesdropping on the user or signaling transmission channel. The attacker can also

Table 7.1 Unauthorized access to data attack

Type of attack	Attack intention	Attack method
Eavesdropping user data	Gain communication contents of users	Eavesdrop on data transmission channel in radio interface
Eavesdropping signaling data	Obtain network management information and other information conducive to active attacks	Eavesdrop on signaling transmission channel in radio interface
Wireless tracking	Acquire the identity and location information of mobile users, realize wireless tracking	Fake the network terminal equipment, then require the mobile station transmit international identity number in plaintext
Passive transport stream analysis	Guess the communication contents and purposes of users	Observe and analyze time, length, speed, source and destination of information transmitted in wireless channel
Active transport stream analysis	Obtain access information	Initiate session communication, and then observe and analyze time, length, rate, source and destination of information transmitted in wireless channel

fake the network terminal equipment and require the mobile station to transmit its international identity number in plaintext in order to obtain the status and location information of users and then realize the wireless tracking. Moreover, the attacker can infer the communication contents among users or initiate a session communication by observing and analyzing the time, length, speed, source, and destination of the information transmitted in wireless channel, and then obtain the access information in the same way. Attack methods and threats are shown in Table 7.1.

Unauthorized access to network service attack: In the unauthorized access to network service attack, the attacker deceives the network terminal by faking a legitimate mobile user's identity, and then acquires authorized access to network services. To evade responsibility, the attacker makes the faked mobile user to take the responsibility. Unauthorized access to network service attack can be realized by many implementation methods. One common method is that the attacker fakes the base station and sends a communication connection establishment request to a mobile user. When the mobile user completes the authentication process successfully, the attacker hijack the communication connection between this mobile user and the base station, and then accesses to network services illegally.

Data integrity threaten attack: The target of this kind of attack is the user data stream and signaling data stream in radio interface. The attacker can achieve the purpose of deceiving the receiver and reaching attack intentions by means of modifying, inserting, deleting, or retransmitting the data stream.

Because of the hidden security threats in radio interface, the attack is not easy to be found, so it is one of the main security threats in GSM-R network.

2. **Security threats on the network terminal**
In GSM-R network, the composition of a network terminal is complex. The network terminal includes many functional units, and the communication media between different units is not the same. Thus, there are some risk factors that cannot be ignored in the network terminal, such as wireless eavesdropping, impersonation identity, data tampering, and repudiation after service. These insecure factors may lead to many different types of attacks. According to the different types of attacks, they can be divided into four types: unauthorized access to data, unauthorized access to network services, data integrity threat, and repudiation after services. Specific analyses are shown in Table 7.2.
3. **Security threats on mobile terminal**
In GSM-R system, the user's handset may be lost, and other mobile devices may be peculated. If the criminal uses those stolen or picked up mobile devices for illegal activities, the serious consequences could be caused. Therefore, it is necessary to enhance the security of mobile terminal.

Table 7.2 Security threats in GSM-R network terminal

Types of attack	Attack intention	Attack method
Unauthorized access to data	Get the user or signaling data that is transmitted between network terminal units	Eavesdrop on communication channel between the network units to obtain data or signaling data, may also fake a network unit to obtain these data
Unauthorized access to network services	Get authorized access to network services or get authentication parameters by faking a legitimate user, then enjoy the network service or do malicious attacks on the network	Fake under the assistance of the network service, home office or a legitimate user, namely hijack access authorization of authorized user to access network services or transfer authentication parameters of legitimate user from a certificate authority by faking home bureau, have access to unauthorized network services and shirk responsibility
Threat the integrity of data	Gain access to network services or interfere communication and the normal operation of the mobile terminal intentionally	Intruder can modify, insert, delete, or retransmit user or signaling data stream transmitted in wireless or wired communication interface, can also modify, insert, delete or retransmit data stored in the network elements, fake a sender which can download applications or data
Repudiation after service	To shirk responsibility, deny after communication	Deny access to web services and send or receive some message

4. **Other security issues**
Usually, the key network element equipment of GSM-R has a redundancy backup mechanism. However, when equipment faults occur or the equipment encounters some irresistible disasters, the failure of equipment could cause serious consequences.

(1) Base station failure

    (1) Natural disasters. In GSM-R network, two base stations of the same site are installed in the same machine room and have roughly the same coverage area. This approach does not consider the disaster recovery issue. If a disaster (fire, flood, lightning, etc.) occurs, two base stations of the same site will be damaged at the same time, which will result in the loss of network coverage in certain sections.
    (2) The base station is damaged. When the base stations in GSM-R network are damaged, the network in a large area cannot work normally.

(2) Core network failures
The reliability of MSC is relatively high. However, once it is damaged, huge losses will be caused and the network in a large area cannot work normally. Thus, in terms of the equipment failure accident, we should take more consideration of the security of devices.

7.1.3 Problems Still Existing in GSM-R

In view of the security threats above, the security measures adopted in GSM-R network, including security algorithm, user identity confidentiality TMSI, user authentication, signaling, and data privacy, have solved some security problems mentioned above. However, there are still some security issues unresolved in GSM-R network.

(1) Pseudo base station problem
GSM-R network only achieves unilateralism authentication, namely only the network can authenticate the user, but the user cannot authenticate the network. Therefore, GSM-R network is faced with the pseudo base station threat [5].
The structure of the pseudo base station is actually relatively simple. It does not have a complete "base station +MSC" physical structure. However, it is composed of several small functional modules, which are used to simulate partial functionalities of the base station and MSC (Fig. 7.2). Figure 7.2 shows the system structure of pseudo base station.
Terminal signal receiving unit is mainly used to obtain the cell location area code and the legal channel frequency; the transmit–receive unit of pseudo base station has a similar structure to the ordinary base station, but with a larger transmission power; the system control unit performs signal modulation, demodulation, and signal processing operations. Report control platform provides services such as

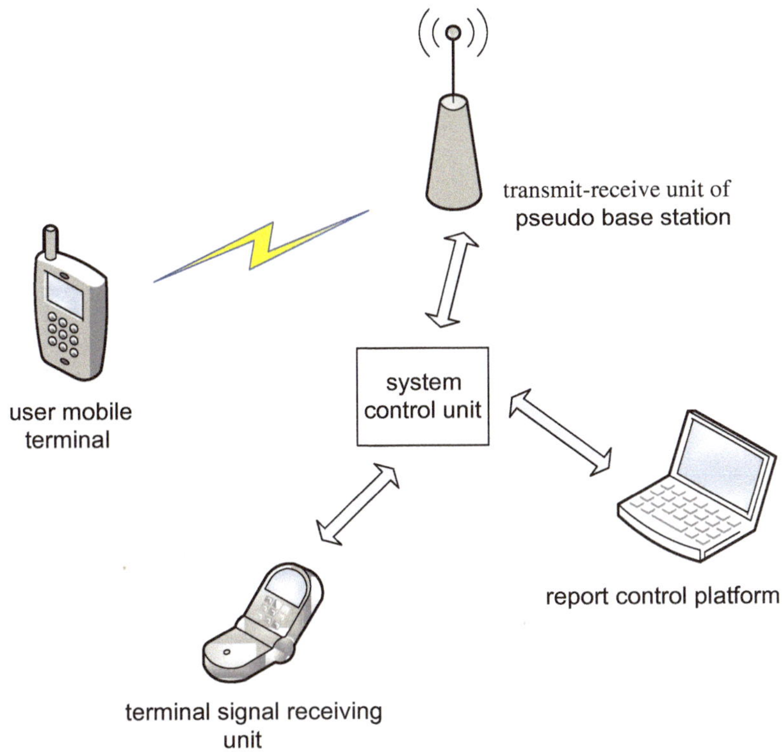

Fig. 7.2 System structure of pseudo base station

user information management, editing and sending short messages, system parameters configuration, cell parameters display, etc.

Generally speaking, only when the user accesses to network, starts up, and updates location,will the user's mobile phone trigger a network access operation. While the pseudo base station makes use of the location update process to attack the mobile terminal, pseudo base station uses the unilateralism authentication in GSM-R communications network to transmit strong power signals, so as to interfere with and shield from base station signals within a certain range. In this way, the cell phone signal is shielded and then the mobile phone will search the base station around automatically. Pseudo base station would figure out the phone number and then push fraud and spam messages to the user's mobile phone. Pseudo base station will lead to mobile phone updating location frequently, which not only affects the normal use of the user, but also makes the wireless network resources of this area scarce and causes the network congestion phenomenon. Countless losses would be brought through the interference of common spectrum resources.

(2) SIM card clone

As SIM card saves the user's important information, it has become the focus of criminals' attacks. Currently, the most typical attack on SIM card is the SIM card

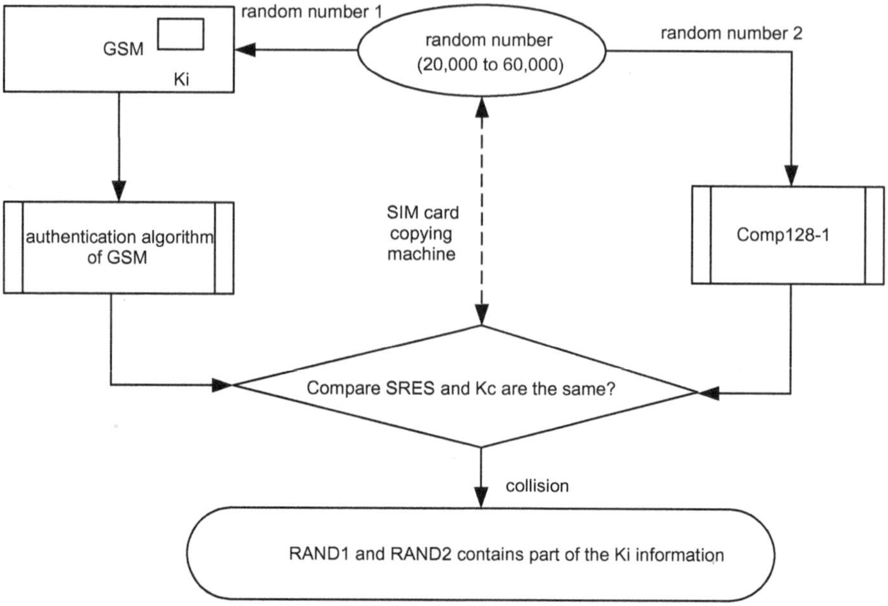

Fig. 7.3 Crack principle of SIM card

clone. SIM card clone attack is mainly aimed at getting the IMSI and Ki, then using the IMSI and Ki to clone a same SIM card.

Ki crack principle is shown in Fig. 7.3. Card reader generates a large number of random numbers to attack on the original SIM card. By comparing the authentication output, the attacker uses the output "collision" to crack. These random numbers are usually continuous and regular in order to get effective attack results. After obtaining the Ki of the original SIM card, the other parameters can be read from the original SIM card, achieving the SIM copy sequentially.

(3) Lack of end-to-end encryption

The data transmission encryption in GSM-R is limited to wireless network, it does not extend to the core network, which means that it only encrypts within the air interface between MS and BS. On the transmission link between two base stations, the user information and signaling data are transmitted in plaintext, which provides the attacker especially the personnel within the network opportunities to attack the network.

(4) User identity leakage

In the case of the user boot his mobile device or VLR data is lost; the user's IMSI has to be transmitted in the network, which causes some risks of being attacked. In railway wireless communication system, the mobile station can easily be lost and also be stolen, so information of SIM card may leak, resulting in SIM card is cloned, which brings security risks to normal operation of railway.

(5) Unavoidable DoS attack
As long as the attacker sends the same channel request to the base station controller repeatedly until the channel is fully occupied, the DoS attack will be caused and make the controller reject service. The attacker can interfere with the correct transmission of the user data, signaling data, and control data on the radio link in the physical or protocol and realize the DoS attack. DoS attacks can effectively destroy the cooperation and service between the nodes in the network and reduce the availability of the network. In addition, it is difficult to detect and prevent DoS attacks because of its good concealment.

(6) Integrity protection
There is no guarantee of the message integrity. If the data is tampered in the transmission process, it will be difficult to find. In the open transmission system of GSM-R, the integrity of message flow is the main consideration to identify the threat. "Message flow" is defined as an ordered set of messages. Under normal circumstances, it is unique to each time window and the receiver in the network. In fact, the received message flow may not be the same as expected for many reasons. There are roughly three particular categories (basic risks): The received message is more than expected; the received message is less than expected; and the received message is out-of-order or modified. There is strict requirement of message integrity in railway wireless communication; the message service will inevitably occur dropout phenomenon that may cause local small range gibberish, and bring security risks to the railway communication. In GSM-R network, it will not be found even if an attacker eavesdropping and tampering with information during transmission without considering the protection of signaling integrity.

(7) Vulnerability of replay attack
The attacker can carry out replay attacks by abusing previous information between the user and the network. Such attacks will continue to be malicious or fraudulent to repeat an effective data transmission. A replay attack can intercept and repeat the data to the destination host by the initiator. Encryption can effectively prevent the plaintext data from being monitored, but it cannot prevent replay attacks. Replay attacks may occur during any network communication. The attacker can use network monitoring or other means to steal authentication credentials, after a certain treatment, and then send it back to the authentication server. In GSM-R system only achieves unilateralism authentication, so GSM-R network cannot resist active attacks and also is difficult to prevent replay attacks.

7.2 Security Enhancement for GSM-R

7.2.1 Security Measures Taken by GSM-R System

Currently, in order to overcome its security issues, GSM-R system has taken many effective measures, including security algorithms, user identity confidentiality, user authentication, signaling and data confidentiality, as well as security guarantee of infrastructures on the ground [6].

1. **Security algorithms**

Both on the network side and in the user's SIM card, a SRES (signed response) is generated by A3 algorithm

$$\mathrm{SRES} = \mathrm{A3}(\mathrm{RAND}, \mathrm{Ki}).$$

And an encryption key is generated by executing A8 algorithm

$$\mathrm{Kc} = \mathrm{A8}(\mathrm{RAND}, \mathrm{Ki}),$$

as shown in Fig. 7.4.

A3/A8 algorithm demands the same inputs, which are a random number RAND and an authentication key Ki. Therefore, in practical applications, they use the same algorithm called COMP128, as shown in Fig. 7.5.

The input parameters of COMP128 are an authentication key Ki of 128 bits and a random number RAND of 128 bits. After the execution of COMP128, a signed response SRES of 32 bits and an encryption key Kc of 64 bits (10 bits zero padding at the end of the output of 54bits) will be generated. So in fact, the input parameter of encryption algorithm Kc has actually 54 effective bits. COMP128 algorithm is a one-way function, which is used in most GSM networks to realize A3 and A8 computations.

2. **User identity confidentiality**

In order to protect the user's privacy and prevent the user's location being tracked, GSM-R employs TMSI (Temporary Mobile Subscriber Identity) to preserve the confidentiality of the user's identity. IMSI will not be used to identify the user unless under special circumstances. Only when the network fails to recognize or connect to the user's HLR/AUC according to TMSI, will the IMSI be used to identify the user by obtaining authentication parameters from HLR/AUC. In

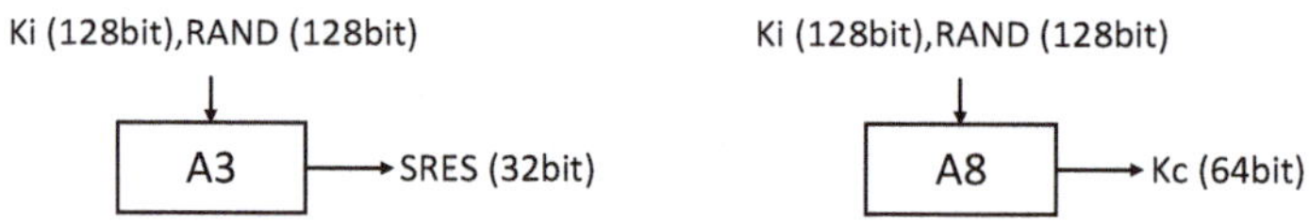

Fig. 7.4 A3/A8 algorithm

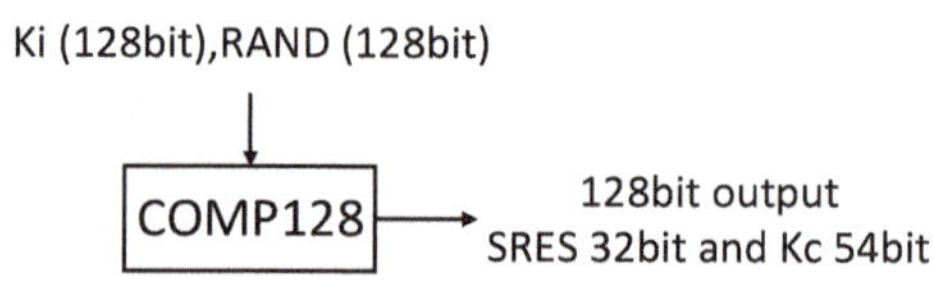

Fig. 7.5 COMP128 algorithm

GSM-R system, TMSI is always associated with some certain LAI (Location area identifier). When the user's LA (Location area) changes, the TMSI reallocation will be realized via the update process of LA. The TMSI reassigned to the user is encrypted and transmitted by VLR after completing the user's authentication and starting the encryption mode, which therefore keeps the confidentiality of TMSI. Meanwhile, VLR will save the new assigned TMSI and delete the old one.

3. **Authentication**

The authentication process is executed by using a group of three authentication parameters (encryption key Kc, random number RAND, signed response SRES). When a user accesses to the network, the user authentication key Ki and IMSI will be assigned to the user simultaneously. Ki will be stored in AUC (authentication center) on the network side and stored in SIM card on the user side. Authentication parameters are generated by AUC trough the execution of corresponding algorithms:

(1) Using random number generator to generate a random number RAND;
(2) Using A3 algorithm to generate a signed response SRES = A3 (RAND, Ki);
(3) Using A8 algorithm to generate an encryption key Kc = A8(RAND, Ki).

Responding the request of MSC/VLR, AUC generates several three-parameter groups (RAND, SRES, Kc) once and stores them in HLR. HLR stores all the three-parameter groups of each user and sends one group to MSC/VLR when MSC/VLR makes a request, so as to ensure that there is at least one unused three-parameter group for each user to access the network. When the user demands an access authentication, MSC/VLR sends RAND to MS and then, MS uses the same Ki and algorithm stored in the SIM card as stored in AUC to calculate SRES. Afterward, MS sends SRES back to MSC/VLR to verify its legitimacy and to decide whether it is permitted to access to the network or not. Specific authentication process is shown in Fig. 7.6.

Authentication is necessary before MS location updating, MOC (Mobile originating call) or MTC (Mobile terminating call) setup, supplementary service activation or deactivation, and registration or deregistration.

4. **Signaling and data encryption**

The network encrypts the data of users to prevent eavesdropping. The encryption process is controlled by the encryption key Kc, which is generated during the authentication process. Kc is generated by the key algorithm A8 and the encryption algorithm A3. Those two algorithms have the same input parameters RAND and Ki, so they can be combined into one algorithm to calculate the signed response and the encryption key. The encryption key Kc is not transmitted at the radio interface, but stored in the SIM card and AUC, where corresponding algorithms will be accomplished respectively, as shown in Fig. 7.7.

Through the process shown in Fig. 7.7, the encryption key Kc is generated on the network side and in MS, respectively.

The encryption process can be described as the following. Taking the encryption key Kc generated by A8 algorithm and the TDMA data frame which carries the user

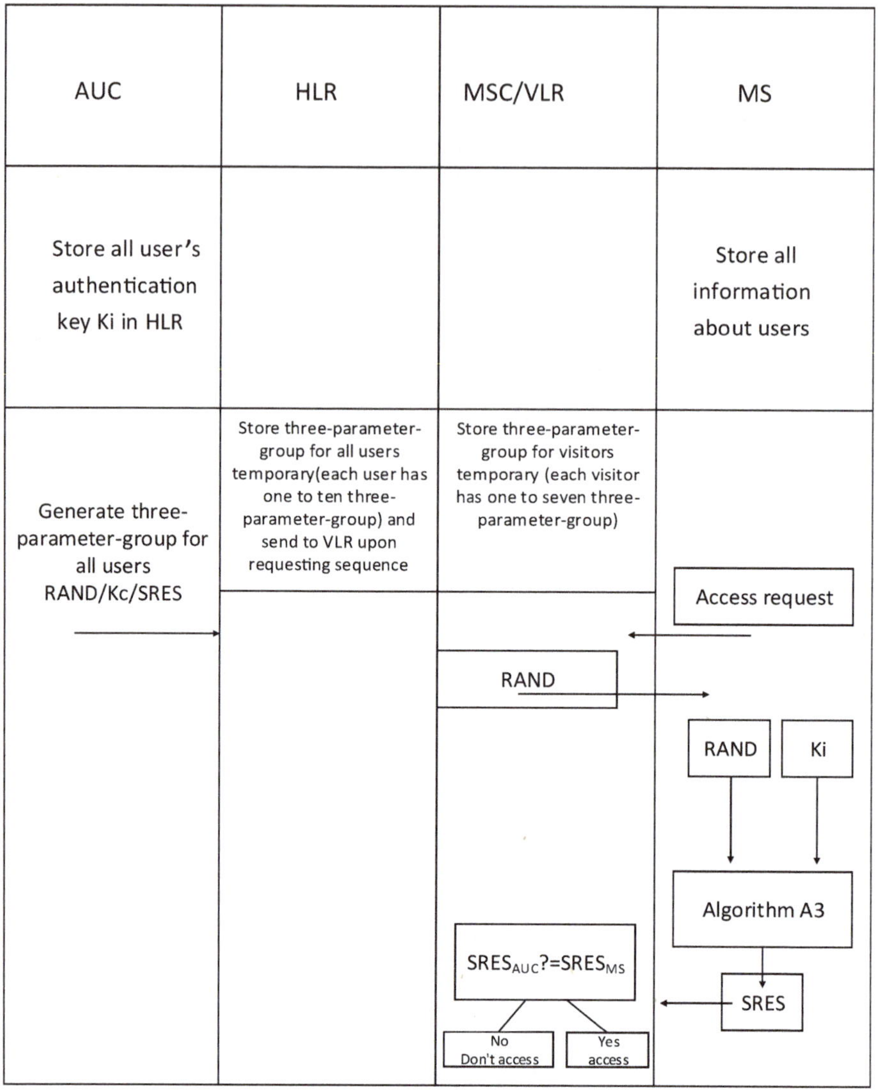

Fig. 7.6 Authentication process

data stream as two input parameters for the encryption algorithm to generate a pseudo-random data stream. Then the pseudo-random data stream is XORed with the unencrypted data stream to obtain the encrypted data stream. The encryption on the network side is completed in BTS, where the encryption algorithm is also stored. The encryption key Kc is transmitted from MSC/VLR to BTS during the authentication process. Specific process is shown in Fig. 7.8.

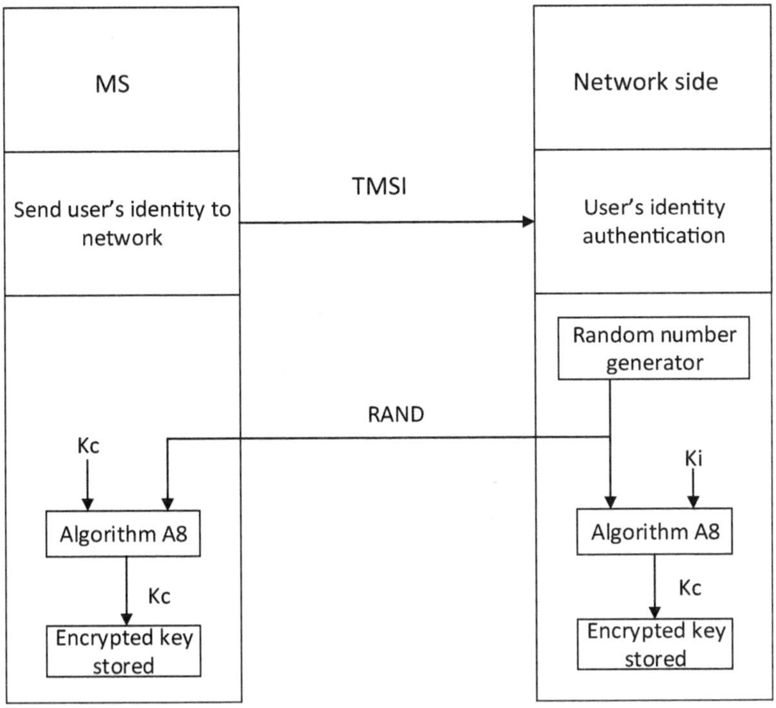

Fig. 7.7 The generation of the encryption key

5. **Security guarantee of ground infrastructures**

Many unexpected factors should be taken into consideration while designing and constructing the network in order to provide backup emergency plans, such as the redundant coverage scheme of base station, which contains two different types, namely the same site covered by double base station and the same site covered by mixed station. Moreover, there are other emergency plans, such as the MSC pool disaster recovery technology and so on.

7.2.2 Bidirectional Authentication

Currently, the authentication of GSM-R network is unidirectional. The network can authenticate mobile stations, while mobile stations do not authenticate the identity of network, so it causes security vulnerabilities [7]. This kind of authentication method cannot resist active attacks, and it is also difficult to prevent the man-in-the-middle-attack (MITMA) and the false base station attack, which may

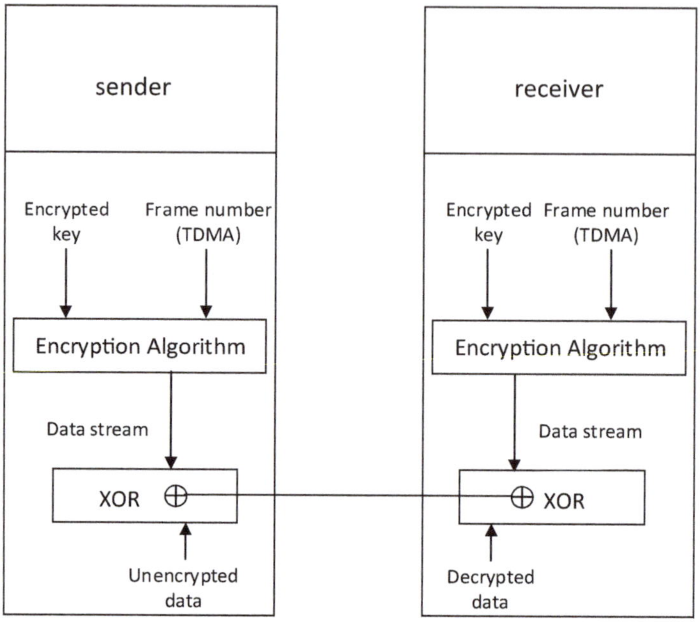

Fig. 7.8 Process of encryption and decryption

lead to the user's sensitive information being stolen or abnormal accesses to network resources.

In order to overcome the shortcomings of unidirectional authentication, we proposed an improved bidirectional authentication method for GSM-R. The improved protocol is described in Fig. 7.9. In the authentication request message shown in the figure below, the network challenges the terminal by using a random number RAND through A8 algorithm and the user's encryption key Ki. The RAND has a corresponding encryption key Kc. IK is the integrity key, which is generated by Ki XOR with Kc. In the authentication response message, the terminal sends back a SRES1 and its own challenging value SEQ rather than only a SRES.

SEQ is a counter stored in the SIM card, which will be incremented by one and stored by the SIM card every time when the terminal sends a SRES1. Supposed that SEQ was big enough (like 32 bits) to not overflow during the terminal survival period or the SEQ value would start counting from zero when it reached the predetermined value if another plan was adopted. There is a same counter on the network side.

When mobile users reach VLR and want to communicate, we should verify their identities and generate encryption keys. The authentication process contains several steps as follows:

(1) MS sends an access request to the network and sends TMSI (or IMSI) to VLR.
(2) When VLR receives the access request from MS, it will send an authentication data request to HLR/AUC and send IMSI to HLR.

Fig. 7.9 The improved bidirectional authentication mechanism in GSM-R

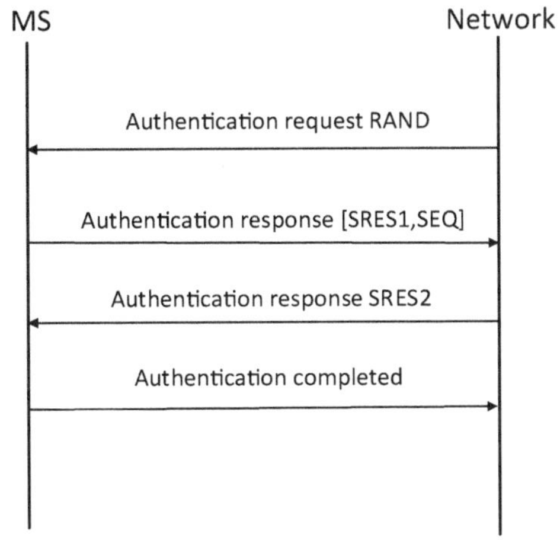

(3) When HLR receives the authentication request from VLR, it will generate a sequence number SEQ and a random number RAND and calculate the authentication vector (RAND,SRESm, Kc,IK,SEQ) and then send it to VLR. In the vector, SRESm = A3(Ki, RAND) is the authentication response expected; Kc = A8(Ki,RAND) is the data encryption key; IK = Ki $\oplus$ Kc is the data integrity key; SEQ is the value calculated by the counter on the network side.
(4) After receiving the authentication vector, VLR will launch an authentication request and send RAND to MS.
(5) After receiving the RAND from VLR, the SIM card in MS will use the authentication algorithm A3 to calculate the authentication response SRES1 = A3(Ki, SRES) and send SRES1 and SEQ to MSC/VLR, meanwhile the counter will be incremented by one.
(6) After receiving the authentication response from MS, MSC/VLR will compare whether the SRES1 received and the SRESm in the authentication vector are identical. If not, the authentication process is failed, and MSC/VLR will send a failure message to MS and end this communication. If they are identical, MSC/VLR will compare the SEQ received and the SEQ on the network side to verify whether they fall in a reasonable range. If not, a synchronization failure message will be sent to MS. Otherwise, the authentication response SRES2 = H(IK, SEQ, RAND) will be calculated and sent to MS. H () is a hash function with a key.
(7) MS also calculates SRES2 by the same algorithm in MSC/VLR. After receiving the authentication response from network side, MS will verify whether the SRES2 in SIM and the SRES2 from MSC/VLR are identical. If they are identical, it means that MS successfully verifies the legality of VLR. If not, MS will send an authentication failure message to the network and stop the authentication process.

Analysis of the improved protocol: this protocol realizes a bidirectional authentication, and MS applies a challenge–response mechanism to execute the network authentication process. It takes a full use of current resources in GSM-R only through some small changes to several algorithms stored in the SIM card and system terminal, rather than making huge changes to current databases. On the basis of the system authentication three-parameter group, this novel protocol adds an integrity protection key IK, forming a new authentication vector.

7.2.3 End-to-End Encryption

Current encryption method in GSM-R system is not realized end-to-end. The encryption function does not extend to the core network, and it is only encrypted within the wireless channel, i.e., the Um interface between MS and BTS. User's information and signaling data are transmitted in plaintext within the link from one base station to another, which offers an opportunity to attackers, especially those employees inside the network [8].

End-to-end encryption uses Asymmetric cryptographic algorithm and WPKI technology, among which WPKI is a system published and updated by a public key authority. WPKI is designed to meet the requirements of mobile systems, and it mainly contains the following parts: certification authority (CA), registration authority (RA), smart cards, and digital certificates.

1. **Certificate management framework**
Figure 7.10 illustrates the GSM-R network-based certification management framework. This framework could provide communication parties with public key certificates as the basis of encryption key exchange. Except the user A of two service entities and the called user B, the network elements involved also include HLR/AUC and a WPKI system (this system identifies user's identity and provides them with certificates).
Before communication, A and B must get in touch with WPKI system. RA will authenticate user entities and then send a certificate signing request to CA. And CA

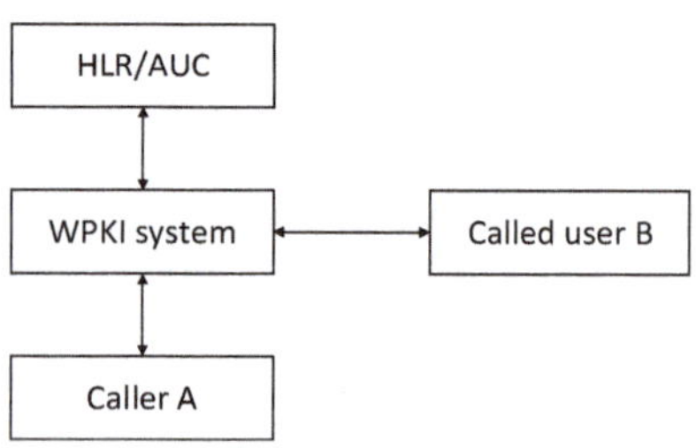

Fig. 7.10 Certification management framework based on GSM-R network

is responsible for certificate making. This certificate contains the user's public key, CA's signature, life cycle information, and so on.

2. **Certificate obtainment and cancelation**

How to obtain the certificates preserved by CA and how to cancel some certificates are two issues necessary to be analyzed.

(1) Certificate obtainment

According to the mechanism of certificate security protocol, MS could not only get certificates from CA, but also from the two sides involved in the security protocol interaction. Here we choose the first scheme in which MS gets certificates from CA.

Certificates are signed by CA, which cannot be forged, so CA can just put them in a directory without other special protections. When the certificates of the two communication users are provided by the same CA, we can get their certificates directly from this directory. But when their certificates are signed by different CAs, it can be realized.

Suppose that there are two users A and B. The certificate of user A is signed and managed by authentication center X, while the certificate of B is signed and managed by authentication center Y. X and Y offer certificates to each other in order to be trusted. Now A wants to communicate with B; in other words, A wants to get B's certificate. Obviously, it is useless to get B's certificate from Y because A does not have Y's public key so it cannot verify the correctness of the certificate. So the right operation order is as follows:

① A gets the certificate signed by X, sends it to Y, and uses X's public key to verify its legitimacy.
② Getting the public key of Y from this certificate.
③ A gets the certificate signed by Y and sends it to B.
④ Using Y's public key to verify the legitimacy of this certificate.
⑤ Getting B's public key from this certificate.

In this way, A could encrypt with the use of B's public key. If B wants to get A's public key, the same procedure should be executed.

(2) Certification cancelation

The cancelation of certificates is managed by CRL. Certificates should be canceled before expiration for some reasons such as that users lose their private keys, users no longer get certificates from this CA or certificates of this CA have been leaked.

These certificates canceled could be issued by different CAs and they are managed by CRL. CRL should paste them in the release directory. So when users get their certificates, they should check CRL to verify whether the certificate is canceled or not.

(3) End-to-end encryption process

① Before using the service that WPKI provides, A and B must register in the RA of WPKI system. Users send their identity authentication requests to RA, and attach their identities in messages.

② RA verifies the legitimacy of user's identities. If they are legal, the certificate signing operation process begins and RA sends certificate signing requests to CA. Otherwise, RA sends back an authentication failure message.

③ CA first verifies whether RA's identity is legal or not and then creates X.509 certificates for A and B by using the public key and some user information submitted by RA. After signing these certificates by using its private key, CA will publish them to the certificate library.

④ After getting the X.509 certificate of each other, users will first verify the legitimacy of the certificate. If the certificate is valid, users will get the public key of each other from the certificate. User A will use user B's public key to encrypt its session key Kc and send it to user B. After receiving the encrypted session key from A, user B will use its own private key to decrypt it to obtain the session key from A. So user B conducts the same operation. If the certificate is invalid, users should resend their requests and demand to update their certificates.

⑤ Users use their own encryption keys to transmit and use the encryption keys of the other party to decrypt and obtain the plaintext message. In this way, an end-to-end encrypted transmission is achieved.

The end-to-end encryption could lead to some delay because of the intermediate steps added, so it should be used according to actual needs. The end-to-end encryption is suitable for some occasions with strict communication security requests. Moreover, the end-to-end encryption mentioned here is limited to the user data encryption because the transmission of signing demands the participation of the network. Therefore, these messages must be transparent to the network and cannot be encrypted end-to-end.

7.2.4 Anti SIM Card Clone

At present, some methods have been proposed to solve the problem of SIM cards being cloned [9]. In the literature, when two SIM cards of the same IMSI are running the Attach process in different SGSN, HLR (Home Location Register) judges whether the SIM card corresponding to this IMSI is cloned by using the location update request from the two SGSNs, as shown in Fig. 7.11. This method will not work when two SIM cards are attached to the same SGSN, because at this time SGSN may not be able to send a location update request to HLR. On the other hand, HLR will send cancel messages to the first SGSN after receiving an update

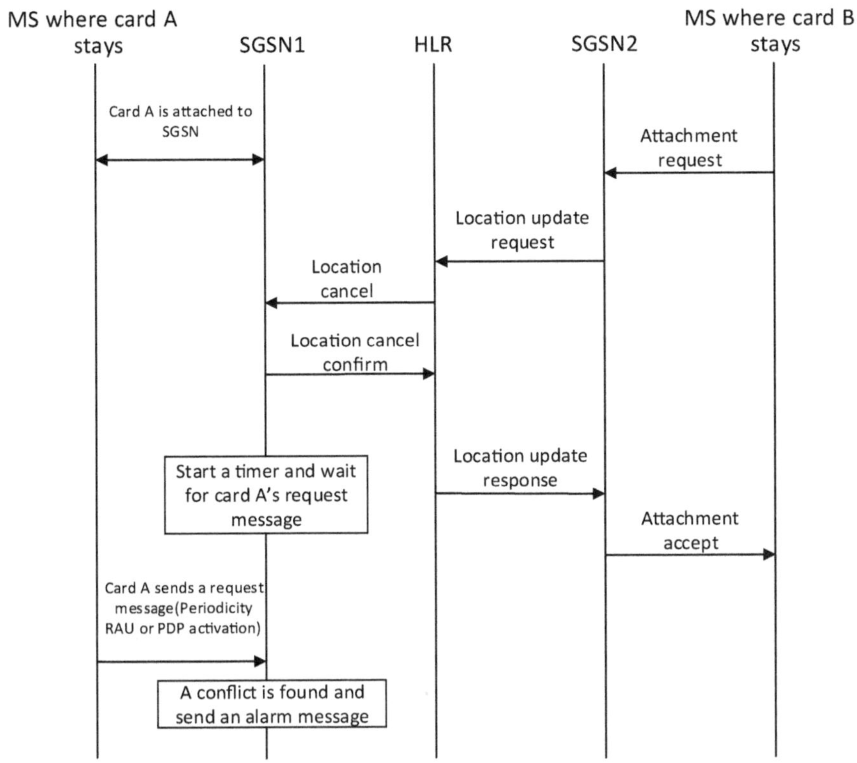

Fig. 7.11 Process of identifying the cloned SIM card

request from the second SGSN, i.e., separating the first user. If the first user is a legitimate user, the practice will lead to the first user being forcibly disconnected to the network, which may bring economic or business losses and deteriorate the user experience. What is more, this solution only tackles the Attach scene, no other scenarios such as that the user initiates a service request.

In this section, a solution based on location estimation will be proposed to recognize cloned SIM cards. We judge whether the USIM card is cloned through the comparison of the location information of two users of the same IMSI. Taking LTE/SAE system as an example, when UE1 and UE2 are within the same range of MME, the specific process is shown in Fig. 7.12.

(1) UE1 sends an attach request message to eNB;
(2) eNB sends Initial UE message to MME, which includes the attaching request message sent by UE1, the location area code of UE, and the community code;
(3) MME sends Initial Context Setup Request message to eNB, which contains the Attach Accept message that will be sent to UE1;
(4) eNB will send the Attach Accept message to UE1 through RRC reconnection configuration message;

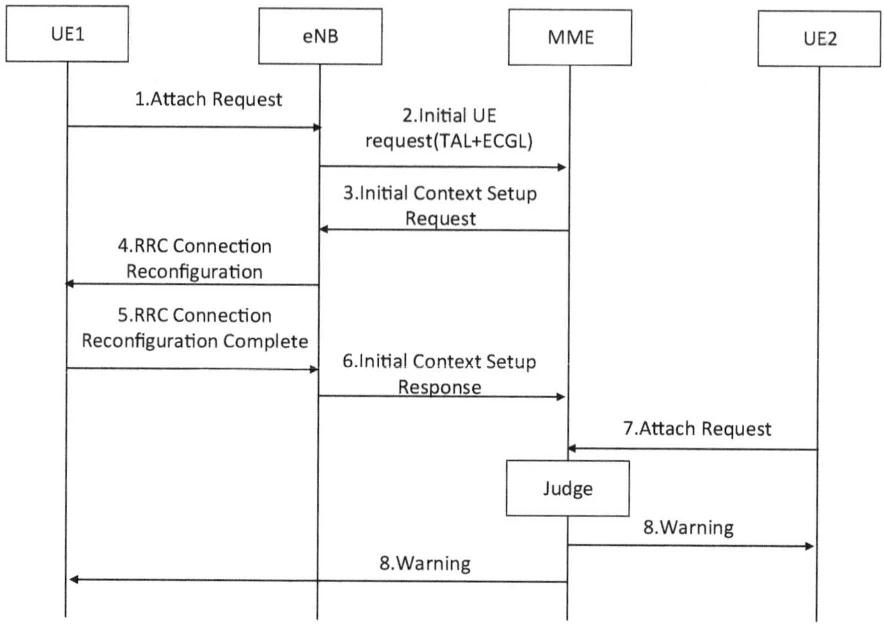

Fig. 7.12 Attach process of UE in the same MME

(5) UE1 confirms the success of the RRC connection to eNB;
(6) eNB confirms to MME that UE1 context is created successfully;
(7) UE2 also sends an Attach request message to MME;
(8) If MME detects that UE1 and UE2 correspond to the same IMSI but the Community Code is not the same, then it sends warning messages to UE1 and UE2, respectively.

For legitimate users and cloned users in different MMEs, Fig. 7.13 describes the process of identifying cloned USIM on the network side.

(1)–(6) The same as Fig. 7.12, UE1 attaches to MME1 successfully;
(7) UE2 sends an Attach request to MME2;
(8) MME2 finds MME1 according to UE2's GUTI, and sends Identification Request to MME1;
(9) After receiving the GUTI sent from MME2, MME1 sends warning messages to MME2 if it detects that the IMSI corresponding to this GUTI is already locally attached;
(10) MME1 and MME2 send, respectively, warning messages to UE1 and UE2 to inform the two users "Your USIM card may be cloned."

This solution is based on the existing attach request process, and adds a MME judgment process, which makes small changes to the existing process. Besides, for

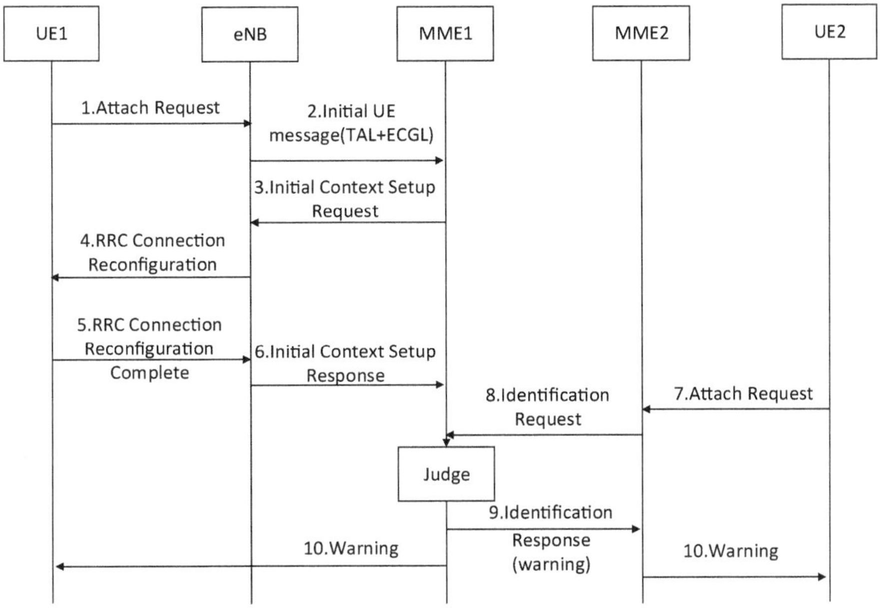

Fig. 7.13 Attach process of UE in different MMEs

abnormal situations detected, the network side will send an alarming message to users. How to deal with this abnormal situation is decided by the user, which reduces the impact on the user experience as much as possible.

Figures 7.12 and 7.13 take the adhesion process, for example, this flow chart also adapts to the Service Request process to identify whether the USIM card is cloned. Moreover, the solution is also applicable to 3G and GSM systems. It is a universal recognition process of cloned mobile phone cards.

Through the scheme proposed in this paper, with the aid of the network, legitimate users can learn whether their USIM cards have been cloned. If legal users receive the same warning message for many times, they can conclude that their USIM cards have been cloned and so, they can replace a new USIM card to solve the potential safety problems.

7.3 Security of Wireless Heterogeneous Networks for Railway

"Mobile broadband, broadband mobile" is the current development features of the railway mobile communication, the coexistence and integration of multi-types of network are the development trend of broadband wireless communications. In order to integrate various access technologies into unified network environment, make full use of all wireless network resources, and provide users with seamless roaming service, International Organization for Standardization and academia have already researched a lot on the integrated heterogeneous wireless technology and scheme.

3GPP proposed six interoperability scenarios of 3G and WLAN fusion in the TS22.934. 3GPP2 focuses on the interconnection between CDMA2000 and WLAN. Though those two standardization organizations are different in proposed schemes, the core idea is using mobile network to achieve authentication certification of WLAN, and allowing the WLAN terminal to use mobile network data service. The IEEE 802.21 Working Group mainly studies on how to provide handover capability which is independent of the media (Handover Media Independent, MIH).

Heterogeneous wireless network refers to deploying a number of small power transmission nodes within the traditional macro-cellular mobile base station coverage as to form a heterogeneous system with different node types. According to the cell coverage area, the cells can be divided into macrocell, microcell, picocell, femtocell, and relay station for signal relay. Heterogeneous network overlaps with different cell ranges and formats heterogeneous hierarchical wireless network.

Application of wireless access technologies such as mobile satellite communication, WLAN (IEEE 802.11), WiMax (IEEE 802.16e), GSM-R, LTE-R, 3G, 4G, and WiMAX-Advanced have made the railway network become a more and more complex heterogeneous network, which brings convenience for users. However, the existing security challenges cannot be ignored. For example, the frequent authentication problem in hot spots is due to deploying a large number of microcells and H (e)NBs, the authentication problem of WLAN access to the EPC, the relay security problem, the mobile terminal trusted security problem, etc.

7.3.1 Fast Re-authentication in Hot Spots

Hot spots refer to the areas where the transit is intensive and the traffic data is large, mainly refer to the large and medium-sized passenger stations in the railway application scenarios. Deploying a large number of H(e)NBs not only can solve the indoor signal coverage problem in high-speed rail stations, but also can provide higher data rate service at lower cost, because H(e)NB has those feathers that is small size, light weight, mobile, etc. However, the coverage of H(e)NB is quite small; the mobile terminal may pass in and out H(e)NB frequently with the user, and may also enter a UMTS or GSM/GSM-R macrocell, which will cause cell handover, and even handover between heterogeneous networks. This frequent handover will cause a great waste of system resources if re-certification is needed every time. Hence, this section proposes a fast re-authentication scheme based on the location lock mechanism; the scheme is based on the security context reuse within a certain time and fast authentication [10, 11].

1. **Handover from H(e)NB to E-UTRAN/UTRAN**

The security context stored in UE and in the core network side has a certain time limit. For the macrocells near UE, they are fixed during most of the time (can be used by the allowance of the operators), an information table of neighboring microcells can be pre-configured in H(e)NB, which contains area information such as Cell ID; H(e)NB gets neighbor cell information at a regular time and report to the core network side through the SeGW (according to the strategy of the operator, there can be many different implementation plans). There are two advantages; on one hand, the core network side can precisely fix the position information of H(e) NB and reduce the threat of illegal use; on the other hand, according to the white list restrictions, the fast re-authentication cell scope can be used at any time. Because H (e)NB cell range is small and the macrocell range is very large, so the fast re-certification scheme is necessary only in the switch between the H(e)NB and its adjacent macrocell, as shown in the following Fig. 7.14.

(1) H(e)NB implements the self-starting process when it is powered on, UE normally attached to the H(e)NB cell and the implements mutual authentication with security gateway. NAS security context and AS security context are saved in ME and MME.
(2) When UE moved to H(e)NB cell edge and detected that BCCH level of adjacent macrocell (E-UTRAN) had exceeded the handover threshold.

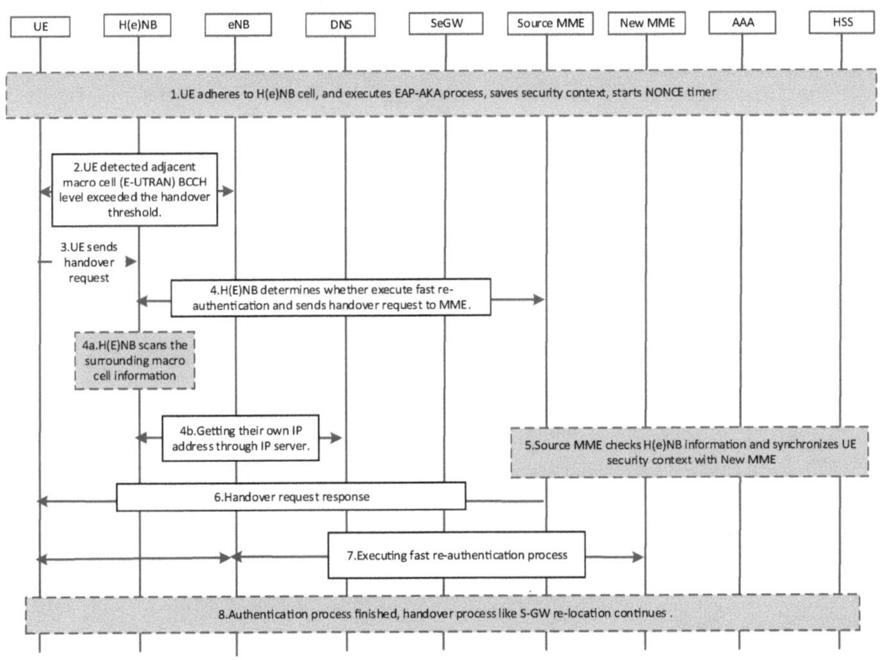

Fig. 7.14 Fast re-authentication when HO from H(e)NB to E-UTRAN

(3) UE sends a handover request to H(e)NB, at meanwhile reports the broadcast information of the target handover cell, such as cell identification number Cell ID, unknown area identification number LAI, the use frequency lists, etc.
(4) H(e)NB maintains a list of trusted adjacent cells that allow for fast re-authentication, H(e)NB compares the target handover cells information in the UE report whether in the trusted list to determine whether to execute fast re-authentication. At the same time, H(e)NB periodically check the location of the MME (the period is determined by the operator strategy).

    (4a) The H(e)NB location check mentioned above mainly has two means, one is scanning the surrounding macrocell information and reporting to MME, this method is suitable for the dense macrocells areas within cities;
    (4b) If H(e)NB is deployed in the area with fewer macrocells, the location check can be completed by getting their own IP address from H(e)NB and the DNS server of the IP server of the core network. Obviously, this strategy, which needs to cooperate with other operators, will increase the cost, and may not be able to achieve the positioning accuracy, so it is an alternative option here. There is another plan which deployed GPS system in the H(e)NB to achieve the precise positioning, although it has a better effect, apparently the cost is higher, which means it is only applicable to a few scenes.

(5) MME checks the location information of H(e)NB based on the operator's strategy. If H(e)NB incidental information agree to perform fast re-authentication, MME will synchronize UE security context saved by itself to MME of handover target; if not allowed, MME does not need to synchronize UE security context, and re-implemented EAP-AKA authentication process after the UE handover attached.
(6) MME returns handover request response messages to UE, and informs whether performs security context reuse and fast re-authentication.
(7) H(e)NB cell and E-UTRAN cell use the same key system, so the security context can be used directly (if timer is not expired), and do not need to key deduction and conversion. In UE and eNB of handover target area, MME implements fast re-certification.
(8) After implementing fast re-certification process, the UE and the new cell network side can use the previous key to protect NAS signaling and RRC signaling, and continue to implement handover process.

The handover fast re-authentication process between UTRAN and H(e)NB is similar to the handover process between E-UTRAN and H(e)NB, which will not be repeated here.

2. **Handover from H(e)NB cell to non-3GPP cell**

Taking CDMA2000 as an example, we design a fast re-authentication scheme, and the key CDMA2000 used is different with 3GPP system, so we must conduct key deduction when reuse security context. Reusing the security context mainly includes the following contents: terminal and network operators' side agreed mutual authentication security algorithm; key materials generated in mutual authentication process, including signaling encryption and integrity protection key and key identifier; valid timer value, key survival, serial number; and temporary identity distributed by service network. LTE/SAE system deduces the various layers' key by implementing the EAP-AKA process, CDMA2000 deduces the various layers' key by implementing the EAP-AKA process. Comparison of the two key systems of the two systems is shown in Fig. 7.15. When H(e)NB cell handover to CDMA2000 cell, the source MME synchronizes the security context to the AAA server, AAA server generates corresponding hierarchical key MK in target network according to the existing KASME, which is labeled as sMK, and the conversion is calculated according to the following formula:

$$\text{sMK} = \text{PRF}(\text{KASME},\quad \text{UE\_ID}|\text{AN\_ID}|\text{NONCEMME})$$

$$\text{sMK} = \text{PRF}(\text{KASME},\quad \text{UE\_ID}|\text{AN\_ID}|\text{NONCEMME})$$

where UE_ID can be IMSI of UE, AN_ID refers to the target network access network name where UE will handover to, and NONCEMME refers to a random number generated by MME, used to resist replay attack. In addition, the calculation of sMK can also be completed by HSS. MME sends the security context to the AAA server, through the HSS route in the middle. The former case pass through the HSS and do not need to deal with, the later situation requires HSS to handle. Because between MME and HSS, as well as protocol between HSS and AAA servers, are based on the Diameter protocol, it modifies protocol a little. AAA server

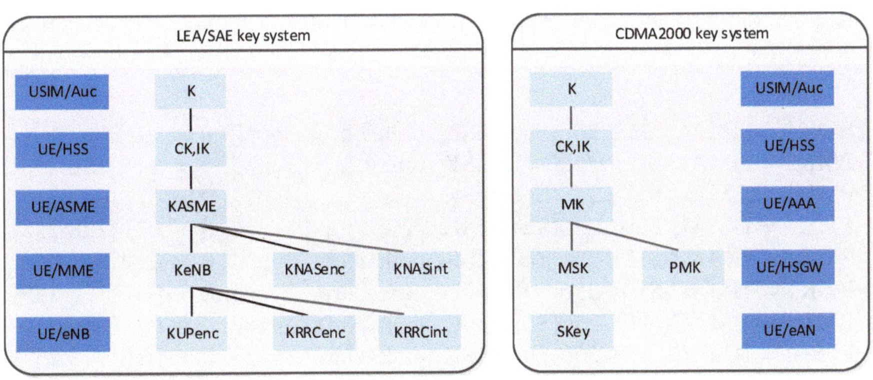

Fig. 7.15 Comparison of LTE/SAE and CDMA2000 key hierarchy

calculates the new root key MK according to sMK, and the calculation formula is as follows:
MK '=PRF' (sMK '0\' |MK in the presence of tag |NONCEAAA|NONCEUE| UE_ID|AN_ID| length)
Where the "MK existing identification" refers to the root key identifier that UE send to the network side, NONCEAAA refers to random number generated by the AAA server, and NONCEUE refers to the random number generated by UE. Then the master session key MSK by MK' can be obtained.
In another way, MME sends security context that contains root key to HSGW directly using S101 tunnel, the prerequisite is that the UE has registered and authorized in the AAA server, and HSGW requests AAA server on the UE authorization and PDN GW information.
When the UE switches from CDMA2000 to H(e)NB cell, AAA server will send secure context that contains CK/IK or EMSK to MME. MME or HSS obtains KASME according to CK/IK and the access network name, and then obtains the key of the access network according to the table above. The formula that can obtain EMSK by CK/IK or KASME is as follows:

$$\text{KASME} = \text{PRF} \quad (\text{CK}|\text{IK}, \text{UE ID}|\text{AN ID}|\text{NONCEAAA}),$$

or

$$\text{KASME} = \text{PRF} \quad (\text{EMSK}, \text{UE ID}|\text{AN ID}|\text{NONCEAAA}).$$

The survival time of the newly generated root key can be given by the initial lifetime of the original key, or re-determine a survival period by the network side. At the same time, the value of the survival time should be the same as the value of survival time in UE.
After obtaining the root key after the handover to the target network, the AAA server sends master session key MSK to network access authentication HSGW. At the same time, UE side also generates MSK; UE and HSGW generate equivalence master key PMK that are used for access network key generation according to MSK. First, dividing the MSK 512 bit into four sub-MSK 128 bits, and respectively generated their PMK according to Sub-MSK, the calculation formulas are as follows.

PMK1 = HMAC-SHA-256(Sub-MSK, pmk@hrpd.3gpp2, 0 × 01), [0:127],
PMK2 = HMAC-SHA-256(Sub-MSK, pmk@hrpd.3gpp2, 0 × 01) [128:255],
PMK3 = HMAC-SHA-256(Sub-MSK, pmk@hrpd.3gpp2, 0 × 02) [0:127],
PMK4 = HMAC-SHA-256(Sub-MSK, pmk@hrpd.3gpp2, 0 × 02) [128:255],
PMK1 = HMAC-SHA-256(Sub-MSK, pmk@hrpd.3gpp2, 0 × 01), [0:127],
PMK2 = HMAC-SHA-256(Sub-MSK, pmk@hrpd.3gpp2, 0 × 01) [128:255],
PMK3 = HMAC-SHA-256(Sub-MSK, pmk@hrpd.3gpp2, 0 × 02) [0:127],
PMK4 = HMAC-SHA-256(Sub-MSK, pmk@hrpd.3gpp2, 0 × 02) [128:255],

where HMAC-SHA-256 function calculates and gets an output of 256 bits. The PMK generated above extracts the corresponding output bit, for example, pmk1 extracts the calculation results from 0 to 127bit. At the same time, UE and HSGW compute a key identifier Pairwise Master Key ID for associating generated PMK. Then, HSGW will send PMK and Pairwise Master Key ID to access network entity eAN/ePCF, to access network key SKey generation.

On the basis of the key deduction mentioned above, the design of the handover from H(e)NB to CDMA2000 based on the location lock fast re-authentication scheme is shown in Fig. 7.16.

(1) H(e)NB implements power-on self-starting process, and UE normally attaches to the H(e)NB cell and implements mutual authentication with security gateway. NAS security context and AS security context are saved in ME and MME.
(2) When UE moves to the H(e)NB cell edge and detects that the BCCH level value of the adjacent macrocell (CDMA) exceeds the handover threshold.
(3) UE sends a handover request to H(e)NB, while reporting the broadcast information of the target handover cell, such as cell identification number Cell_ID, unknown area identification number LAI, the use frequency lists, etc.

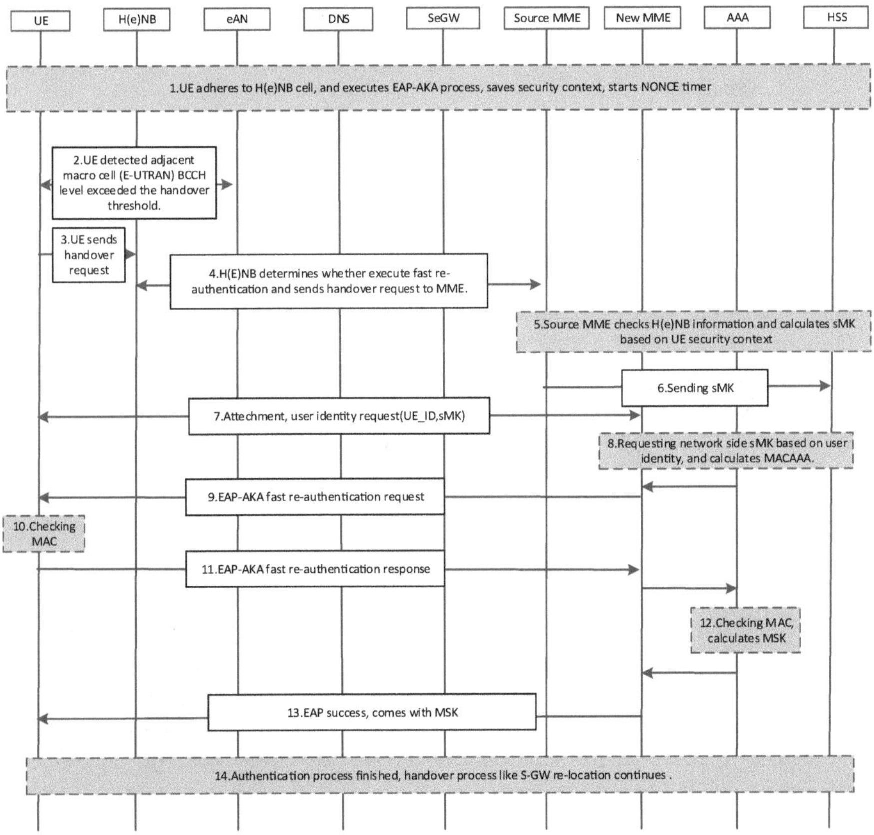

Fig. 7.16 Fast re-authentication when HO from H(e)NB to CDMA2000

(4) H(e)NB maintains a list of trusted adjacent cells that allow for fast re-authentication. H(e)NB checks whether the target handover cell information in the UE report are also in the trusted list to decide whether to execute the fast re-authentication or not. At the same time, H(e)NB periodically executes MME location check (the period is determined by the strategy of operators).

  (4a) There are two means to realize the H(e)NB location check mentioned above. The first is to scan the surrounding macrocell information and report to MME, which is suitable for dense macrocell areas in the city.
  (4b) If H(e)NB is located at fewer macrocell areas, the location check can be completed by getting its own IP address from the DNS server between H(e)NB and the core network. Obviously, this method needs to discuss strategies with other operators in advance, which will bring additional costs, and may not be able to achieve the positioning accuracy. Hence, it is only an alternative option here. Other program is to deploy a GPS system in H(e)NB to achieve a precise positioning, which performs the best but apparently costs the most, only be applicable to a few scenes.

(5) MME checks the H(e)NB location information based on the operator's strategy. If the H(e)NB incidental information agrees to perform the fast re-authentication, MME calculates sMK based on the KASME and NONCEMME of UE security context saved by itself; if not perform the fast re-authentication, then it does not need to deduce the key, and the UE re-performs the authentication process after handover attachment. At the same time, the UE side will calculate sMK using KASME and NONCEMME.
(6) MME will send sMK and the corresponding user's UE_ID (IMSI) to HSS and save.
(7) UE connects to the HSGW attachment in CDMA2000 cell; the HSGW sends an identity request to UE; UE responses to the request, and informs the root key MK on the network side of the IMSI or the pseudonym which carries the identification.
(8) HSGW transmits the response message to AAA server. AAA server requests the root key sMK of EAP-AKA to HSS and calculates the MACAAA, and generates a random number NONCEAAA and a counter Counter.
(9) AAA server sends a fast re-certification request to UE, which contains the MACAAA, NONCEAAA, and Counter.
(10) After receiving the message, UE calculates the MACAAA on the user side by using the parameters in the message to see whether it is identical to the MACAAA in the received message, so as to verify the legality of the network side. Moreover, UE compares its own Counter to verify the freshness of security context. If the condition is satisfied, MACUE is calculated to generate NONCEUE, and the sMK calculated before is used to calculate MK ' after the deduction for fast re-authentication, and MK' is used to generate the master session key MSK as well as the extend master session key EMSK in the lower layer.

(11) UE sends a fast re-certification response message to HSW, which contains the parameters generated in the last step, such as the MACUE, NONCEUE, Counter, and so on.
(12) HSGW sends the fast re-authentication message to AAA server. AAA server receives the message and checks the correctness of MACUE to verify the legitimacy of UE. If the validation is successful, MK ‘ is calculated, and the master session key MSK as well as the extend master session key EMSK is obtained.
(13) AAA server sends to UE the fast re-certification success message and sends the master session key MSK to HSGW for generating the access network session key.
(14) After the completion of the fast re-certification, the cell handover process continues.

3. **Security analysis**

Taking use of the fast re-authentication of previous certification is a good method to deal with the waste of resources caused by frequent handover in hot spots. Operators need to configure the key deduction algorithm and the fast re-authentication trusted white list in advance. This list has been saved in H(e)N on the network side, also in MME and H(e)MS (used for the location check when software updates) and it can be changed by the operator or H(e)MS strategies. So, the UE does not need to re-execute a complete EAP-AKA process. It just uses the existing security context, selects the matched key deduction conversion algorithm according to the target handover cell, so as to establish the updated security context and complete the fast re-authentication process.

7.3.2 Wlan and Cellular Authentication

With the rapid development of wireless communication, it is an inevitable trend that the wireless access mode WLAN will integrate with 3GPP system and form the interworking network. How to realize the fast and authentic handover scheme of WLAN-3GPP interworking network becomes one of the focus issues in this field. At present, there is no good way to solve the problem of handover security. This section proposes an improved fast and authentic handover scheme. The basic idea is adding a management server (MS) to the existing architecture, which can enhance the security and efficiency of handover. There are two scenarios of handover procedures.

1. **STA handover between APs in the same ESS**

In the WLAN-3GPP interworking network, the STA handover procedure between APs in the same ESS is described as follows:

(1) STA sets up link with current AP.
(2) STA sends handover requests or notification messages to the current AP. The handover request message includes a pre-choice target AP list.

(3) MS eliminates counterfeit APs and filtrates APs through formulation and contention strategy. Then it outputs the choice target AP list. If MS disagrees with handover, then it turns to step (4), otherwise it turns to step (5).
(4) MS does nothing or sends a rejection response of the handover request to STA, and the handover is ended.
(5) MS sends a pre-handover notification to current AP, which will send STA information to each AP in the choice target AP list (called target AP). Meanwhile, current AP sends the pre-handover notification, which includes the choice target AP list, to STA transparently.
(6) Current AP sends STA information to target AP.
(7) Target AP updates STA state to handover state.
(8) Target AP delivers the response to current AP. If STA information is delivered successfully, the deliver–success–response is sent to current AP and MS. If information delivery is failed, it turns to (4).
(9) Current AP determines the final target AP list. After receiving the response message of each target, current AP deletes the APs which are failed in information delivery from the choice target AP list, and then it acquires the final target AP list.
(10) Current AP sends handover notification which includes the final target AP list to STA.
(11) Current AP updates the local two-layer delivery table.
(12) Current AP declares a two-layer delivery table updating broadcast.
(13) Target AP updates the local two-layer delivery table and changes the STA state into handover state.
(14) STA sends reassociation request to the target AP in the final target AP list.
(15) Target AP sends the reassociation response to STA.

Additional explanations about some special cases are as follows:

(1) If MS detects that the load of some AP overweight and its neighboring AP has idle resources, it will start the handover procedure. Then it carries out step (2), and the STA sends notification information instead of handover requests, the other steps are identical.
(2) After receiving the pre-handover notification, and if the weak signal results in communication failures with current AP, STA may attempt to associate with the APs in the choice target AP list from the first one to the last one. If all attempts do not succeed, it shows the handover failure. During the attempts, if STA receives the handover permission response sent by current AP, it turns to step (14).

Figure 7.17 shows the handover procedure.

2. **STA handover between APs in different ESSs**

In the WLAN-3GPP interworking network, the STA handover procedure between APs in different ESS is described as follows. MS needs to detect the AP's validity belonging to its own management scope. If MS detects that APs are not controlled by local MS, it needs to detect which neighboring MSs are controlling the nonlocal

APs by using the pre-stored neighboring AP list. After the detection, MS will send these APs' information to its neighboring MS and request handover to the neighboring MS. The neighboring MS performs the same procedure as current MS. Figure 7.18 shows the handover procedure.

3. **Security analysis**

A management server (MS) is added to the existing architecture, which is responsible for the establishment and maintenance of the AP information list, detecting counterfeit AP, and providing the AP legitimacy query to terminal users.

7.3.3 *Relay Security*

There is a strict requirement of capacity in LTE-R system. In LTE-R, a relay scheme is suggested, which brings many benefits to the system, e.g., improving spectrum efficiency, increasing system capacity, and a guarantee of user's requirements for high-speed data transmission. However, because of its special characteristics, the deployment of relay node equipment will bring some new problems. For instance, with the emergence of the relay node equipment, there are

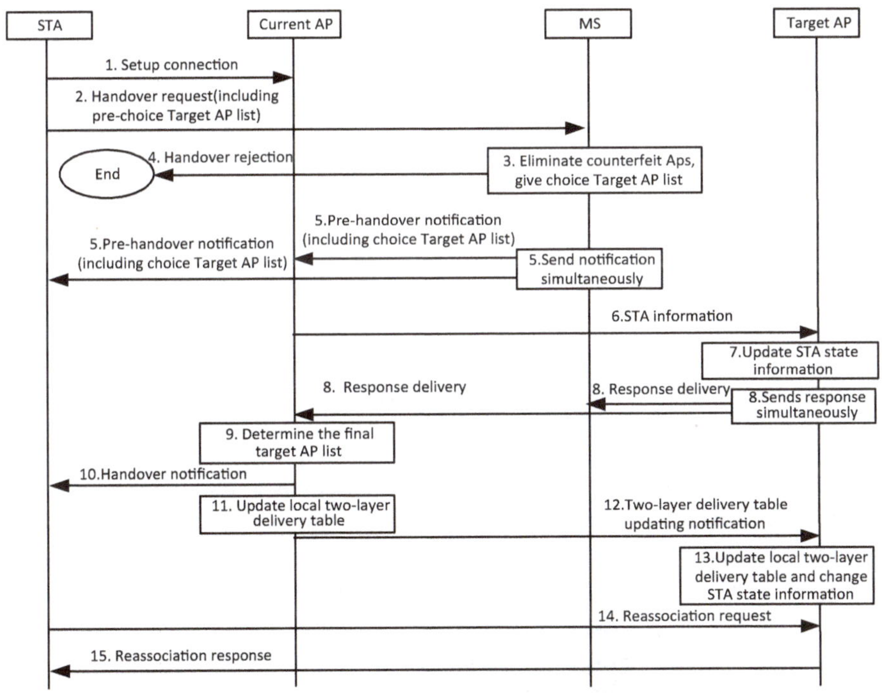

Fig. 7.17 STA handover flow between APs in the same ESS

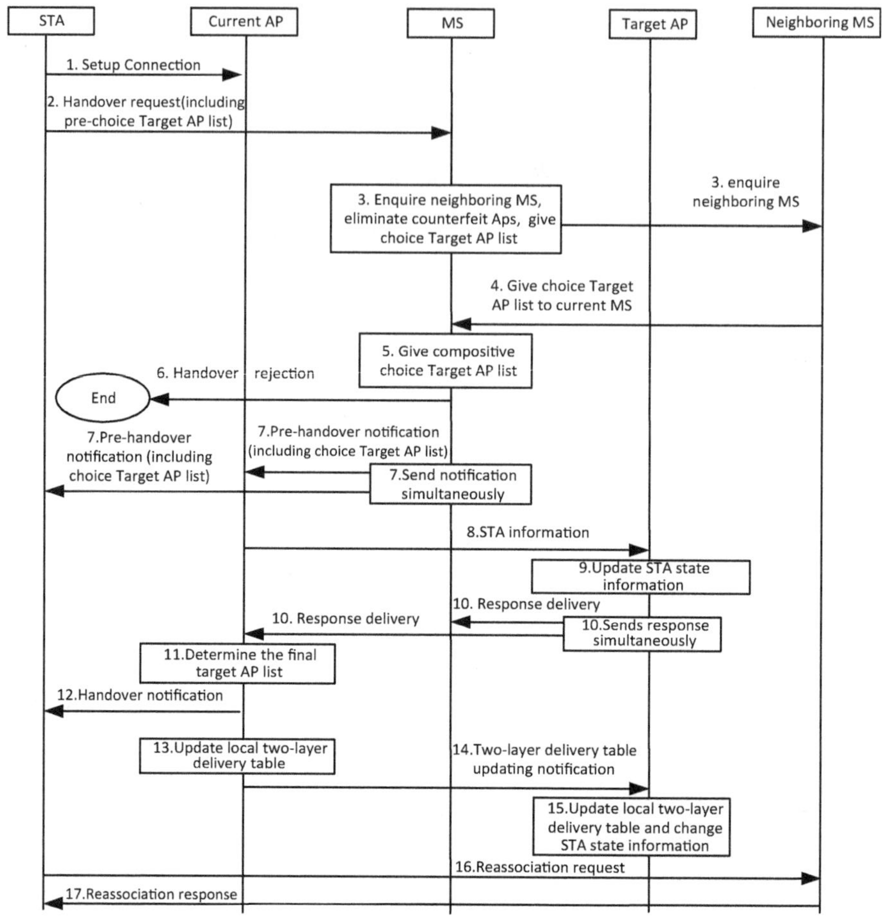

Fig. 7.18 STA handover flow between APs in the different ESSs

new types of handovers when users move through the network. The traditional security scheme is not suitable for the new scenario any more. Therefore, a new handover procedure should be proposed to ensure the reliable communications during handover.

In this section, we will analyze the security issues emerging when the user's terminal executes a handover process under the deployment of relay nodes. Then we give a solution that the source node and destination node have synchronization security signals, thus keeping the communication normal after handover.

1. **Relay network infrastructure**

In the E-UTRAN network of LTE-A, relay nodes are introduced to support the relay communication. The relay nodes are usually connected to eNB via wireless channel. The eNB which provides service to the relay node is called donor eNB (DeNB). The

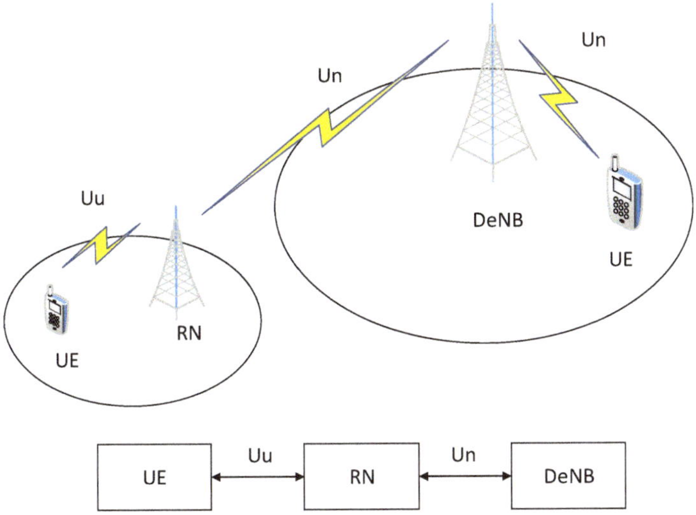

Fig. 7.19 The LTE-A network with relay nodes

air interface between RN and DeNB is Un. Un is based on the Uu between UE and eNB. The LTE-A network with relay nodes is shown in Fig. 7.19.

Seen from Fig. 7.19, with the deployment of relay nodes, the air interface between user equipment and donor eNB is divided into two parts, namely the Uu interface between UE and RN, which is also called the access link, and the Un interface between RN and DeNB, which is also called backhaul link.

There are some handover scenarios for user equipment under relay deployment situation.

Scenario 1: The user equipment connected with RN switches to the DeNB with the same RN via X2 interface.

Scenario 2: Handover between two RNs in the same DeNB via X2 interface. The MME connected with DeNB remains unchanged.

Scenario 3: The user equipment connected with RN switches to another DeNB. There are two conditions. If the DeNB connected with the RN and the destination DeNB are linked with the same MME, the handover is via X2 between DeNBs. Otherwise, the handover is via S1 between DeNB and MME.

Scenario 4: The user equipment connected with RN switches to a RN connected with another DeNB. There are two conditions. If the source and destination DeNBs are linked with the same MME, the handover is via X2 between DeNBs. Otherwise, the handover is via S1 between DeNB and MME.

Scenario 5: The user equipment connected with eNB(or DeNB) switches to the RN via X2 between DeNB and RN.

2. **Case of handover process and security enhancement**

Because of the limitation of length, we take scenario 3 as a case to introduce handover process and security enhancement.

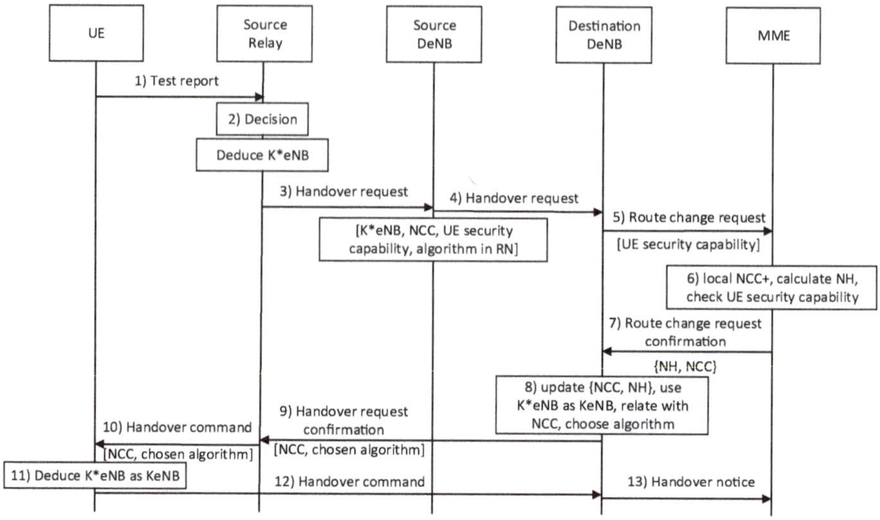

Fig. 7.20 UE switches from RN to another DeNB via X2

When the user equipment connected with RN switches to another DeNB via X2, the secret key used by the destination DeNB can be generated in the RN or source DeNB. The handover procedure is listed below.

In Fig. 7.20, NCC stands for next hop chaining counter, and NH stands for next hop key. This is the secure protection for handover and key update. KeNB is the key used for handover. It is also used for the generation of RRC layer key and user encryption key. K*eNB is the key generated by mobile equipment and eNB in the vertical and horizontal key deduction. If the source DeNB and destination DeNB are not the same one, it is S1 handover including MME redirection. The key used in the destination DeNB can be calculated in RN or source DeNB. The detailed process is shown in Fig. 7.21.

In the above, we introduce two handover processes. When the source and destination node are connected to the same MME, the handover is via X2. Otherwise, the handover is via S1. There is consistent security information with the handover, thus which ensures the reliable communication.

3. **Security analysis**

As the relay nodes are deployed, there are more handover scenarios. To ensure the reliable communication during the handover, all of the keys and algorithms have to be calculated in the destination node. In this session, we talk about handover processes and security measures in different scenarios. The destination node has consistent security information with the user equipment, thus keeping the communication normal.

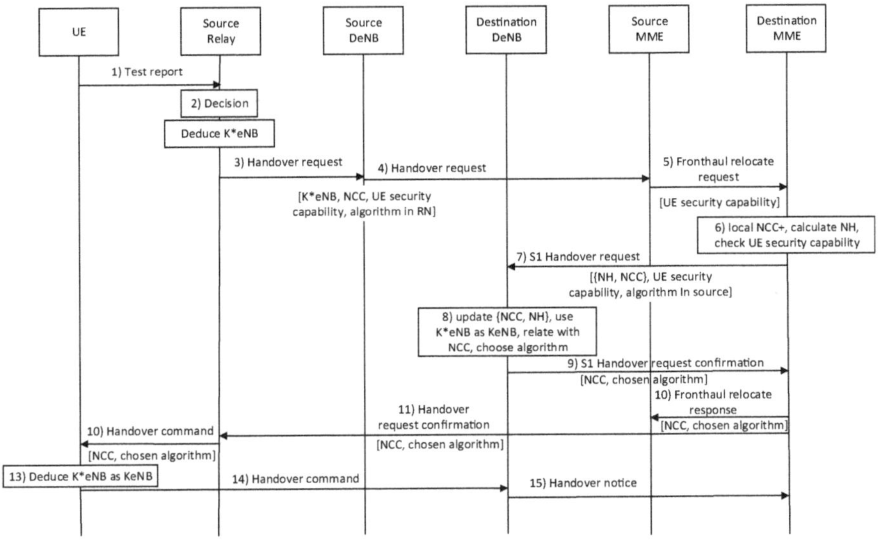

Fig. 7.21 UE switches from RN to another DeNB via S1

7.3.4 Access Authentication for Mobile Trusted Computing

With the spread of wireless network, the railway mobile communication network is over IP. The 3GPP, WLAN, and WiMAX protocols have provided security insurances for user access authentication and data transmission. However, the service provider and IP network are open and have many security holes, which leads to various security threats, such as virus, hacker attack, and user information embezzlement, in application layer. The tradition security mechanism is not able to solve these threats [12–14].

It is easy to supervise the inner threats from core network of railway mobile communication. However, the supervision of mobile terminals is very hard. The mobile terminals are less protection due to limited abilities. They are vulnerable to virus when the system holes or security applications are not immediately updating. Moreover, the mobile terminals have extensive numbers, wide distributions, and high mobility. The infected terminal can be new source of virus transmission. In this session, we introduce an access authentication method for mobile terminals. The method can monitor threats effectively and enhance the security of mobile terminals. Ultimately, it improves the security of the whole railway mobile communication network.

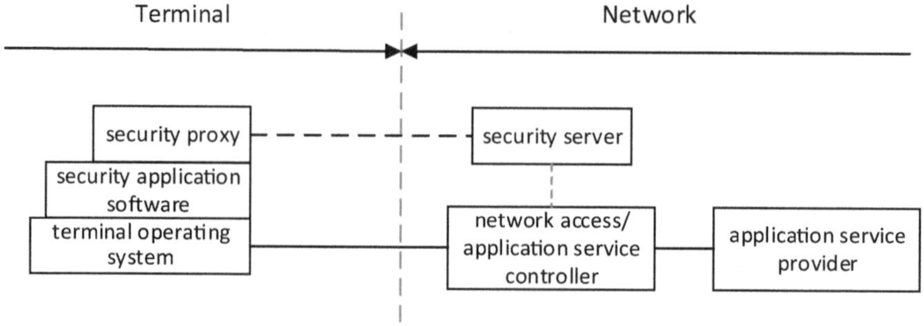

Fig. 7.22 Trusted access authentication model for mobile network

1. **Trusted Access authentication model for Mobile network**

In this session, we introduce a trusted access authentication model for mobile network, which is aimed at unsafe mobile terminals. In the model, we make assessment of the security condition of mobile terminals. The result is used to guide the network access control of mobile terminals. According to the result, it reminds the mobile terminals to update their safety application. The model provides security enhancement to the network and mobile terminals and prevents the virus from fast spreading. The model is shown in Fig. 7.22.

From the figure above, we can see there are four parts in trusted access authentication model for mobile network: security proxy of the mobile terminal, security server of the network, network access controller, and application service controller of the network. The core of the model is security proxy and security server.

The security proxy is responsible for collecting security information of mobile terminals and communicates with security server. The security server makes assessment according to the collected information and gives security level that demines whether the mobile terminal is admitted to the network and what services are permitted. Above all, the security proxy is information collector and the security server is decision-maker.

The security server makes assessment, when it receives security status information from security proxy. If the security server thinks the mobile terminal is not safe, it will give instructions to the network access controller and application service controller. The instructions are also sent back to the security proxy.

If there are patches or components to update, the security server will inform the security proxy to assist updating. The updating packets are saved in the security software server and mobile terminal's system server.

The mobile terminal is controlled by the security server, thus the network access and application service is controlled by the security server.

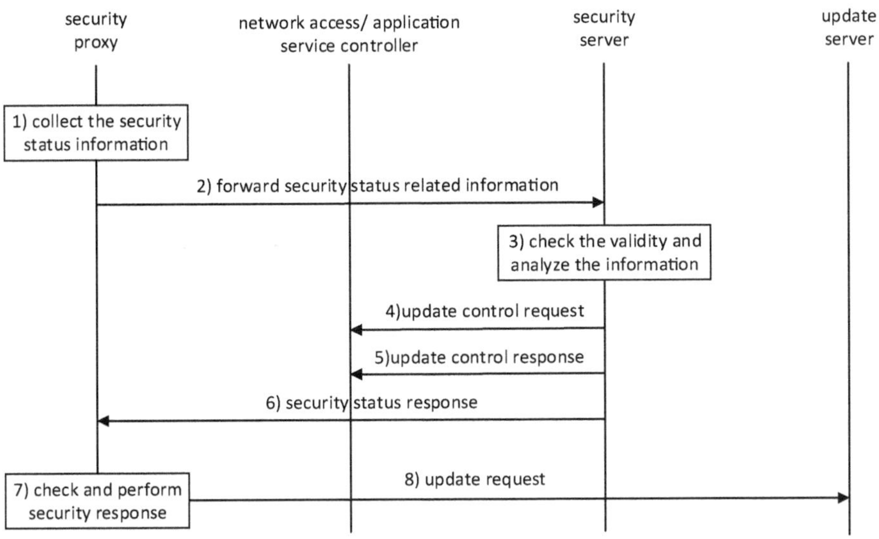

Fig. 7.23 Assessment process

2. **Assessment of mobile terminals**

When a mobile terminal tries to join the network, it needs to perform authentication with the network. After successful authentication, the security proxy will actively collect the security status information of mobile terminals and then forward the information to the security server. The security server makes assessment and decides whether the mobile terminal is admitted to the network. The process of assessment is shown in Fig. 7.23.

From Fig. 7.23, we can see the detailed process of assessment is as follows:

(1) The security proxy collects the security status information of mobile terminal.
(2) The security proxy forward information to the security server.
(3) The security server checks the validity and analyzes the information according to security strategies. The security server gives a security status level.
(4) If the mobile terminal needs updating, the security server will inform the network access controller and application service controller.
(5) If step 4 performs, the network access controller and application service controller will send back updating results.
(6) The security server sends responses to the security proxy and tells the assessment results. The responses include the strategies that the security

proxy collects information, and generates and transmits information about the state of security. If the mobile terminal needs security update, the responses will also include the address to get update and the strategies to update. The security proxy checks the validity of responses and updates local information and strategies.

(7) If the security status information assessment response indicates that the mobile terminal needs security update, then the security proxy assists the mobile terminal to finish the security update according to the information and strategies stored in the response. Here, the security update includes of the security proxy's own upgrade and update.

Among all the steps, steps (4), (5), and (7) are optional. The mobile terminals connect with the network after successfully updating and passing the authentication.

The trusted access authentication model gives instructions to control the network access and application service by making assessment of the mobile terminals. The terminal updates the patches and components in time. Ultimately, it provides security enhancements to the network and mobile terminals and stops the virus from fast spreading.

7.4 Future and Challenges

The GSM-R is turning into LTE-R. Many technologies in 4G systems can be used for reference for security of railway mobile communication. Meanwhile, the railway communication network is tightly integrated the computer network. Apart from the traditional security threats, there are more network security problems, for example, the security of SDN network framework. Moreover, with the research of cyberspace security in recent years, related research has been raised in railway mobile communication. The cyberspace security is the advanced development stage of information security and network security, including confidentiality, integrity, availability, and facticity. The cyberspace security is one of the most serious potential risks in transportation system. Therefore, the cyberspace security is an important research direction in the future.

References

1. Xu S, Ma W, Wang X (2003) Security technology in wireless communication network. Posts and Telecom Press, Beijing, pp. 36–41
2. Vines RD (2002) Wireless security essentials: defending mobile systems from data piracy. Wiley Publishing, New York

3. Maxim M, Pollino D (2002) Wireless security. McGraw-Hill, Maidenhead
4. Liu Y, Fan Z (2004) Analyses and countermeasures of GSM—R network security. Railw Signal Commun Eng 6:27–30
5. Song Y, Zhou K, Yao B, Chen X (2010) A GSM/UMTS Selective Jamming System. International Conference on Multimedia Information Networking and Security
6. Zhang M, Fang Y (2005) Security analysis and enhancements of 3GPP authentication and key agreement protocol. IEEE Trans Wirel Commun 1(12):734–742
7. Wu H, Shi X, Gu Y (2009) Research and improvement on security of GSM-R system. J Beijing Jiaotong Univ 33(2):127–1301
8. Zeidler HM (1986) End-to-end encryption system and method of operation. US, US 4578530 A
9. Al-Fayoumi MA, Shilbayeh NF (2014) Cloning sim cards usability reduction in mobile networks. J Network System Management 22(2):259–279
10. Wu H, Gu YH, Zhong ZD (2013) One type of GSM-R fast handover algorithm for high-speed railway. In: IET, international conference on wireless, Mobile and Multimedia Networks, IET, pp.175–178
11. Black J, Halevi S, Krawczyk H, Krovetz T, Rogaway P (1999) UMAC: fast and secure message authentication. In: Advances in cryptology - CRYPTO '99, international cryptology conference, Santa Barbara, California, Usa, August 15–19, 1999, Proceedings (Vol.1666, pp.216–233), DBLP
12. Asokan N, Ekberg JE, Kostiainen K, Rajan A, Rozas C, Sadeghi AR et al (2014) Mobile trusted computing. Proceed IEEE 102(8):1189–1206
13. Madlmayr G (2008) A mobile trusted computing architecture for a near field communication ecosystem. In: iiWAS'2008 - The tenth international conference on information integration and web-based applications services, 24–26 November 2008, Linz, Austria (pp.563–566), DBLP
14. Chengdu (2006) Trusted computing based user authentication for mobile equipment. Chinese J Comput 29(8):1255–1264

Chapter 8
Channel Simulation Technologies for Railway Broadband Mobile Communication Systems

8.1 Simulation Approaches

Nowadays, the performance evaluation approaches can be classified in three groups: computer simulation, field trial, and Hardware-in-Loop (HIL) simulation.

Computer simulation is pure numerical simulation and is the most cost-effective and flexible approach. We can model very complicated systems in Computer with the simulation software, such as Matlab, NS2, OPNET, etc. By thus, single variable and multivariable systems, linear and nonlinear systems, continuous, discrete and hybrid systems, time-invariant and time-varying systems, engineering and nonengineering systems can all be simulated with low cost. However, the models in simulation are the simplification and approximation of the real system and environment, and the accuracy of the simulation strongly relies on the accuracy and validity of the numerical models used. In addition, hardware performance evaluation cannot be carried out in computer simulation.

Field trial (or on-road testing) evaluates performance of devices in a realistic system and environment. The accuracy of field trial is very high. However, field trial is tedious, time-consuming, and expensive, even inacceptable. Furthermore, parameters' tuning in performance evaluation is very frequent; unfortunately the adjustment is very difficult in field trial. In addition, the testing environment cannot be fully repeated in field trial, which is very important in failure investigation.

HIL simulation is an intermediate solution between computer simulation and field trial. In HIL simulation, testing is executed in a virtual test scenario, instead of on-road or in real devices. Much of the test environment is replaced by mathematical models, so components can be inserted into a closed loop. Compared to field trial, HIL simulation can substantially lower the cost and risk, and is able to adjust system and environment parameters at ease. Furthermore, HIL simulation can fully control the system and environment parameter, which makes for tests that are reproducible, systematic, and more reliable. So HIL simulation is a trusted, cost-effective alternative for field trial. HIL is a powerful tool in the design and test

Z.-D. Zhong et. al., *Dedicated Mobile Communications for High-speed Railway*, Advances in High-speed Rail Technology, DOI 10.1007/978-3-662-54860-8_8 of complex equipment and system. In recent years, HIL simulation has applied more and more widely in research, design and development of automobiles, airplanes, missiles, satellites, rockets and locomotives, and so on.

In high-speed railway, the period of field trial is very limited, and the risk is very high. In recent years, the HIL simulation has attracted more and more attention.

8.2 Simulation Scenario for Railway

High-speed railway has many scenarios, such as viaduct, cutting, tunnel, stations, and so on. The network deployment and channel model change from different scenario. In 3GPP RAN4, several typical scenarios are discussed. At last, the four scenarios were agreed, and could be used as baseline scenarios for the performance evaluation of HSR.

8.2.1 *Scenario 1: Open Space SFN*

In 3GPP TSG-RAN WG4 #74 meeting, the new Study Item (SI) "LTE performance enhancement under high speed scenario" had been discussed. Firstly, many new high speed deployment scenarios had been proposed. In which, the single-frequency network (SFN) deployment scenario with remote radio head (RRH) along high-speed train is the most important, shown in Fig. 8.1.

In SFN deployment scenario, multiple RRHs are connected to one BBU with fiber and share the same cell ID. With the SFN deployment, the coverage of cell is enlarged, the frequency of handover is reduced, and the multipath gain can be obtained. Thus the SFN deployment is an appropriate network deployment in viaduct.

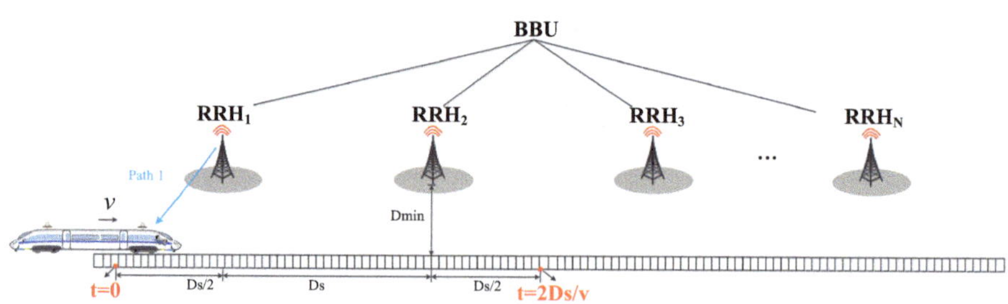

Fig. 8.1 Sketch map of single-tap SFN channel model

Table 8.1 Parameters for scenario 1

Parameter	Value
RRH Railway track distance	50–300 m
Distance between RRH	1–1.5 km
Cell ISD	Maximum: 6 km (6 RRHs connect to 1 BBU) Minimum: 2 km (2 RRHs connect to 1 BBU)
RRH height (compared to railway track)	15–25 m

The descriptions of Scenario 1 are as follows:

- RRHs are connected to one BBU with fiber
- Multiple RRUs share the same cell ID
- No repeaters installment (Table 8.1).

8.2.2 Scenario 2: Tunnel Environment

Scenario 2 is for tunnel environment. For different network deployment, the Scenario 2 can be cataloged in 5 sub-scenarios, 2a–2e, described as follows.

(1) Scenario 2a:

- RRHs or RAUs is deployed through fiber in **tunnel** environment
- RRHs or RAUs share the same cell id
- Repeaters are installed on the carriage and distribute signal inside the carriage

(2) Scenario 2b:

- RRHs or RAUs are deployed through fiber in tunnel environment
- RRHs or RAUs share the different cell id
- Repeaters are installed on the carriage and distribute signal inside the carriage

(3) Scenario 2c:

- Leaky cables are used to extend the signal through the **tunnel** environment
- Repeaters are installed on the carriage and distribute signal inside the carriage

(4) Scenario 2d:

- RRHs or RAUs are deployed through fiber in tunnel environment
- RRHs or RAUs share the same cell id
- Repeaters are **not** installed on the carriage

Table 8.2 Parameters for scenario 2

Parameter	Value
RRH Railway track distance	Closest: 1 m, farthest: 9 m
Distance between RRH	500 m
RRH height (compared to railway track)	Lowest position: 2.5 m

(5) Scenario 2e:

- RRHs or RAUs are deployed through fiber in **tunnel** environment
- RRHs or RAUs share the different cell id
- Repeaters are **not** installed on the carriage

The parameters of Scenario 2 are shown in Table 8.2

8.2.3 Scenario 3: Open Space ENB to RP

The penetration loss of HST train is 20–25 dB, thus the most reliable method for the wireless communication is to install repeater (RP) onboard. The communication link between eNB and RP in open space is defined as Scenario 4. By the way, the scenario in the train, i.e., the link between RP and user terminal, is just a traditional scenario.

The description of Scenario 3 is shown as follows:

- In a portion of the high-speed outdoor coverage, eNB are installed through the railway on same frequencies as public network coverage
- In the remaining cases, the railway is covered with public network only.
- Repeaters are installed on the carriage and distribute signal inside the carriage through leaky cables.

The Parameters of Scenario 3 are shown in Table 8.3

8.2.4 Scenario 4: Public Network

The original description of Scenario 4 is outdoor eNB installed through the railway on same frequencies as public network coverage.

The parameters of Scenario 4 are shown in Table 8.4

Table 8.3 Parameters for scenario 3

Parameter	Value
eNB Railway track distance	10 m
Distance between eNB	5 km
eNB height (compared to railway track)	20 m

Table 8.4 Parameters for scenario 4

Parameter	Value
eNB Railway track distance	300 m
Distance between eNB	3 km
eNB height (compared to railway track)	25 m

8.3 Channel Model in Simulation

Traditional fading channel models, such as COST 207 models [8], ITU 3G models, 3GPP models, are designed for public mobile communication system. However, radio propagation on high-speed railway has some properties, such as environment, high mobility, coverage, etc. In order to evaluate the system in high-speed railway, 3GPP RAN4 have proposed HST channel models for SFN.

Since the channel environment of high-speed railway in viaduct is open, with less scatterers and no obstacle. Thus, the multipath cased by the scatterers is out of consideration. Assuming that there is only line-of-sight propagation, UE will observe superposition of signals coming from multiple RRHs. If we further assume transmission timing is perfectly synchronized in RRHs, UE will observe multipath channel due to different propagation delay between UE and RRHs. Since UE is moving along the railway, signal properties such as Doppler shift, time delay and power will dynamically change with time. In our view, channel model for SFN deployment should capture dynamic nature of multipath component to allow proper evaluation of demodulation performance in real network.

8.3.1 *Single-Tap HST Channel Model*

In the single-tap HST channel model, only the strongest path is taken into consideration. Single-tap HST channel model is simplified and non-fading propagation channel model, which only captures the dynamic Doppler shift while assuming static delay and power of multipath. The Doppler shift is calculated as follows:

$$f_{d\max} = f_c \cdot \frac{v}{c} \cdot \cos\theta_n(t) \tag{8.1}$$

To maintain the continuity of the frequency offset and avoid the alternation of Doppler shift when handing over, the cosine of angle $\theta_n(t)$ of the nth path can be expressed as

$$\cos\theta_n(t) = \frac{-\left(n+N-\frac{1}{2}\right)D_s + vt}{\sqrt{\left[-\left(n+N-\frac{1}{2}\right)D_s + v\right]^2 + D_{\min}^2}}, \quad n \in [1, N] \tag{8.2}$$

Fig. 8.2 The Doppler shift trajectory (D_s = 300 m, $D_{\min}$ = 10 m, f_c = 2690 MHz, v = 300 km/h)

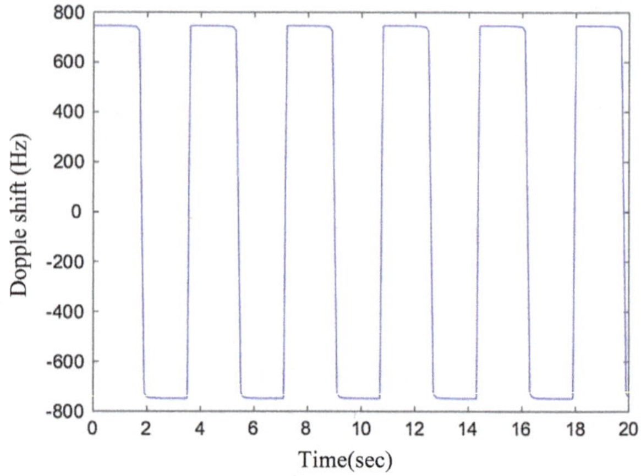

When $t \geq 2ND_s/v$,

$$\cos\theta_n(t) = \cos\theta_n(t \bmod (2ND_s/v)) \tag{8.3}$$

Obviously, it is very complicated taking N paths into consideration. One-tap high-speed train channel model is denoted as equitation.

$$\cos\theta(t) = \begin{cases} \frac{0.5D_s - vt}{\sqrt{D_{\min}^2 + (0.5D_s - vt)^2}}, & 0 < t \leq \frac{D_s}{v} \\ \frac{-1.5D_s + vt}{\sqrt{D_{\min}^2 + (-1.5D_s + vt)^2}}, & \frac{D_s}{v} < t \leq \frac{2D_s}{v}, \\ \cos\theta(t \bmod (2D_s/v)), & t > 2D_s/v \end{cases} \tag{8.4}$$

The Doppler shift trajectories are shown in Fig. 8.2

8.3.2 *Two-Tap HST Channel Model*

Qualcomm Incorporated had proposed two-tap HST channel model with low modeling complexity. The channel model takes the two strongest paths, corresponding to two nearest RRHs, into consideration, and capture dynamic propagation condition, including dynamic Doppler shift, channel tap delay, and channel tap power.

When UE passes wrap around point, RRH for weaker tap is replaced with newly approaching RRH. For example, when UE passes RRH2, UE stops receiving signal from RRH1 and starts receiving signal from RRH3 (Fig. 8.3).

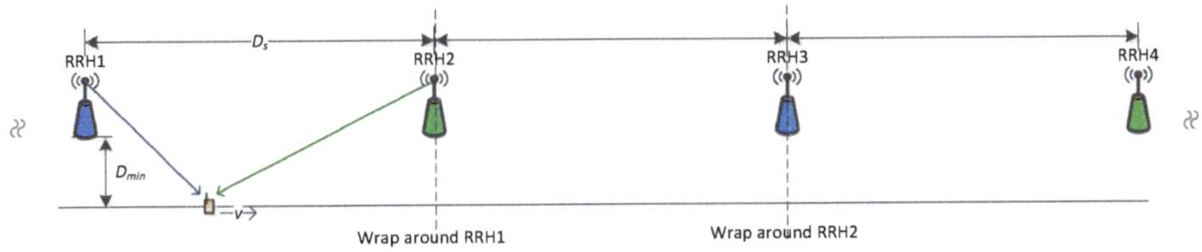

Fig. 8.3 SFN channel model with wrap around

In the model, the Doppler shift, delay and power of two channel taps are calculated as below

$$f_{s,1}(t) = f_s\left(t + \frac{0.5D_s}{v}\right)$$
$$f_{s,2}(t) = f_s\left(t + \frac{1.5D_s}{v}\right) \tag{8.5}$$

where $f_s(t) = f_d \cos\theta(t)$. Here, $\cos\theta(t)$ is given by

$$\begin{aligned}
\cos\theta(t) &= \frac{0.5D_s - vt}{\sqrt{D_{\min}^2 + (0.5D_s - vt)^2}}, & 0 \le t \le \frac{D_s}{v} \\
\cos\theta(t) &= \frac{-1.5D_s + vt}{\sqrt{D_{\min}^2 + (-1.5D_s + vt)^2}}, & \frac{D_s}{v} < t \le \frac{2D_s}{v} \\
\cos\theta(t) &= \cos\theta\left(t \bmod \frac{2D_s}{v}\right), & t > \frac{2D_s}{v}
\end{aligned} \tag{8.6}$$

Note that this Doppler shift model is exactly same as existing HST channel model.

Time-varying tap delay for blue RRH and green RRH is given by

$$\begin{aligned}
d_1(t) &= D_1(t)/c \\
d_2(t) &= D_2(t)/c
\end{aligned} \tag{8.7}$$

Here, $D_1(t)$ and $D_2(t)$ are distances from UE and two RRHs and given by

$$D_1(t) = \begin{cases} \sqrt{D_{\min}^2 + (vt)^2}, & 0 < t \le \frac{D_s}{v} \\ \sqrt{D_{\min}^2 + (2Ds - vt)^2}, & \frac{D_s}{v} < t \le \frac{2D_s}{v}, \\ D_1\left(t \bmod \frac{2D_s}{v}\right), & t > \frac{2D_s}{v} \end{cases}$$
$$D_2(t) = D_1\left(t + \frac{D_s}{v}\right) \tag{8.8}$$

Assuming that transmit power from RRHs are same, signal power received by the UE from each RRH is given by

$$P_1(t) = \left(\frac{D_{\min}}{D_1(t)}\right)^2$$
$$P_2(t) = \left(\frac{D_{\min}}{D_2(t)}\right)^2 \quad (8.9)$$

where received power is normalized with respect to the received power when UE is $D_{\min}$ away from RRH. Note that free space attenuation exponent 2 is used in the equation.

Figure 8.4 shows Doppler shift, delay and received power of the two-tap SFN channel model. It is assumed that carrier frequency is

8.3.3 *WINER Channel Model*

The WINNER project has begun in 2004 and developed in the course of the ISTWINNER, which focuses on channel modeling for Beyond 3G (B3G) system. WINNER channel models support the frequency range 2–6 GHz and maximum 100 MHz bandwidth. WINNER model defines 16 propagation scenarios, including indoor office, large indoor hall, indoor-to-outdoor, urban microcell, bad urban microcell, outdoor-to-indoor, stationary feeder, suburban macrocell, urban macrocell, rural macrocell, and rural moving networks.

The WINNER model is a so-called geometry-based stochastic channel model. As shown in Fig. 8.5, the cluster is the basic component, which is made up of N (N = 20) rays reflected by the closed scatters. The number of clusters N_c varies with scenarios (in D2a, N_c = 8).

The scenario for D2a is for high-speed network, and the parameters of the WINNER D2a scenario was compared with that in real high-speed railway environment as Table 8.2.

Two critical assumptions in WINNER D2a model are reasonable. The environment around high-speed rail is open. This means that very little scattering, relatively far from the MS. The WINNER D2a model is assumed by the 20 clusters. And each path is assumed to be included in the traditional unlimited-ray channel model, such as COST207 model. Since the MS is high-speed, the channel is dynamic. In other words, the channel parameters, such as the latency and amplitude, depend on the position. Besides, WINNER D2a is randomly assigned with a certain channel parameters generated. So we think WINNER D2a model is an appropriate channel model for high-speed railway environment (Table 8.5).

Fig. 8.4 Doppler shift, delay, and power for two-tap SFN channel model ($D_s = 2000$m, $D_{min} = 50$m, $fc = 2.69$ GHz, $v = 350$ km/h)

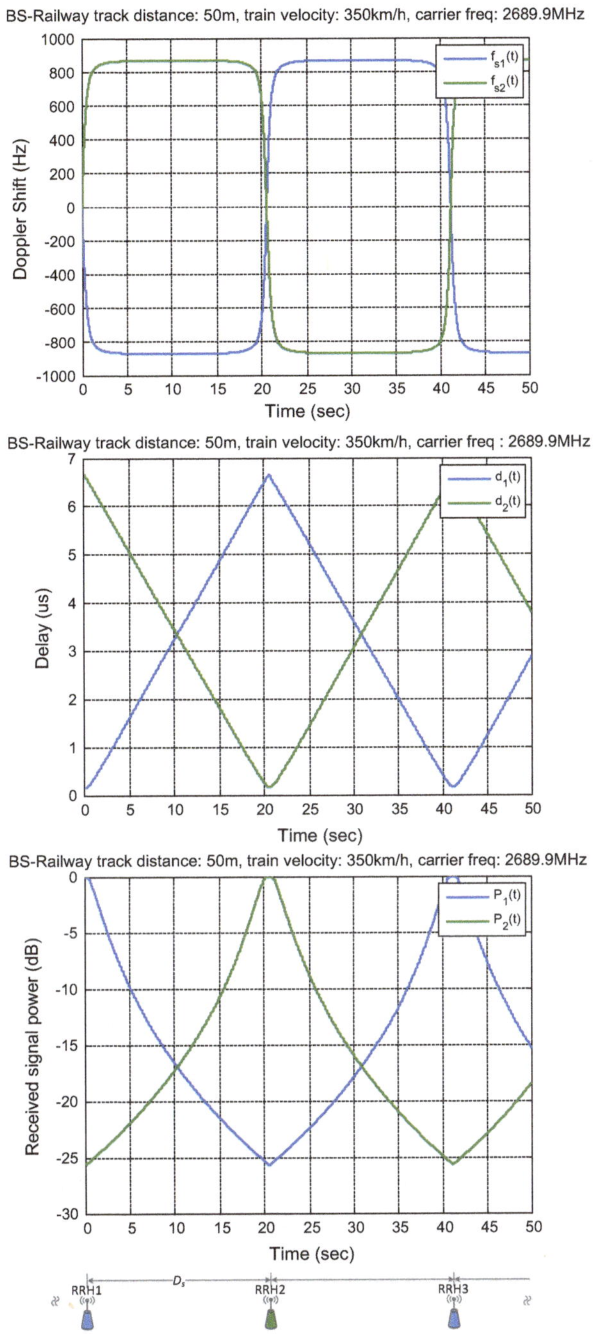

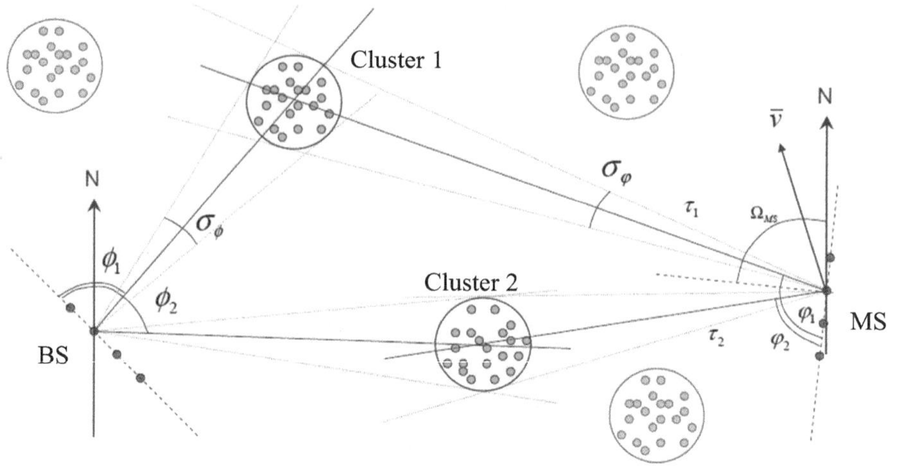

Fig. 8.5 The WINNER fading model

Table 8.5 WINNER D2a scenario versus real high speed railway environment

	WINNER D2a secnario	High-speed railway environment
Distance between adjacent BSs (m)	1000–2000	3000
The antenna heights of MS (m)	2.5	3.8–4.2
The antenna heights of BS (m)	30	20–45
The location of BS	50 m away from the rail	30–50 m away from the rail
The speed of MS (km/h)	120–350	200–350
LoS/NLoS condition	LoS	LoS

The computation of channel impulse response is described as below.

Step 1 Delays

The delay of the *n*th cluster satisfies the exponential distribution

$$\tau_n' = -r_\tau \sigma_\tau \ln(X_n), \tag{8.10}$$

where X_n is uniformly distributed over the interval (0,1), r_τ is the delay distribution proportionality factor and σ_τ is rms delay spread (r_τ = 3.8 and σ_τ = 40 ns in D2a model).

Subtract τ_n' with minimum delay to normalize the delays and sort to descending order.

$$\tau''_n = \mathrm{sort}(\tau'_n - \min(\tau'_n)) \tag{8.11}$$

Due to existence of Light-of-Sight (LOS), a modification of τ''_n is required. Thus, the delay of the nth cluster is obtained by

$$\tau_n = \tau''_n / D, \tag{8.12}$$

where $D = 0.7705 - 0.0433\ K + 0.0002\ K^2 + 0.000017\ K^3$, K is Rice factor in dB.

Step 2 Power
The power of the nth cluster is determined by

$$P'_n = \exp\left(-\tau_n \frac{r_\tau - 1}{r_\tau \sigma_\tau}\right) 10^{\frac{-Z_n}{10}}, \tag{8.13}$$

where $Z_n \sim N(0,4)$ is shadowing term in dB, $N(0,4)$ represents Gaussian distribution with zero mean and standard variation 4.

Since the sum power of all cluster should equal to one, the power of nth cluster is

$$P_n = \frac{P'_n}{\sum_{n=1}^{N} P'_n} \tag{8.14}$$

Step 3 Angle of Arrival (AoA) and Angle of Departure (AoD)
The AoA of the nth cluster with respect to north direction is

$$\varphi_n = (X_n \varphi'_n + Y_n) - (X_1 \varphi'_1 + Y_1) + \varphi_{LOS}, \tag{8.15}$$

where $\varphi'_n = \frac{2\sigma_{AoA}\sqrt{-\ln(P_n/\max(P_n))}}{C_{LOS}}$, $C_{LOS} = C \times (1.1035 - 0.028\ K - 0.002\ K^2 + 0.0001\ K^3)$, $\sigma_{AoA} = \sigma_\varphi / 1.4$, σ_φ is AoA rms spread, and $\log_{10} \varphi'_n \sim N(1.5, 0.2)$, φ_{LOS} is the LoS direction between MS and BS, $Y_N \sim N(0, \sigma_{AoA}/5)$.

Clearly, $\varphi_1 = \varphi_{LOS}$, and the directions of other clusters are distributed around the LoS direction.

The AoA of the mth ray in the nth cluster is determined by

$$\varphi_{n,m} = \varphi_n + c_{AoA} \alpha_m, \tag{8.16}$$

where $c_{AoA} = 3$ is the cluster-wise rms azimuth spread of AoA, and α_m is the angle offset factor of the mth ray shown in Table 8.6.

Table 8.6 Ray offset factor (in degree)

M	α_m	m	α_m
1, 2	± 0.0447	11, 12	± 0.6797
3, 4	± 0.1413	13, 14	± 0.8844
5, 6	± 0.2492	15, 16	± 1.1481
7, 8	± 0.3715	17, 18	± 1.5195
9, 10	± 0.5129	19, 20	± 2.1551

The computation of AoD is analogous to that of AoD with $c_{AoD} = 2$. Finally, the AoD and AoA of each ray will pair randomly.

Step 4 Initial phase

The initial phases $\left\{\Phi_{n,m}^{vv}, \Phi_{n,m}^{vh}, \Phi_{n,m}^{hv}, \Phi_{n,m}^{hh}\right\}$ of the mth ray in the nth cluster for four different polarization vertical-vertical, vertical-horizontal, horizontal-vertical, horizontal-horizontal are uniformly distributed over the interval (−180°, + 180°]. The initial phases of LoS vertical-vertical and horizontal-horizontal satisfy the same distribution.

Step 5 Cross-polarization power ratios (XPR)

The XPR of the mth ray in the nth cluster

$$\kappa_{n,m} = 10^{X/10} \tag{8.17}$$

where $X \sim \boldsymbol{N}(12, 8)$.

Step 6 Channel Impulse Response

The channel coefficients for the nth cluster in the link from the sth transmitter element to the uth receiver element is

$$\begin{aligned}
\boldsymbol{H}_{u,s,n}(t) = {} & \sqrt{\frac{1}{K_R+1}}\sqrt{P_n}\sum_{m=1}^{M}\begin{bmatrix} F_{tx,s,V}(\varphi_{n,m}) \\ F_{tx,s,H}(\varphi_{n,m}) \end{bmatrix}^T \begin{bmatrix} \exp\left(\mathrm{j}\Phi_{n,m}^{vv}\right) & \sqrt{\kappa_{n,m}}\exp\left(\mathrm{j}\Phi_{n,m}^{vh}\right) \\ \sqrt{\kappa_{n,m}}\exp\left(\mathrm{j}\Phi_{n,m}^{hv}\right) & \exp\left(\mathrm{j}\Phi_{n,m}^{hh}\right) \end{bmatrix} \\
& \cdot \begin{bmatrix} F_{rx,u,V}(\phi_{n,m}) \\ F_{rx,u,H}(\phi_{n,m}) \end{bmatrix} \exp\left(\mathrm{j}d_s 2\pi\lambda_0^{-1}\sin(\phi_{n,m})\right)\exp\left(\mathrm{j}d_u 2\pi\lambda_0^{-1}\sin(\varphi_{n,m})\right)\exp\left(\mathrm{j}2\pi\upsilon_{n,m}t\right) \\
& + \delta(n-1)\sqrt{\frac{K_R}{K_R+1}}\sum_{m=1}^{M}\begin{bmatrix} F_{tx,s,V}(\varphi_{LOS}) \\ F_{tx,s,H}(\varphi_{LOS}) \end{bmatrix}^T \begin{bmatrix} \exp\left(j\Phi_{LOS}^{vv}\right) & 0 \\ 0 & \exp\left(\mathrm{j}\Phi_{LOS}^{hh}\right) \end{bmatrix} \\
& \cdot \begin{bmatrix} F_{rx,u,V}(\phi_{LOS}) \\ F_{rx,u,H}(\phi_{LOS}) \end{bmatrix} \exp\left(\mathrm{j}d_s 2\pi\lambda_0^{-1}\sin(\phi_{LOS})\right)\exp\left(\mathrm{j}d_u 2\pi\lambda_0^{-1}\sin(\varphi_{LOS})\right)\exp\left(\mathrm{j}2\pi\upsilon_{LOS}t\right)
\end{aligned} \tag{8.18}$$

where $F_{tx,s,V}(\varphi_{n,m})$, $F_{tx,s,H}(\varphi_{n,m})$, $F_{rx,u,V}(\phi_{n,m})$ and $F_{rx,u,H}(\phi_{n,m})$ are the field pattern of transmitter and receiver antennas on vertical and horizontal;

φ_{LOS} and ϕ_{LOS} are the angles between LoS direction and north direction on BS and MS;

$\upsilon_{n,m} = \frac{\|v\| \cos\left(\varphi_{n,m} - \theta_v\right)}{\lambda_0}$ is the Doppler shift of the mth ray in the nth cluster, $\|v\|$ and θ_v are the speed and direction of MS, λ_0 is the wavelength;

υ_{LOS} is the Doppler shift of the LOS;

$\delta(\bullet)$ is the Dirac's delta function, K_R is Rice factor in linear scale;

d_s and d_u are spacing between transmitter and receiver antenna elements.

Thus, the CIR between the sth transmitter element and the uth receiver element is

$$\boldsymbol{H}_{u,s}(t,\tau) = \sum_{n=1}^{N_c} \boldsymbol{H}_{u,s,n}(t - \tau_n)\delta(\tau - \tau_n) \tag{8.19}$$

The CIR will be used in HIL simulation.

8.4 Hardware-in-Loop Simulation Testbed

8.4.1 Architecture

We build the HIL simulation platform in the laboratory as shown in Fig. 8.6, which contains radio channel emulator, test mobile terminal (MS), base station (BS), LTE core network and QoS test software.

The test mobile terminal is ZTE ME3760 supporting the frequency of 2.330 GHz under TDD duplex mode, and it achieves the function of network registration and data reception. The radio channel emulator C8 is used to emulate the radio propagation, which both has the flexibility of software simulation and the

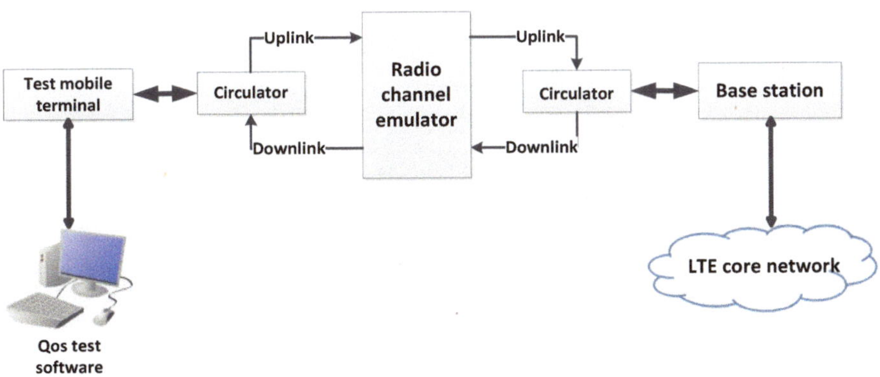

Fig. 8.6 HIL simulation platform for LTE system performance evaluation

accuracy of the hardware emulation. As for the radio channel, the classical channel —Rician channel—is used, what's more, the K-factor,the angle between LOS path and velocity and moving speed can be set flexibly on the platform, which provides the feasibility for the evaluation. The QoS software can test the UDP delay and throughput of the system.

In order to evaluate the performance, some environmental parameters should be determined, such as configuration of base station, performance indicator, and received power.

The test mobile terminal is ZTE ME3760 supporting the frequency of 2.330 GHz under TDD duplex mode, and it achieves the function of network registration and data reception. The radio channel emulator C8 is used to emulate the radio propagation, which has both the flexibility of software simulation and the accuracy of the hardware emulation. The QoS software can test the UDP delay and throughput of the system.

The components in HIL simulation testbed are identical to those on high-speed rail except radio channel emulator. Therefore, we can evaluate the performance on high-speed railway propagation environment in the laboratory, and study the correlation between channel environment and performance.

The radio channel emulator C8 is the core of the HIL simulation platform, which is technology-independent and supports all mobile communication systems working on 350 M–3 GHz frequency band with bandwidth no more than 70 MHz, such as GSM/UMTS/WiFi/WiMax/LTE. The radio channel emulator has 8 independent physical channels (see Fig. 8.2), and the radio channel characteristics, such as frequency, multipath delay, attenuation, noise, interference, and shadowing, can be set independently on each physical channel.

8.4.2 HIL Simulation Results

In this subsection, we conduct the HIL simulation. The parameters are shown as follows:

Parameter	Value
MCS	13
Duplex	FDD
Frequency	Uplink:1920–1980 MHz Downlink:2110–2170 MHz
Width	10 MHz
MIMO	2 × 2
Channel model	WINNER D2a

As shown in Fig. 8.7, BLER increases with the speed. When speed is 0 km/h, BLER is zero, which indicates the performance is good and stable. When speed is

Fig. 8.7 The performance with various speed

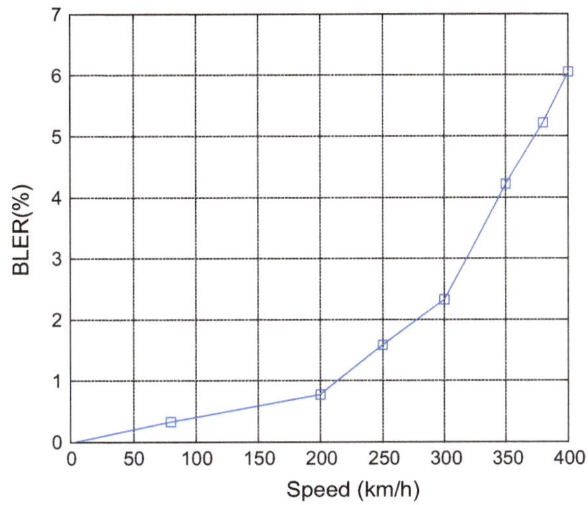

(a) BLER Vs speed (RSRP=-85dBm)

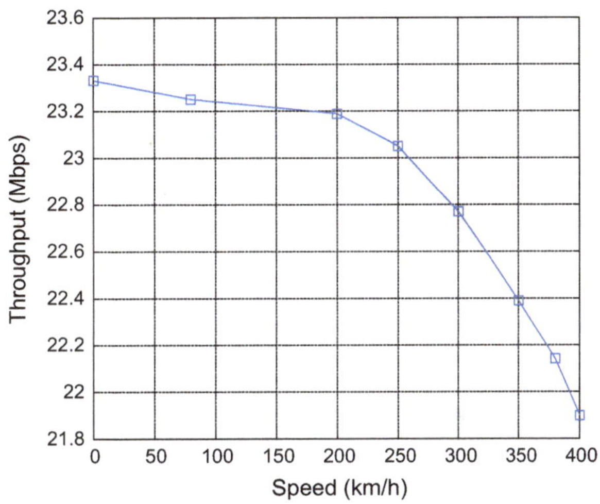

(b) Throughput Vs speed (RSRP=-85dBm)

up to 400 km/h, BLER will exceed 6%. The throughput will deteriorate with the increase of speed. When speed is up to 400 km/h, the throughput is only 94% of the peak throughput.

Fig. 8.8 The performance with various RSRP

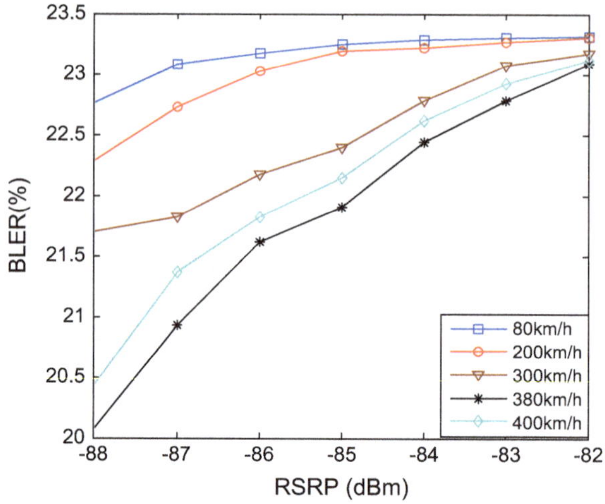

(a) BLER Vs RSRP

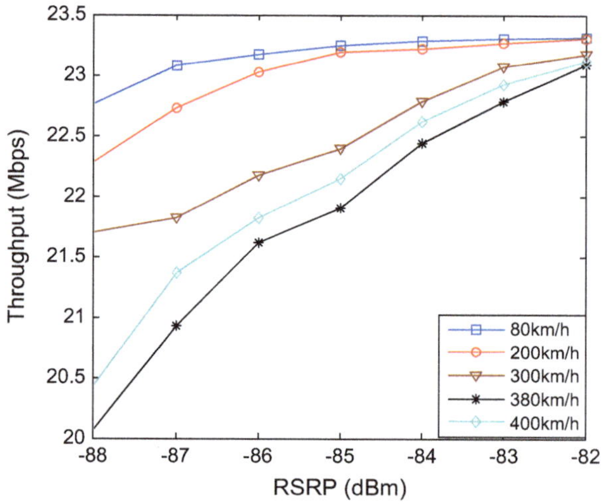

(b) Throughput Vs RSRP

As shown in Fig. 8.8, the BLER and throughput will improve with the increase of RSRP, especially in the high speed. Thus, we should increase the coverage 3–4 dB in high-speed railway to improve performance.

FSC
www.fsc.org
MIX
Papier | Fördert gute Waldnutzung
FSC® C083411

Kolibri 360 GmbH
Ferdinand-Jühlke-Straße 7
99095 Erfurt, Deutschland
produktsicherheit@kolibri360.de